LAITHWAITE, E. R.
A HISTORY OF LINEAR ELECTRIC
MOTORS. ERIC R. LAITHWAITE.
621.462   54-341191

A History of Linear Electric Motors

Other Macmillan titles of related interest

W. A. Atherton *From Compass to Computer—A history of electrical and electronics engineering*
J. D. Edwards *Electrical Machines—An introduction to principles and characteristics*

A History of Linear Electric Motors

Eric R. Laithwaite

PhD, DSc, CEng, FIEE, FIEEE

Professor of Heavy Electrical Engineering
Imperial College, London

MACMILLAN

© Eric R. Laithwaite 1987

All rights reserved. No reproduction, copy or transmission
of this publication may be made without written permission.

No paragraph of this publication may be reproduced, copied
or transmitted save with written permission or in accordance
with the provisions of the Copyright Act 1956 (as amended).

Any person who does any unauthorised act in relation to
this publication may be liable to criminal prosecution and
civil claims for damages.

First published 1987

Published by
MACMILLAN EDUCATION LTD
Houndmills, Basingstoke, Hampshire RG21 2XS
and London
Companies and representatives
throughout the world

Distributed in the USA and Canada by
San Francisco Press, Inc.
Box 6800, San Francisco
CA 94101-6800

Typeset by
TecSet Ltd, Sutton, Surrey

Printed in Hong Kong

British Library Cataloguing in Publication Data
Laithwaite, E. R.
A history of linear electric motors.
1. Electric motors, Linear—History
I. Title
621.46'2 TK2537
ISBN 0-333-39928-5

Contents

Preface   ix

Acknowledgements   x

1 Introduction—the first age of topology   1
   1.1 Design and evolution   2
   1.2 The earliest rotating machines   3
   1.3 An explanation of some terminology   4
   1.4 'Magnetic' and 'electromagnetic' machines   5
   1.5 *De mortuis nil nisi bonum*   6
   1.6 Fundamental shapes   8
   1.7 A classification of machine shapes   9
   1.8 The early inventions classified   10
   1.9 An engine should look like an engine!   12
   1.10 Steps forward and backward   21
   1.11 Shape becomes an end in itself   26
   1.12 1888—the biggest milestone since 1831   27
   1.13 References   29
   Appendix: Chronological order of rotating machine inventions   30

2 The early inventors and their patents   31
   2.1 The pre-induction motor period   31
   2.2 A.c. reciprocating motors   34
   2.3 The start of the induction motor era   35
   2.4 Boucherot, Zehden and Birkeland   36
   2.5 Bachelet and World War I   38
   2.6 Japolsky   41
   2.7 Liquid metal pumps   43
   2.8 World War II   45
   2.9 References   50

3 The contributions of the textile men — 52
- 3.1 'Action at a distance' is the 'bait' — 52
- 3.2 How a power loom weaves — 53
- 3.3 The amazing Monsieur Jacquard — 55
- 3.4 Classifying shuttle drives — 56
- 3.5 A topological giant — 57
- 3.6 Classification using electrical considerations — 58
- 3.7 A curiosity — 59
- 3.8 Milestones revealed by the patent history — 61
    - 3.8.1 Concerning Monsieur Bachelet — 61
    - 3.8.2 The men who had it didn't know they had it — 62
- 3.9 Duplication of effort — 66
- 3.10 Ingenuity continues — 67
- 3.11 An amateur textile engineer tries his hand — 68
- 3.12 A naturally self-oscillating motor — 70
- 3.13 Series and parallel worlds — 76
- 3.14 The 'DNA' molecule of machine engineering — 76
- 3.15 References — 78
- Appendix: Chronological order of patents on electromagnetic textile devices — 79

4 'Fashions' in engineering — 84
- 4.1 The task of the engineer — 84
- 4.2 The importance of 'fringe subjects' — 84
- 4.3 Ignorance really is bliss! — 85
- 4.4 Digression on a theoretical dilemma — 85
- 4.5 The bad old days — 87
- 4.6 A 'fashiongraph' — 87
- 4.7 Extrapolation of the fashiongraph — 88
- 4.8 The great pollution bandwagon — 89
- 4.9 The danger of breeding a race of robots — 90
- 4.10 Contrasts in industrial outlook — 90
- 4.11 'We must simplify the theory' — 90
- 4.12 Useful theory or formal discipline? — 91
- 4.13 Unification for its own sake — 91
- 4.14 Industrial 'pros and cons' — 92
- 4.15 Automation — 92
- 4.16 Design or evolution? — 93
- 4.17 Accountants become fashionable as dictators — 93
- 4.18 References — 94

5 Electromagnetic levitation — 95
- 5.1 Different worlds — 95
- 5.2 Obsession with cylindrical geometry — 97
- 5.3 A theory 'brought out and dusted' — 98

5.4	Levitation melting	102
5.5	Levitation with a degree of freedom	104
5.6	References	106

6 Academics and industrialists — 107

6.1	Background	107
6.2	Portrait of an academic	108
6.3	The importance of a great industrialist	110
6.4	The new generation	112
6.5	John Lowe	112
6.6	The 'clean-up' men	113
6.7	The blending of electronics and 'machines'	116
6.8	A lot depended on communication	120
6.9	Series connection	122
6.10	References	123

7 The high-speed transport game — 125

7.1	What to leave out, not what to put in!	125
7.2	The railway experiments	125
7.3	The 'Aérotrain'	130
7.4	The formation of Tracked Hovercraft Limited	134
7.5	The pace makers	136
7.6	The long pole pitch problem	140
7.7	The 'darkest hour before the dawn'	144
7.8	The three-part problem concept of the mid '60s	149
7.9	Global competition 1966–1972	153
7.10	Transpo 72	160
7.11	The stop-go phase	166
7.12	The Select Committee	167
7.13	References	168

8 The world-wide game — 169

8.1	New readers begin here!	169
8.2	The Toronto Urban Transit Scheme	170
8.3	Earith and after	172
8.4	The Consortium of British Universities	175
8.5	'Landspeed'	178
8.6	The world scene up to 1976, country-by-country	179
	8.6.1 USA	180
	8.6.2 Canada	187
	8.6.3 Japan	188
	8.6.4 Europe	190
	8.6.5 USSR	194
8.7	References	195

CONTENTS

9	The second age of topology	196
	9.1 A voice from the past	196
	9.2 Can no-one be right?	197
	9.3 The lateral dimension	197
	9.3.1 Concern about stability and about power factor	200
	9.4 Tubular motors	202
	9.5 The vertical dimension	205
	9.5.1 Electromagnetic levitation	206
	9.5.2 The topology of track joints	209
	9.6 The longitudinal direction	212
	9.6.1 Surface current topology	213
	9.6.2 Superimposed skewed windings	215
	9.7 Linear synchronous motors	220
	9.7.1 Claw-pole motors	220
	9.7.2 Combined cryogenic Maglev and propulsion	222
	9.8 References	223

10	A continuing story	224
	10.1 General	224
	10.2 'Stop-press' on HSGT	225
	10.2.1 USA	226
	10.2.2 Canada	226
	10.2.3 Japan	226
	10.2.4 West Germany	226
	10.2.5 UK	226
	10.3 What have we learned?	229
	10.3.1 Wisdom	229
	10.3.2 Courage	229
	10.3.3 Theory, design and evolution	230
	10.3.4 Invention	231
	10.4 References	233

Bibliography   *234*
   General   235
   Industrial applications   255
   Levitation   277
   Transport   296
   Theory   349

Index   *385*

Preface

The manuscript of this book was first prepared at the request of the Institution of Electrical Engineers in 1971. This coincided with a period of great activity in one particular aspect of linear motors, that of high-speed ground transport (HSGT) applications. It was a time when the Tracked Hovercraft project showed such great promise, when new inventions grew like mushrooms and hopes for a world-adopted British system ran high, only to be dashed by the Government's closure of the project in 1973 and demolition of the full-scale track in 1974. It was a time when personal emotions ran high.

The original draft contained a whole chapter on the rise and fall of Tracked Hovercraft, some of which was found unacceptable in that it might have given grounds for libel action. The whole manuscript was in fact revised to avoid giving any offence whatsoever. Meanwhile, new developments in HSGT were taking place in several countries from the USA to Japan and more papers were flooding in to swell the bibliography. By 1975 new papers on linear motors were appearing at the rate of between 14 and 16 per month and it was decided to close the bibliography at that date on the grounds of space and of the escalating task of cross-referencing to check that the bibliography after 1975 was complete.

Originally it had been intended to publish in 1983 with the final chapter bringing everything up to date, but the final manuscript was declined by the IEE's publishers in 1984 and a re-revision took place for the present publishers.

Some updating has been carried out on the last chapter in view of the time that has elapsed since the original target date for publication. Perhaps future historians may record that not until 1986 could the linear motor be said to have really become industrially accepted, for the author was invited to address the IEE in London on January 15, 1986 on the subject of 'Linear motors—a new species takes root', which was then to be published in *Electronics and Power*. In newspaper language and in the context of this book, this paper must surely be classed as 'Stop Press'!

Eric R. Laithwaite

Acknowledgements

This work might never have been completed had it not been for the untiring efforts of my secretary, Miss Elizabeth Boden, who not only typed and re-typed the manuscript several times but corrected spelling and grammar and did a tremendous job on the bibliography in respect of accuracy and attention to detail. While neither of us pretends that it is now perfect, all I would say is that as the result of her efforts, 'It's a lot better than it was!'

I am also much indebted to Dr Brian Bowers of the Science Museum, South Kensington, firstly for help with research on the early days of the subject and the patents of Wheatstone in particular, and secondly for reading the whole manuscript and advising me on the production of a revised version that would offend no-one.

I am also grateful to the publishers for picking up the pieces when, at one time, all seemed a lost cause.

The diagrams were drawn by Mr E. Lawler.

1 Introduction — the first age of topology

The greatest difficulty in compiling a history of almost any specialist subject is to know where to begin. There is a popular saying—'There is nothing new under the sun'—and certainly it is often impossible to credit the invention of the most useful of devices to any known single person. In *The Children's Encyclopaedia* (1922) a verbal picture of a stone-age man is painted with the use of phrases such as "He lifts the stone, raises it above his head, and flings it from him. The others imitate him. They chuckle and grin. The first game of cricket is being played."

At first sight this appears to be exaggeration taken to extremes. 1.7 million years were to elapse before anything like the modern game of cricket began to emerge, but since no single man invented it and since the fascination of cricket, in common with that of most other ball games, comes from appreciating a sphere in motion, perhaps the encyclopaedia has a most valid point.

Linear motors are undoubtedly a product of technology and there is much confusion in the modern world between science on the one hand and engineering and technology on the other. John Lenihan (1969) in his Kelvin lecture of 1968 put technology into its proper perspective when he said:

"It is often supposed that technology is the product of science, emerging unpredictably as a reward for untrammelled intellectual curiosity. Two of the most significant and spectacular advances in the history of civilisation were accomplished by technology alone. Until the eighth century, a mounted soldier had the advantage of a better view but had to rely on his own muscle power and was helpless when unhorsed. Once provided with stirrups, however, his fighting power was greatly increased.

"The horseman now merely held the lance between his arm and his body, allowing the muscle power of the horse to deliver the blow. The Franks absorbed and exploited the new technology of war. The Anglo-Saxons did not take it seriously and in 1066 they paid the price of their neglect."

He continues with his second example as a dual invention, the two occurring at about the same period but with no precise origin in time.

". . . the ox was commonly used for ploughing and other agricultural tasks. Horses were not very serviceable, partly because their hooves were easily damaged and partly because their tractive effort was very limited. The ox yoke was inefficient when applied to the horse because, as soon as the animal took the strain, the neck strap pressed on the windpipe with discouraging results.

"The nailed horse-shoe appeared during the ninth century and the modern type of harness, with a padded collar allowing the animal to exert full effort was developed almost simultaneously. The rise of the middle classes, the growth of commerce, industry and education and many other features of modern civilisation may be traced back to the agricultural revolution—technological but not scientific—of nine or ten centuries ago."

[The scientist's view of the great landmarks in the history of civilisation is very different from that of the architect (Clark, 1969).]

1.1 Design and evolution

In the history of electric motors similar influences will be seen to have come together to produce the rotating machines of today, perhaps as much an *evolution* as a process of logical design. These developments may be referred to as 'the First Age of Topology' for in the nineteenth century, particularly from Faraday's discoveries in 1831 to Tesla's induction motor of 1888, the study of shape dominated electric motor development. So far as *linear* electric motors are concerned, we may well be only at the middle or even the beginning of what is called in chapter 9 'the Second Age of Topology'. As with the horseshoe and the stirrup it is almost impossible to say where linear motors began, for linear motion, as such, plays a much greater part in human life than does rotary, except perhaps in centres of very high industrial content. In Nature there is little rotary motion, except that of the planets and solar systems themselves, but in plant-life and animals there are virtually no wheels and one could argue that linear motion is 'natural' motion.

A great pioneer in electrical machines—no less a person than the great French engineer Paul-Marie-Joachim Boucherot (1869–1943), inventor of the double-cage rotor—had something very emphatic to say on precisely this point (Boucherot, 1908).

"Man has adapted his machinery to rotate because he found it the easiest to produce artificially. To conclude that rotary motion is *natural* motion is begging the question. Stars have rotary motion, electrons also, but animals only move and act by oscillating movements. Snakes move on a miniature system and we know with just what exotic speed. I am of the opinion that there are, in animal mechanics, profound fundamentals which have been analysed far from completely."

Linear motors can be regarded, in their most general form, as the physical result of the splitting and unrolling of rotary machines, and there are therefore at least as many types of linear motor as there are of rotary, for each one (induction, synchronous, commutator, reluctance, hysteresis etc.) has its linear counterpart. It is therefore profitable, by way of introduction to the importance of topology in the development of linear machines, to study the First Age of Topology (1830-1890), for many types of linear motor are no more than one simple topological exercise from their rotary equivalents.

1.2 The earliest rotating machines

Figure 1.1 shows the earliest and certainly one of the simplest shapes of electric motor, although in its original form it was used as a generator—in fact it is the basic arrangement used by Faraday for his dynamo of 1831. It should be made clear at the outset that all electrical machines are *reversible*, so far as power flow is concerned. A d.c. motor may, without a change in its basic physical shape, be used as a d.c. generator—that is, the power can flow from shaft to electric terminals and vice versa. As it is with rotary machines, so it is with linear, but so far the use of a linear machine as a motor has far exceeded its use as a generator. Nevertheless there are periods in the duty cycles of commercial linear motors when they may operate as generators, plugging induction motors etc.

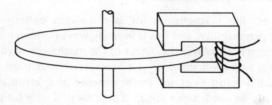

Figure 1.1 Schematic arrangement of Faraday's disc dynamo, 1831

Following only months behind Faraday's disc came Hippolyte Pixii's machine shown diagrammatically in figure 1.2. This machine had all the essential elements of the modern alternator in that

(i) The field-generating magnets, rather than the output coils, comprised the *moving* part
(ii) The output was *alternating* current rather than 'battery-like' current
(iii) The airgap could be made quite large
(iv) The system could be made multi-polar.

The principal difference between Pixii's generator and the modern machine was that the working surface was a plane rather than a cylinder. The use of the term 'working surface' needs some further explanation here for, like some other phrases not in common use, it will be used throughout the book as a convenient way of expressing some of the theoretical arguments which follow.

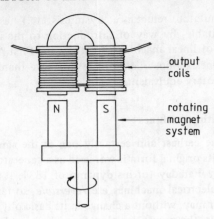

Figure 1.2 The arrangement of Pixii's a.c. generator, 1831

1.3 An explanation of some terminology

In a transformer, the magnetic circuit is completed entirely in ferromagnetic material of high permeability, and this is in many ways (although not all) ideal for an electromagnetic machine. Where the object of the machine is to convert energy from an electrical form to a mechanical or vice versa, relative motion between different parts of the machine must exist and in the absence of a ferromagnetic fluid which could conveniently be used, some airgap is necessary, if only for purely mechanical reasons, between one ferromagnetic part of the machine and another. Because the relative permeabilities—that is, magnetic 'conductivities', of certain steels and air are of the order of 1000, the reluctance of the airgap often dominates the magnetic circuit design. To this extent it can be argued that the airgap of a machine is 'where the action takes place', even though we know that electromagnetic phenomena have little meaning unless considered in relation to interlinked, closed circuits (electric and magnetic). So we speak of the available tangential thrust of a rotating machine 'per unit of pole area', but since the extent of poles is not always easy to define (slot openings being convertible into effective pole area) the idea of a working surface in which slots and teeth alike are regarded as homogeneous is often a convenient one.

A second philosophical outlook, not found in the usual textbooks on electrical machines, should perhaps also be explained here, for in this case it is not only a 'convenient' way of describing a machine but sets out a most *fundamental* difference between certain types. Various inventors (both in the field of linear and rotary machines) have discovered this difference for themselves, yet it does not, even at the present time, form a part of formal courses in electrical machines. The fact that it is related to the 'working surface' makes it convenient to explain it here.

1.4 'Magnetic' and 'electromagnetic' machines

Electricity and magnetism are said to be manifestations of a fourth dimension, which is perhaps another way of saying that we shall never appreciate them to the extent that we appreciate three-dimensional objects. To say that we shall never 'explain' electromagnetic phenomena is probably true but unhelpful. To say that one particular, so-called, 'explanation' is *better* than another however is unworthy of a scientist, for he should know that a particular concept (and it is generally agreed that it can be no more than a concept) can be much more profitable than another for one particular person. Nevertheless, a history of any aspect of electrical phenomena will record the general acceptance of particular concepts which go in and out of 'fashion' as do designs of human clothing. Perhaps one of the simplest examples, although not an electrical engineering one, is the corpuscular theory of light, followed by the wave theory, followed by the quantum theory. In many cases, unfortunately, the teaching of a new 'gospel', especially by those who first propounded it, inhibits the next generation to a large extent, often inferring even that 'this is the *only* way' a phenomenon may be explained or a device manufactured. It is impossible to estimate the extent of the delay in the development of linear machines which was occasioned in particular by the handing on of the dogma—'only machines with small airgaps are *good* machines'—especially during the period 1920-1950, as is discussed more fully in chapter 2.

Another aspect of the writing of any history of a subject is that the writer can see both the wisdom and the folly of his ancestors, at least in part, and benefit therefrom; but it then becomes necessary to relate, in the introduction, pieces from the end of the story in order that the reader can appreciate, and enjoy the more, the work of the early pioneers.

The action of most electrical machines may be explained on the basis of a set of magnets on the stationary member reacting with a second set of magnets on the moving part. The word 'magnets' in this context however is too broad for the student who would make a closer study, for these magnets may be of several types, thus:

(a) permanent magnets, which are perhaps the only real magnets in the true sense of the word
(b) electromagnets energised by windings whose current may be controlled
(c) steel in which magnetism is induced in one set of pieces, simply as the result of the proximity of another set, whether the latter be of (a) or (b) type.

In the context of this book, classes (a) and (b) will be referred to as 'Electromagnetic Machines'. Class (c) will be called 'Magnetic Machines'. The fundamental difference between these two groups lies in the fact that any electromagnetic machine, when simply scaled up in size, becomes a 'better' machine, while magnetic machines become better as they are scaled down. Formal proof of these statements and a definition of what is meant by 'better' can be found in other publications (Laithwaite, 1966, 1971). It will suffice at this point to say that a factor of 'goodness' can be defined which is based only on the product of firstly, the resistance of the electric circuits of a machine, secondly the reluctance of the magnetic circuit and thirdly the speed of the machine. This quantity is, to a first approximation with leakage ignored, the ratio of the full load current of a machine to its no-load current. Since the latter is largely determined by the airgap, attention is bound to be focussed on this point of the machine, which brings us back to the idea of a 'working surface'.

Another concept which can be useful in some thought processes is that the presence of what we normally call 'magnetic poles' is due only to stepped changes in reluctance of the magnetic circuit. If, for example, a piece of electric wire of high resistance were connected to a battery via thick conductors of low resistance and the whole circuit were then to be immersed in a brine bath and probes used to detect the current flow lines in the liquid, the pattern would appear similar to that shown in figure 1.3. Had we been told that the flow lines were lines of magnetic force, we would at once have labelled the points A and B as 'magnetic poles'. This is not a new concept, rather it is a neglected one, for Faraday himself was aware of it.

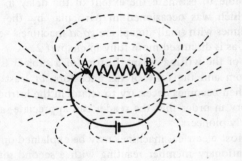

Figure 1.3 Electric analogue of a bar magnet showing that poles arise only as the result of a change in impedance

1.5 *De mortuis nil nisi bonum*

To return to Pixii's machine of figure 1.2; an original machine was, in 1971, lent by the Smithsonian Institution to the Science Museum at South Kensington, London where an excellent replica is on display to the public. The caption under the name-plate reads

> "Faraday produced the first continuous current in 1831. Later in that year Pixii made a machine incorporating an iron-cored multi-turn bobbin. This was a great advancement in design, and a year later, at Ampère's suggestion, a simple commutator was also added."

Since the only man-made electric currents before Faraday's discoveries were generated chemically, there was much praise for anyone who could turn alternating current (which we now know to be the most convenient and economical to use) into 'battery-like' current, and I have no doubt that Ampère's suggestion of fitting a simple reversing switch driven by the moving part of the machine itself received much applause at the time. Yet we now know that the struggle to make d.c. set back the course of electrical engineering history for perhaps as much as several decades.

History, of course, is a source of wisdom as opposed to pure science which is a source of knowledge. In history we can read of great men *and* of their mistakes, for it is often in the making of mistakes that the experimenter learns. Indeed it is hoped that this volume itself may, by recording retrograde, as well as forward steps in rotating machinery, enable the linear motor designer of the future to avoid the analogous pitfalls.

The generation of alternating current is perhaps one of the most fundamental of all topological considerations, for when a bar magnet is pulled out of a coil (as Faraday demonstrated) a pulse of current can be made to flow in the circuit of the coil and this will be unidirectional. But the process cannot be repeated with the same, or any other magnet, unless a current flowing in the opposite direction is first produced or the coil is at some point disconnected and reconnected. Of course, it must have been incredibly difficult for Ampère to have seen his simple reversing switch as the forerunner of the multi-segment, modern commutator of d.c. machines, and even more difficult to see Faraday's disc dynamo as the ultimate limit of multi-segment commutators in which the size of each segment could be said to be zero and the number of segments infinite. But these ideas, which are simple enough for us to appreciate now, are examples of the Wisdom which comes only after a wealth of experience (a history) has been accumulated.

What is even more fascinating is the next *direct* development of Pixii's machine which occurred in 1846 and was the result of an invention by Emil Stöhrer of Leipzig. His machine is shown in figure 1.4. This is claimed to be the first multi-polar machine, carrying three horseshoe magnets. The extra weight of the magnets led the inventor to interchange the stationary and moving parts. This in itself demonstrates great appreciation of Faraday's experiments, for when you know 'nothing' and someone tells you that an electric current (which you know to be only the result of a chemical action) can be generated by pulling a magnet from inside a coil, it is by no means to be *assumed* that pulling the coil from the magnet produces the same effect.

The moving coils could now only pass current to an external circuit, where it could be put to work via slip rings and brushes, and this machine therefore represents the second major retrograde step in generator design. We now see that Pixii had the modern turbo-alternator right in the palm of his hand and yet let it slip away through the advice of men whom he no doubt considered greater authorities than himself. The Pixii machine has no commutator or slip rings to carry the output current. Apart from the replacement of permanent magnets by electromagnets, the only difference in arrangement between the modern alternator and Pixii's machine was the relatively simple topological re-arrangement of making the working surface that of the surface of a cylinder rather than of a disc. It is interesting also that, for certain purposes, modern machines with disc surfaces can be profitable. They are

Figure 1.4 Six-pole generator developed by Emil Stöhrer of Leipzig (*courtesy of Trustees of the Science Museum*)

now usually described as 'axial-flux machines'—a term we shall meet again in linear motor terminology. Perhaps foremost in 're-discovering' the usefulness of the disc surface was the firm of Fairbanks–Morse in the USA.

1.6 Fundamental shapes

The pressure on the early pioneers to produce d.c. was increased with the rapid developments in the technology of electro-plating. From early experiments by Jacobi in 1838, commercial-sized plating baths were in use in the 1840s and in 1844 Messrs Prime and Son of Birmingham built, to a design by Woolrich, what was probably the earliest industrial electrical machine of any magnitude. This generator was first described in *The Electrician* (1889). The four compound horseshoe magnets had like poles on the same side. The moving armature had eight coils with a common connection to a slip ring on one side and to separate commutator segments on the other. Topologically the arrangement was basically the same as Faraday's disc as shown in figure 1.5 but the rotor configuration was essentially a *new* topology at that time.

A few years ago one could have said that the complete exploitation of a Faraday disc dynamo would involve the use of magnetic circuits which would be very difficult to make—structures perhaps in the form of the plates of a quadrant electrometer—in order to utilise the whole surface of the disc. Yet the appearance of a new material in the form of niobium–tin (superconductor) caused the whole topological

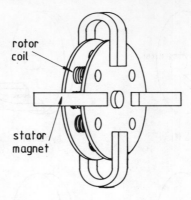

Figure 1.5 Physical layout of Woolrich's design, 1844

concept of a machine to be changed overnight. The first large-scale superconducting generator to be built at Newcastle returned basically to Faraday's disc, using virtually no ferromagnetic material and a layout as shown in figure 1.7, type (1a).

1.7 A classification of machine shapes

At this point it is interesting to simplify each invention of the nineteenth century into its basic topological arrangement, regarding the primary member as a stationary permanent magnet system and the secondary as a disc (elongated to a cylinder whenever the *edge*, rather than the face of the disc is used as working surface). With a single horseshoe magnet and a disc, six fundamentally different arrangements are possible, as shown in figure 1.6, (1) to (6). These arrangements make no assumption as to how the magnetic circuits are completed, nor even whether the disc is of (i) solid conductor, (ii) conducting windings only, (iii) slabs of steel or (iv) a combination of steel and copper. Since there are in general three planes in which the magnetic circuit can be completed and these are entirely independent of the plane in which the primary horseshoe lies, there are $6 \times 4 \times 3 = 72$ *basically* different topological arrangements for rotating machines without involving all the many ingenious variations in each group and the variations in winding arrangements which, in part, overlapped the inventions of the nineteenth century but were, in the main, invented after that period.

From the basic arrangements of figure 1.6, (1) to (6), other secondary topologies can emerge. Figure 1.7 shows a modified form of (1), labelled (1a), two arrangements (2a) and (2b) derived from (2), (4a) derived from (4), and a (5a) from (5). Figure 1.8 shows four arrangements developed from the basic type (3) of figure 1.6. In discussing the inventions of the first age of topology in strict chronological order, the object is to illustrate that, while the various machines may now be classified as in figures 1.6 to 1.8 in a logical manner, no such logic existed, nor can we expect it to have so existed in the minds of the inventors of the 1800s, and there-

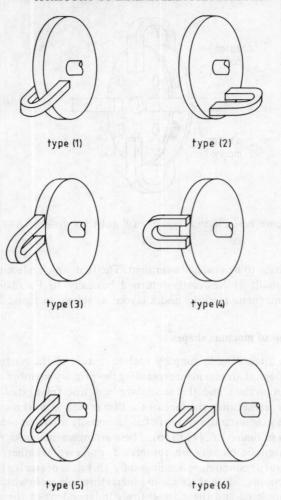

Figure 1.6 Fundamentally different topologies

fore some of the ideas were clearly 'ahead of their time' while others were just as clearly retrograde steps. Thus the forward steps of technology are invariably stumbling ones. The inventors of linear motors, having the history of rotary machines to study, should have been able to develop the best machines much more rapidly. Yet they failed to do so, as subsequent sections will show.

1.8 The early inventions classified

Faraday's disc was type (1) of figure 1.6, while Pixii's was type (2a) of figure 1.7. The fact that most writings of the 1800s were illustrated by pictures of magnets

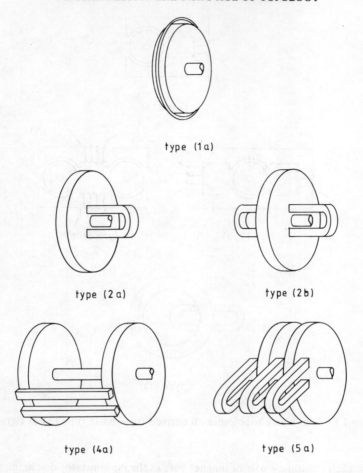

Figure 1.7 Secondary topologies seen as extensions of the basic types

with the flux coming 'out of the end' of the poles, rather than from their sides, makes Faraday's machine all the more remarkable. Types (2) and (3) make use of the pole ends, while (4) and (5) use the sides. That the early pioneers did not 'brainwash' each other to anything like the extent that the engineers of 1900–1920 did their sons is evidenced by the fact that Pixii's machine was followed by Woolrich's arrangement which clearly used magnet pole 'sides', as in type (4a).

Another interesting machine was built by W. Sturgeon in 1832 and is shown in figure 1.9. It differed from Pixii's machine, not only in that it was designed to run as a motor, rather than a generator, but that it also ran on d.c. and by modern notation could be said to be 'inside out', for the discs of type (4a) were simply bar magnets which were pulled into line with the bar-type electromagnets whose coils were fed from a commutator, driven by the rotor. Rotor inertia served to carry the

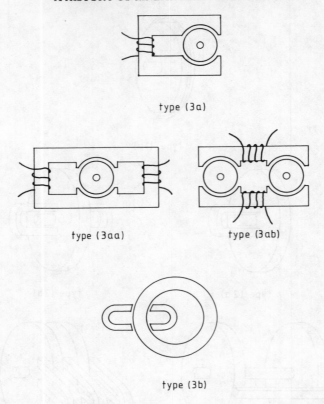

Figure 1.8 Secondary topologies all derived from basic type (3) of figure 1.6

bars past each stationary pair of magnet bars as the commutator de-energised them. It will be noted at this point that type (4a) is very nearly the same as type (6).

The full significance of types (4a) and (6) will not emerge until their linear counterparts have been discussed and this does not occur historically until 1967 (see chapter 7 in particular). At this point it is only necessary to notice that these two types of rotary machine could be defined simply by saying that the axis of the primary m.m.f. is parallel to the axis of rotation, although even this definition requires some stretch of the imagination if the horseshoe magnet of figure 1.6 (type (6)) were replaced by an electromagnet with its exciting coils on the pole faces.

1.9 An engine should look like an engine!

Trying to place oneself mentally in the position of an early inventor and seeing his difficulties is itself an extremely difficult thing to do. In later chapters, however, I can recall my own feelings at a time when I did not know the usefulness of a three-

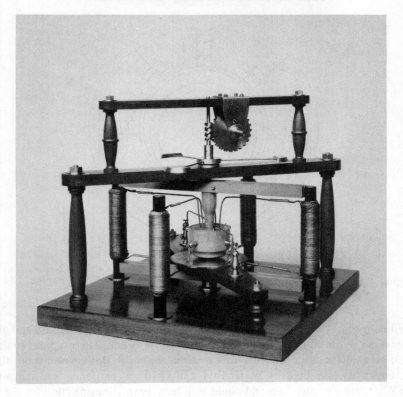

Figure 1.9 Sturgeon's machine of 1832 was virtually a d.c. motor 'inside out' (*courtesy of Trustees of the Science Museum*)

dimensional approach, of magnetic equivalent circuits, the transverse flux idea and other discoveries in which, subsequently, I had a hand. It is therefore perhaps justifiable that a history of a subject should be written by one himself 'immersed' in it, for the difference between not knowing a concept on one day and being fully aware of it on the next is the equivalent of climbing the North Face of the Eiger and much more so for the earlier pioneers.

In 1840, there were many engines in use—steam engines mostly, and it was not unnatural for electrical inventors to use the steam engine as their first guide to general layout. Couple with this the fact that the properties of magnets had been known for 5000 years and that the principles of electromagnets were now fairly well known, and their efforts to harness these forces become clearly justified for us at the present time.

Wheatstone (1841) devised a series of four electromagnetic engines, three of which are described in his patent. In a fascinating article Brian Bowers (1972) of the Science Museum, South Kensington, London, describes these developments in detail. All are reluctance machines, the first of which was a type (6) in which the

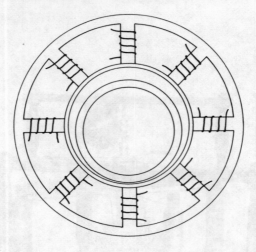

Figure 1.10 Reproduction from Wheatstone's patent of his first engine, 1841

horseshoe magnets were electrically energised and could be switched on and off in sequence so as to pull the 'keepers' on the rotor into line. The machine could be said to be an elaboration of Sturgeon's machine, although Bowers points out that a very similar machine built by W. H. Taylor in 1840 was exhibited less than a mile from Wheatstone's workshop and could well have been the inspiration for the first of Wheatstone's four reluctance motors. Although Taylor's machine no longer exists it is described in the *Mechanics Magazine* of 9th May, 1840. Bowers himself 'unearthed' the Wheatstone machines only a few years ago, thus contributing a valuable link in the chain of electromagnetic machine discovery. Figure 1.10 shows the patent drawing of the first of Wheatstone's 'eccentric' engines and figure 1.11 a photograph of the actual motor. The rotor was rolled inside the bore of a ring of electromagnets each of which could be switched on and off in turn to attract the steel cylinder rotor and cause it to rotate. This machine is therefore a reluctance motor and its place in a history of linear machines is assured by Wheatstone's concept of the machine being 'unrolled' to give linear motion.

To quote directly from Bowers' article, describing Wheatstone's patent specification: "Essentially it says that whereas other electromagnetic engines had electromagnets and armatures with their active faces [my term—'working surface'] arranged in concentric circles (as is the case also with modern rotating electrical machines) these engines had their armature surfaces in a circle which was not concentric with the circle linking the surfaces of the electromagnets. The two circles could have either parallel or intersecting non-parallel axes. It then goes on to suggest that one of the circles might be unrolled into a straight line so that the other circle would roll along it and produce a reciprocating machine. This is by far the earliest reference to the idea of a linear motor."

Figure 1.11 Wheatstone's first 'eccentric' engine, 1840 (historically his second machine, his first being a simple concentric type (6)) (*courtesy of Trustees of the Science Museum*)

The non-parallel axis idea is extraordinarily ingenious and could well be unique in the whole history of electromagnetic machines. The basic layout was a type (2) (figure 1.6) but the rotor, instead of spinning on a fixed axis, performed precisely the motion of a dinner plate which is set in motion on a table such that its axis gyrates in conical fashion while the edge of the plate contacts the table at one point only, a point which rotates in a circle. I agree of course with Dr Bowers' conclusion about the earliest reference, provided a linear motor fits the modern American term for an electrical machine, namely an 'electromechanical energy converter'—that is, it converts electrical energy into mechanical energy or vice versa. If it were not so we would be back with the lodestones!

At the time of writing it appeared that Wheatstone had tossed in the idea of linearisation as an afterthought, almost as a chef would add a pinch of herbs to a soup, but during this period Brian Bowers made a further search of the same locality in which he found the eccentric engines and pulled out a row of coils made by Wheatstone which, although there was no secondary, could only have been used as the primary of a linear motor (figure 1.12).

Figure 1.12 The first linear motor of all time, built by Wheatstone between 1841 and 1845

Invention, for me, is not mere observation of a phenomenon. Although it is possible that a schoolboy of 1820 had pulled a small iron rod along a thin table by means of a magnet underneath the table, such an arrangement was not proposed for direct application to a practical system until 1844 (see chapter 3) and this is therefore the proper date to record the system as having been invented. The true inventor is a man of vision, not the man who, like a child on its way home from school, sees a diamond, picks it up because it shines, loses it on the way home and never knows what it was worth. What is interesting is that the linear system using *electromagnets* —that is, a true electric motor—preceded by three years the more obvious system using *permanent* magnets (which is properly described as a 'clutch').

To classify Wheatstone's eccentric machines, his first was a type (3), his second, shown in figure 1.13, a type (6) and, as just described, his third a type (2). I can only comment that in the light of all this work it took my co-workers and me a disgracefully long time to achieve such flexibility in thinking in three dimensions, as is described in chapter 9 on the 'Second Age of Topology'. Wheatstone's third eccentric machine is shown in figure 1.14.

INTRODUCTION—THE FIRST AGE OF TOPOLOGY

Figure 1.13 Wheatstone's second eccentric engine (*courtesy of Trustees of the Science Museum*)

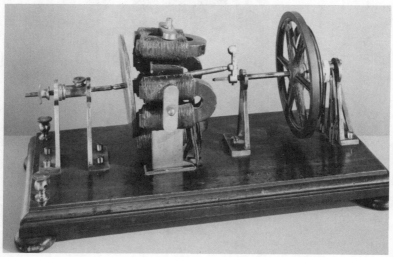

Figure 1.14 Wheatstone's 'rolling plate' eccentric engine (*courtesy of Trustees of the Science Museum*)

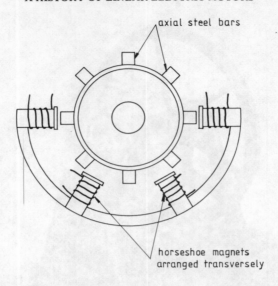

Figure 1.15 Layout of Froment's electromotor, 1845

Figure 1.16 Apps' second machine (perhaps an electromagnet version of Froment's engine) (*courtesy of Trustees of the Science Museum*)

Bowers points out that this third machine of Wheatstone bears a remarkable similarity to an unusual design of steam engine developed by E. and J. Dakeyne and patented in 1830.

Perhaps the next significant machine was due to Froment in 1845. Using a similar layout to Taylor's machine for the stator, the rotor was also composed of steel bars of which over 50 per cent were in use at all times. Froment's machine layout is shown in figure 1.15 and is clearly a type (6).

In addition to Froment's 'electromotor' which resides in the South Kensington Science Museum there are two machines, due to Apps, whose dates are difficult to trace but which are certainly later than 1866 and therefore a long time after Froment. The first is simply a Pixii machine with the permanent magnet replaced by an electromagnet and this machine could lay claim to be the first electromagnet-electromagnet system. Apps' second machine (figure 1.16) is an electromagnet version of Froment's electromotor and this alone suggests that Apps made it after 1845, at the earliest.

If Wheatstone's 'unrolled' eccentric rotor motor was the first invention in linear motors, the second was undoubtedly a machine patented in 1852. The inventor's name will no doubt come as a surprise for it was William Henry Fox Talbot, better known for his pioneering work in photography. Not only was he the first man to take a paper photograph (1835) but the first man to make a negative and use it for printing. There is no doubt that from time to time Nature produces a 'superman' capable of switching his fertile brain from one subject to another, so apparently remote as to stagger the mind of even the better-than-average scientist.

The description of Fox Talbot's linear motor in his patent of 1852 reads: "This Invention is a new kind of electro-magnetic machine. A heavy cylinder is made to roll upon a long but narrow metallic table or plate, close beneath which a long row of horse-shoe electro-magnets is placed. These magnets stand vertically with their poles uppermost, so that as the cylinder rolls along the plate, it unites the two poles of each magnet consecutively. If the magnets are placed close enough together and on a level with each other, their summits will form a sufficiently firm surface for the purpose required. The machine is so contrived that the cylinder is always attracted forwards till it reaches the end of the row of magnets; the action is then reversed, and it returns in the opposite direction till it reaches the other end, and so on. The whole distance traversed by the moving cylinder makes one stroke of the engine. The cylinder communicates its force and motion to the rest of the machinery, by its axis being attached to a crank and fly wheel. [Note still the insistence that dynamo-electric machines are 'engines' that make 'strokes' which are converted to rotary motion by cranks.] The motion of the cylinder itself sets in action the contrivance (sometimes called a commutator, sometimes a rheotome), which magnetises the magnets that are before the cylinder, and unmagnetises those behind it."

For a man of Fox Talbot's imagination the building of a linear motor was clearly a sideline, and I am sure that had he given this topic the attention he gave to photography, it is likely that fewer than 45 years would have elapsed before anything like his linear motor was given an application.

It is also very important to record at this point that a row of electromagnets which are switched on in a sequence (only one or two at a time being energised) in order to propel a single and small piece of steel, by reluctance change, does in fact produce a dominant travelling field component moving in the *opposite* direction to the sequence of coil switching. This field is capable of propelling a secondary, consisting of a simple sheet of conductor in the reverse direction. A complete explanation of this phenomenon in terms of conventional concepts is given by Laithwaite (1967).

By 1853, electromagnetic machines had not made any real impact on industry and the thought was still being pursued that to generate rotation *per se* from magnets was basically wrong. Electric 'engines' should be designed like steam engines with pistons, cranks being used to convert what was essentially linear motion into rotation. Allan's Electromagnetic Engine of 1853 (figure 1.17) har-

Figure 1.17 Allan's magnetic piston engine (*courtesy of Trustees of the Science Museum*)

nessed the attractive force between an electromagnet and a steel block; but realising that magnets could only exert a really effective pull over about $\frac{1}{4}$ inch, he put four coils and keepers in tandem, switching on each magnet at the appropriate time so that coil (1) lifted not only its own keeper but those of coils (2), (3) and (4), after which coil (2) was energised and lifted armatures (2), (3) and (4), and coil (3) then lifted (3) and (4), so that when coil (4) had 'fired' the total lift transmitted to the crankshaft was about 1 inch. With four such stacks phased at 90° to each other, his four-stroke engine used 'magnetic' forces quite effectively.

He was not to know, of course, that because he had used purely 'magnetic' forces, his machines could never be successful in large sizes. However, Allan was not by any means the last inventor to try to use the very 'promising' B^2 compressive forces of magnetism as opposed to the weaker BJ shearing forces of electromagnetism, and the pronouncements of Boucherot (1908) on this subject are evidence of the continuing study of this apparently frustrating feature of the Laws of Nature, for creatures of the size of men! Allan's machine could, quite properly, however, be described as an early 'linear' motor.

It is interesting also that Allan's engine, which undoubtedly 'looked like an engine', was preceded by both Fox Talbot's and Wheatstone's linear machines, which looked much less like mechanical engines and which effectively used a linear-to-rotary device the opposite way round—that is, to convert rotary motion, by rolling friction, into linear traverse.

1.10 Steps forward and backward

About 1857 Professor Holmes' 'Magneto-Electric' machine used a multiplicity of type (5) machines (type (5a)). One such machine was assessed and approved by Faraday himself. In a later version (1871) Holmes used six discs and seven rows of horseshoe magnets. Each row consisted of eight horseshoes and the whole machine therefore carried 56 horseshoes of such a size that the overall diameter was over 6 feet and the length 7 feet. From such a machine which is now seen to have inherited the worst features of types (3) and (5) an output of less than 2 kW was obtained. Nevertheless it was a milestone in its own way for it was largely responsible for the introduction of arc lamps in lighthouses and was used for Souter Point lighthouse for many years (see figure 1.18).

Overlapping the 'lighthouse era' and following it there was a period of concentration on electric circuits and in 1856 Siemens wound a rotor like the shuttle in a loom, a cross-section of which is shown in figure 1.19. These shuttle armatures were to survive even to the present time (in very small machines) but did not lend themselves readily to subdivision for the purpose of heat distribution. They represent, in one sense, the 'opposite end of the scale' from Faraday's disc in which infinitesimal subdivision of rotor conductor obtained. Large modern machines with commutators find an optimum arrangement between these extremities. The first recorded use of this type of rotor which is perhaps the first 'slotted rotor' is that it was incorporated in a dynamo by Ladd in 1856 which had a permanent horseshoe magnet stator.

22   A HISTORY OF LINEAR ELECTRIC MOTORS

Figure 1.18 The Souter Point Lighthouse generator installed in 1871 (*courtesy of Trustees of the Science Museum*)

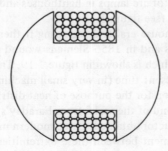

Figure 1.19 Siemens' shuttle rotor, 1856, in cross-section

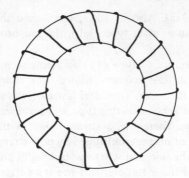

Figure 1.20 Pacinotti's ring-wound armature, 1860

A new concept of electric windings for rotors was presented by Pacinotti in 1860. His ring-wound armature shown in figure 1.20 made possible semi-infinite distribution of rotor current with its inherent uniform spreading of rotor heat. This type of rotor was also to find much favour, even to the present day. It is surprising that Faraday's induction experiments of the 1820s in which he used precisely this type of construction did not warrant a second look until 1860.

In 1870 a Belgian named Gramme produced a machine using the ring winding for the rotor and somehow *his* name, rather than Pacinotti's became associated with these ring windings. While not *fundamentally* different from the later 'surface' or 'drum' windings so far as the electromagnetic action on the working surface is concerned, their linear equivalents have a profound topological difference from surface windings. In the rotary machine, all return or 'non-useful' portions of the windings (the term 'non-useful' being often misleading, for it insists that an e.m.f. is induced in one *part* of a circuit and not in another, and this concept can at times prove to be a very false friend) are bunched together in the bore of the rotor and, since 'what goes *in* must come *out*', the net current in the bore of an a.c. rotor of this type is zero. In linear machines, the return parts of a 'Gramme-ring' winding are not bunched and hence give rise to high leakage reactances. Incidentally Gramme's machine had its rotor axis turned through 90° from the layout shown in type (3aa) so that the shaft lay in the plane of the diagram.

In 1867 Wilde revived Stöhrer's type (2) by using a double-sided magnet system, type (2b), and a rotor which consisted of a co-axial set of coils with iron cores. In the same year Dickenson, no doubt trying to avoid the very large diameter, short length machines which almost inevitably resulted from the use of a multi-pole type (3), reduced the number of poles to two and extended his horseshoe electromagnet to that of type (3a). These machines were usually mounted with the back of the horseshoe horizontal and field windings on the side limbs (this last feature having the effect of reducing leakage). If the rotor was above the field coils, the machine was known as an 'over' type, or if the rotor was below, an 'under' type. Siemens used a horizontal (3a) type but fed the working surface magnetically from both sides, as in type (3aa). Also in 1867, Ladd used the flux from a type (3a) twice, having twin armatures as in type (3ab), the one providing the excitation for the other.

The large diameter feature of type (3) also seems to have inspired the arrangement of type (4) with multiple magnets 'folded over' to make the overall diameter little more than that of the rotor. The latter however was still of short axial length. A Gramme-ring armature version of this type was produced in France in 1881 (see figure 1.21). At this time there was a strong belief in the potency of 'magnetic moment'—that is, length of magnet multiplied by pole strength—and the fashionable philosophy was to take the line that what was lacking in pole strength (the machine world was still waiting for the metallurgist to give it a better ferromagnetic material) could be compensated by use of long magnets. Field winding cores over four feet long were not uncommon and it is on record in the *Journal IEE* (1922) that it was a good idea to have long poles "to give the flux a good run at the airgap"!

1878 is another landmark in that the Brush dynamo, using a type (2b) with bobbin coils like those of a Wilde alternator, removed the iron cores from the rotor coils and, for the first time, used laminated iron in the stator part of the magnetic circuit. It is recorded as an almost *casual* comment that: "In larger machines the

Figure 1.21 A permanent magnet stator system in which the magnet system was not radial and co-planar (*courtesy of Trustees of the Science Museum*)

armature* was laminated, an improvement which increased the output by 50%." What publicity would be made today of a commercial machine with an output of 1.5 times its predecessor?

Siemens, in 1880 made a type (2b) using a Wilde stator and Gramme-ring rotor with steel core. The first age of topology was now virtually at its height; many of the 72 possible topological arrangements listed earlier with reference to figures 1.6 to 1.8 became real pieces of hardware rather than merely ideas. For example, a dynamo by Pilsen and Joel in 1885 was employed to light Bethnal Green museum — a type (2b) machine, the rotor being iron-cored and Gramme-ring wound, and therefore differing from the Brush, Gramme and Ferranti machines.

One has the impression on reading the literature of the period that it was all being done 'for fun' and in the spirit of 'mine's better than thine' — of friendly rivalry and infinite fascination. Pioneers of what was to become one of the most universal businesses of the modern world, they were as yet unhindered by the need to make their machines 'fit the system' for there *was* no system fixing the voltage or frequency (if a.c.) with which a machine design must comply. They were making their own rules and jargon as they went along. Much of design technique was empirical, pure technology rather than anything mathematically inspired.

I feel entitled to include, by way of illustration of the previous paragraph, a few quotations from the IEE's Proceedings at Commemoration Meetings (*Journal IEE*, 1922).

My first extract was related by Crompton and refers to Kapp:

"Having seized on this I went back to my own designer (Kapp) and told him we must cut down the air-gap. "Oh," he said, "you will diminish the space needed to get the necessary number of turns on the armature, and you will not get the electromotive force you require." I urged that if we made our magnets and our armature cores of largely increased cross-section we should get such an improved magnetic field that one layer of winding on the armature would give us the E.M.F. we required. But Kapp would have none of it. Luckily, Kapp's holiday was then due and during his absence I made a dynamo with huge field magnets of Swedish iron, with a huge armature core-ring of soft-iron plates pushed on to spiders and with only one layer of armature winding, so that the air-gap on which so much depended was reduced in this machine to a little over $\frac{1}{8}$ inch on each side, the double air-gap being thus only $\frac{1}{4}$ inch. This machine was finished and on the bed-plate when Kapp returned. When the test commenced, owing to the fact that the large mass of iron had but little residual magnetism the magnetic field took some time to build up, so that at first there was only a very small E.M.F. in the circuit. Kapp said quite triumphantly, "I told you so, you have wasted your money; you have nobody but yourself to thank." But even as he spoke the field began to build up and the E.M.F. mounted

---

*The word 'armature' has been used throughout the history of machines to indicate stator, rotor, primary or secondary almost at random. The word is therefore avoided as much as possible in this history, so as not to confuse; but in this case, since the machine was a generator, the stator is clearly the armature, for the rotor contained no steel.

rapidly, the testing resistance became red hot and the belt came off; but the modern dynamo was born. I only knew later from what Hopkinson himself told me and from what Wordingham will tell those present here, that Hopkinson had been working on identically the same lines as I had, neither of us knowing what the other was doing. I showed a dynamo of the new form at the Inventions Exhibition and the judges gave me a Gold Medal."

The second quotation is from a speech by Sir Oliver Lodge:

"The electrified rail was about one foot off the ground, on little posts by the side of the railway, and the railway ran by the side of the road. Of course, this rail sometimes went past gates and once it killed a cow. I forget the voltage, but quite a nice, harmless shock could be got from it. There was nothing to prevent people from touching the live rail, and FitzGerald and I amused ourselves by taking these shocks. Lord Kelvin, however, could not feel them at all. He asked us what we were jumping about for. We said, "You try it." He tried it, gravely pressing both hands on the rail. After a pause he remaked, "I think I do feel something." I suppose his skin was so dry that a moderate number of volts had no effect on him. It reminded me of the old charwoman in 'Three Men in a Boat'. Many will remember the incident about a cheese, which was so strong that it made the cab-horse gallop when they were taking it away to bury it, and the charwoman said yes, she did notice a something."

Finally, this gem from Mr G. W. Partridge:

"To give another example of accidents; at one time an arc of 10 feet length was started up between the terminals of a transformer and the iron roof of the building. One of our men threw his hat at the arc and thus put it out."

1.11 Shape becomes an end in itself

One of the great men of his time was Sebastian Ziani de Ferranti. An alternator which he designed with A. Thomson in 1882 was the prototype of the 1000 kVA alternator for the Deptford scheme built in 1889. This was a type (2b) with steel in the rotor. The rotor coils consisted of a single continuous ribbon of insulated copper. The machine generated at a frequency of 80 Hz.

In 1885 an arc-lighting dynamo by Thomson and Houston introduced a new dimension to large machine design, *literally*—it was the third dimension (axial), for the rotor of their machine was spherical. Not since Froment's machine of 1845 had this dimension been exploited and not until 1957 (see Williams *et al.*, 1959) was it to be so again. The strange thing about this design however, is that it is difficult to see what benefit was derived from the spherical shape, except that of a satisfaction in elegance in that the sphere was the 'perfect' shape for a solid body—topology for its own sake perhaps? The machine now in the Science Museum, South Kensington, has been cut away to show the construction (figure 1.22).

Figure 1.22 The spherical Thomson–Houston dynamo of 1885. Part of the stator has been cut away to reveal the shape of poles and rotor (*courtesy of Trustees of the Science Museum*)

1.12 1888—the biggest milestone since 1831

A Crompton generator of 1888 marked the end of the Gramme-ring wound machines, its rotor being in effect a half-way stage between ring and drum windings, but this year was far more notable for the achievement of Nikola Tesla, who constructed his first successful induction machine. Tesla was different from the rest. While others enjoyed their inventing and the friendly rivalry which went with it, Tesla was in deadly earnest. He neither married nor had friends, declaring both to detract from a man's performance. He did however enjoy a very special mental relationship with his mother and it is said that they practised the art of conversation by telepathy very successfully. At times he was both poverty stricken and despondent about his ideas. He worked as a road digger for a time. At one stage he deduced that a rotating secondary could only produce a perfect rotating field (that is, circular not elliptic) when rotating in exact synchronism, by which time there was no torque developed, by definition, so he felt it impossible that induction motors could ever be useful. (We know now that sizable torques are developed by the crudest of elliptic fields.)

Tesla was apparently unaware that in 1879 Professor Walter Baily had propounded an experiment in which four electromagnets were arranged at 90° intervals below a conducting disc which was free to rotate. Although a.c. was not used, the electromagnets were fed from a battery via a hand-operated rotary switch and undoubtedly produced a two-phase rotating magnetic field. The device was first demonstrated at a meeting of the Physical Society in London on 28 June 1879.

When asked about its use by one of the audience, Baily replied that he regarded it only as a scientific toy. Tesla's intention was always that a brushless machine should dominate the world of electric drives as indeed it does, and nowhere less than in the most recent developments in high-speed ground transport where the vehicles are required to free themselves entirely from contact with the ground (chapter 7).

The importance of Tesla's invention can be judged by the fact that the Crompton-Brunton alternator being developed in 1890 was seen to be capable only of generating single-phase current and was therefore abandoned because Tesla's later induction motors of 1889 required three-phase current. The first three-phase power transmission was demonstrated between Laufen and Frankfurt in 1891 when Oscar von Miller transmitted 240 kW at 15 kV over 110 miles. The first three-phase public supply was from the Wood Lane generating station in West London, which began operating in October 1900.

By this time, type (3) topology had been firmly established, even for magnetic machines, and a Pyke-Harris reluctance alternator of 1892 was undoubtedly the forerunner of the aircraft alternators of 1950 (figure 1.23). The first half of the twentieth century was to see the abandonment of long poles, the use of axially longer machines and a most ingenious array of machine windings. But the magnetic circuit was to remain unchanged, co-planar, short, fat, laminated (a.c.) and using as small an airgap as economics allowed, especially in the case of the now all-dominant induction motors. (It has been estimated that of the total horsepower of all electric drives in the world today, over 95 per cent is supplied by induction

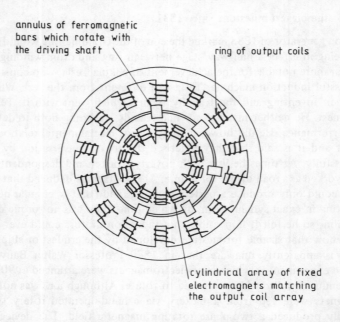

Figure 1.23 Pyke-Harris reluctance alternator schematic, 1892

motors.) The first age of topology had ended. Fashion in engineering turned to 'squeezing the last drop out of the lemon' in terms of efficiency, and in the new fashion of increased power/weight ratio. This last was the concern of metallurgists (who invented better steel alloys), chemists (who developed better insulating materials—notably varnishes) and mechanical engineers skilled in fluid flow (who developed vastly superior cooling systems). The electrical engineer, by comparison, stood aside and watched.

1.13 References

Boucherot, P. (1908), 'Appareils et machines à courant et mouvement alternatifs', *Bulletin de la Société Internationale des Électriciens*, Vol. 8, pp. 731-755

Bowers, B. (1972), 'The eccentric electromagnetic engine—a chapter from the very early history of the electric motor', *Electronics and Power*, Vol. 18, pp. 269-272

Clark, Sir Kenneth (1969), *Civilisation—A Personal View* (British Broadcasting Corporation and John Murray).

Journal IEE (1922), 'Proceedings at the meetings held in commemoration of the first ordinary meeting of the IEE in 1872', Vol. 60, No. 308, pp. 377-500

Laithwaite, E. R. (1966), *Induction Machines for Special Purposes* (Newnes)

Laithwaite, E. R. (1967), 'The moving window motor', *Electrical Review*, Vol. 181, pp. 278-279

Laithwaite, E. R. (1971), *Linear Electric Motors* (Mills and Boon)

Lenihan, J. M. A. (1969), 'The triumph of technology', *The Philosophical Journal*, Vol. 6, No. 1, pp. 12-18

The Children's Encyclopaedia (1922), Vol. I, p. 47 (Amalgamated Press)

The Electrician (1889), 'The Birmingham Exhibition—Woolrich's magneto-plating machine', Vol. 23, p. 548

Wheatstone, C. (1841), British Patent No. 9022

Williams, F. C., Laithwaite, E. R. and Eastham, J. F. (1959), 'Development and design of spherical induction motors', *Proc. IEE*, Vol. 106A, No. 30, pp. 471-484

Appendix: Chronological order of rotating machine inventions

Date	Inventor and invention	Basic topology (figures 1.6 to 1.8)	Form of secondary
1831	Faraday's disc dynamo	1	Copper disc
1831	Pixii's a.c. generator	2a	Iron-cored, axially-wound coils and yoke
1832	Sturgeon's electromagnetic engine (motor)	4a	Bars of steel (reluctance motor)
1840	Taylor's eccentric rotor motor	3	Steel 'keepers'
1841	Wheatstone's eccentric rotor motors (i)	3	Steel cylinder
	(ii)	3	Steel cylinder } all reluctance motors
	(iii)	6	Steel disc
1844	Woolrich's electro-plating generator	2	Iron-cored, axially-wound coils and yoke
1845	Froment electromotor	1	Iron-cored, radially-wound coils and yoke
?	Apps (i)	6	Iron-cored, axially-wound coils and yoke
	Apps (ii)	2a	Iron-cored, radially-wound coils and yoke
1846	Stöhrer	6	Iron-cored, axially-wound coils and yoke
1853	Allan's magnetic 'piston engine'	Linear with cranks	Iron keepers
1856	Holmes' lighthouse generator	5a	Iron-cored, axially-wound coils
1856	Ladd's dynamo	3a	Shuttle type armature (first slotted rotor)
1860	Pacinotti's dynamo	3a	'Ring-wound' armature
1867	Wilde alternator	2b	Iron-cored, axially-wound coils
1867	Dickenson 'over' and 'under' types	3a	Gramme-ring, iron-cored
1867	Ladd double armature dynamo	3ab	Shuttle type, iron-cored
1870	Gramme's machine	3aa	Ring-wound rotor but with shaft in plane of magnetic circuit
1878	Brush dynamo, laminated stator	2b	Air-cored, axially-wound coils
1880	Siemens double-sided dynamo	3aa	Surface-wound, iron-cored
1881	de Méritens, Paris dynamo	4	Gramme-ring, iron-cored
1882	Ferranti alternator	2b	Air-cored, strip-wound, axial coils
1885	Pilsen and Joel's dynamo	2b	Iron-cored, Gramme-ring
1885	Thomson and Houston arc-lighting dynamo	Spherical	Surface-wound rotor, iron-cored
1888	Crompton generator	3aa	Half Gramme-ring, half drum
1888	Tesla	3	Short-circuited, wound rotor, iron-cored
1890	Crompton-Brunton alternator	3a	Air-cored, strip-wound, axially-mounted coils
1892	Pyke-Harris reluctance alternator	3b	Salient steel blocks
1900	Parsons alternator (the first commercial three-phase generator)	3	Drum-wound with three slip rings

2 The early inventors and their patents

2.1 The pre-induction motor period

Linear motion is common in Nature and in the motions of early steam, gas and other internal combustion engines. In the mechanical devices the 'natural' linear motion is converted to rotation by other mechanical means. The history of linear electrical machines is concerned with the influence of such reciprocating machines only to the same extent as that to which rotary electric motors left their mark, as discussed in chapter 1.

Clearly there can be no electromagnetic machines before Faraday's discovery of the laws of induction in 1831. To this extent it is surprising that the earliest linear electric motor emerged as early as 1838, for it then took over a century for any linear machine to make a substantial commercial profit. In another sense however, it is perhaps *not* so surprising that linear motors were tried so early, for their direct ancestors were beam engines and the like with pistons and cylinders.

A historian of the 1920s (although I doubt very much whether he ever saw the introduction to his paper as a potted history of 80 years of linear motor development) was undoubtedly P. Trombetta. His paper to the Spring Convention of the American Institution of Electrical Engineers held in Chicago in 1922 is one of my favourite references on the subjects of engineering in general and linear motors in particular (Trombetta, 1922). He begins his historical section thus:

"The writer invented the electric hammer early in 1915 when still at school. After coming out of school the war was in full swing and it was thought inadvisable to start developments of any kind." (How many good ideas are lost because their originator was too young to have the confidence to proceed?) He continues: "... the earliest attempts to make electric motors were mainly along the lines of imitating the reciprocating steam engine and the first patent for a reciprocating motor to be applied to a pump was issued in 1852." This patent contains the drawing reproduced in figure 2.1.

Tomlinson's *Cyclopaedia of Useful Arts*, first published *c.* 1852, however, describes earlier reciprocating electromagnet-driven engines, figs 838 and 839 of that work being reproduced in figure 2.2 (Tomlinson, 1852–54). Of this machine it is

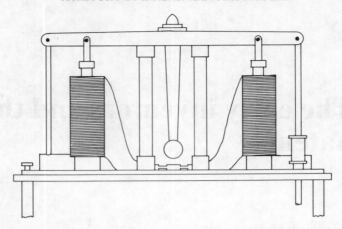

Figure 2.1 The first patented reciprocating motor applied to a pump (1852)

written: "By means of an engine on this principle, Professor Jacobi*, of St. Petersburg, in 1838 and 1839, succeeded in propelling a boat upon the Neva at the rate of four miles an hour. This boat was 28 feet long, about 7 feet wide, and drew nearly 3 feet of water. It contained 10 persons. The engine was worked by a voltaic battery of 64 pairs of plates." Later it is said, "The most powerful engine of this kind yet made has been equal to one horse." (Not in every respect, one hopes!)

The same volume goes on to mention a Mr. Llewelyn who, in 1848 "exhibited to the members of the British Association a similar experiment on a lake at his residence near Swansea." It also records that "In 1842 Mr. Davidson constructed an electro-magnetic engine, which was tried on the Edinburgh and Glasgow Railway." Tomlinson's *Cyclopaedia* concludes:

"Various forms of electromagnetic machines have been invented by Wheatstone, Talbot, Hearder, Hjorth and others, but it is not likely that such machines will meet with much encouragement, while coal, the food of the steam engine, is abundant."

The last word on the replacement of steam engine cylinders by electromagnets must surely rest with Trombetta, however, for he wrote without any bitterness, I feel sure: "The idea of plunger reciprocating motors prevails until the present time and it would be useless to recount here the innumerable patents that have been issued on the subject. It may be said however, that the idea remains unchanged and the only thing that changes is the inventor."

In the pre-induction period, dominated as it was by pistons and cylinders, it was natural that the 'electromagnetic engines' (which we now recognise as reluctance

---

*Jacobi is perhaps better known for the mark he left on the world of mathematics, the 'Jacobian matrix' consisting of partial derivatives of two functions with respect to two other functions symmetrically arranged in a 2 × 2 matrix.

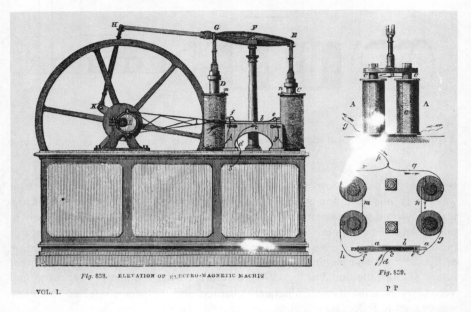

Figure 2.2 An early reciprocating electromagnet engine from Tomlinson's *Cyclopaedia of Useful Arts*

motors of a crude kind—in modern reluctance motor parlance they had but a single 'tooth') should retain piston-in-cylinder construction. They were little more than electromagnets with holes through their centres so that the iron armatures (or 'pistons', if you prefer) could be embraced completely by the primary coil and thus form a reasonably good magnetic circuit without the use of primary iron.

Nowadays, with our wide variety in electric motors, it is necessary to resort to the biologists' solution to a complexity and *classify* them. Thus the earliest types of linear motor are not to be seen as the linear equivalent of a rotating machine which has been split along a radial plane and unrolled, the change from (a) to (b) in figure 2.3, for this exercise produces the now almost *standard* linear motor which is *flat*. The re-rolling of the flat structure about a perpendicular axis as in (c) produces a class of linear motors which have had cylindrical symmetry restored but which produce their travelling fields and secondary motion axially, as it were 'down the barrel' of an electromagnetic gun. Such machines have now been classified as 'Tubular Motors', but let this not detract from the fact that although the earliest linear motors undoubtedly fall into such a class, they were inventions in their own right, based only on their true ancestors, the piston and cylinder.

When I published my first paper on linear motors in 1957 I found myself unable, for a time, to decide on whether to call the linear equivalent of the rotor of a rotating machine a 'secondary member', thereby effectively succumbing to the patent agent's necessarily (to avoid ambiguity) cumbersome language, or whether

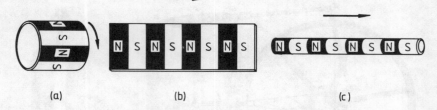

Figure 2.3 Topological developments of the two basic linear motors from a rotating machine

to retain the word 'rotor' on the grounds that no confusion could possibly arise when only linear motion was involved. I was flattered that other writers, in the main, followed my example and new words such as 'linor' are to be found only in one or two remote points in the literature. But with the moving primary motor becoming a natural choice for high-speed transport applications and the invention of hybrid 'rack and pinion' motors in which one member was linear and the other rotary (Laithwaite and Hardy, 1970), a return to the terminology 'primary' and 'secondary' was inevitable. Subsequent chapters and the papers referred to therein will be seen to reflect these trends.

2.2 A.c. reciprocating motors

Trombetta describes a linear asynchronous polyphase induction motor for a rock drill, appearing in a patent of August 1901 issued to R. D. Mershon. This patent states that earlier linear motors have been proposed in which reversal was effected by means of a switch connected to the generator. Such systems are condemned by the inventor as being able to start only if the natural period of the moving part is related to the generator speed, and Mershon prefers instead to have the switch operated by the rotor. This patent puts the date of the first linear induction motor at least prior to 1901.

Trombetta goes on to describe a patent issued to P. Centener in 1905, in which: "each stator is fed with two separate sources of alternating current of different frequencies and in this way the motion of the field is reversed at the beats occurring in the two frequencies, so the moving part would reverse its motion without the necessity of reversing the current by means of contactors or switches." This patent is undoubtedly the forerunner of the West and Jayawant Oscillator described later. In typical style, Trombetta concludes on this aspect of the subject: "The success of these inventions in rectilinear motors is apparent from the fact that none of them are at the present time on the market."

His philosophy on the subject of special-purpose motors could well serve as a model to certain sections of heavy industry in the 1980s, for he wrote: "It may be said further that in the early stages of electrical development, the total demand for motors was very small and of this total demand, the percentage which required an ordinary standard motor was far greater than that which required a special straight-line motor, so that it was altogether out of the question for any manufacturer to undertake the development of as many special motors as there were special cases to be taken care of. At the present stage of electrical industries, the conditions are precisely reversed in that there are now many special cases which require a much larger number of motors than was required for the total electrical industry in the earlier days of the art and certainly anybody who is acquainted with the industry will agree that many manufacturing enterprises are now existing, and in fact prospering, in the manufacture of only one or two special appliances. It is for this reason that it is felt justifiable to state that the application of standard rotary motors to special cases has gone beyond its limits." A man ahead of his time. Forty-four years were to elapse before the present author, then unaware of Trombetta's philosophy, produced a book devoted almost exclusively to linear machines under the title *Induction Machines for Special Purposes*. I hasten to add that this is *not* the reason for Trombetta being one of my favourite authors!

Between 1850 and 1888, nothing strikingly significant was invented in the way of linear motors. There was a proposal to use one in a sewing machine, there were two patents on a solenoid rock drill in 1880 and a patent by Van Depoele in 1885 in which he carried the exciting winding of a d.c. solenoid on the moving part, and this last idea seems to have created some excitement at the time. It was described by Trombetta as "a well-developed idea on straight-line motion".

2.3 The start of the induction motor era

On March 6, 1891 Professor J. A. Fleming delivered a most important Discourse at the Royal Institution which was subsequently published as a paper in *The Electrician* (Fleming, 1891). In this he introduced subjects which were to have a major influence on research into high-speed vehicles 80 years later and this aspect of it properly belongs in the chapter on that subject. The fact is that his apparatus consisted of objects of great curiosity at the time. One piece in particular has aroused many an undergraduate from slumber in an otherwise unillustrated lecture, which assures it of a place in this chapter also. I refer, of course, to the famous 'jumping ring' experiment, the basic tubular motor, in which a conducting ring is thrown off the extended straight core of an a.c. magnet on switching on. Among other things the jumping ring represents the first tubular induction motor. Fleming's lecture was a landmark in the study of induction, despite his modest and generous recognition of the work of others, declaring the whole to be based on the work of Ampère and attributing the invention of the jumping ring, electromagnetic levitation and that most important phenomenon of the shaded pole effect to Elihu Thomson, who was clearly as great a hero and inspiration for Fleming as Faraday is for me.

There were many great principles and fundamentals disclosed in novel and simple ways in this Discourse, which indicated the depth to which Fleming appreciated the subject of induction. Thus, one reads such sentences as: "The secondary coil is, in fact, an electromagnetic eye which can *see through* an inch of wood, but to which a sheet of zinc is semi-transparent, and a thick sheet of copper quite opaque."

The language in which people of those days wrote their findings was never dull. Would that many of our modern authors could drop their scientifically correct 'suits of impenetrable armour' if only for a few sentences in a paper and say such comparable things as Fleming's "All good conducting rings will execute this gymnastic feat, and rings of copper and aluminium are found to be the most nimble of all." His use of analogue was free-flowing and brilliant. A resilient mechanical model to illustrate the nature of self-inductance to a not entirely scientific audience —"If rings of different metal and equal size are placed on a tube, they float at different levels like various specific-gravity beads in a liquid"—and the following quotation are both examples of this: "There is a symmetry in the relations of magneto-motive force and the magnetic induction and electromotive force and electric current, and we can, as Faraday pointed out long ago, make the symmetry complete, if we suppose the two interlinked magnetic and electric circuits immersed in an imperfectly conducting medium." Why, I might reasonably ask, was I not taught these easy-to-appreciate analogues and duals? In what I regard as my most important paper 'The goodness of a machine' (Laithwaite, 1965) I was re-creating the same concept of that circuital interconnectedness which is surely the very bedrock of machine studies and was first put together in this paper by Fleming, which was unknown to me until after I had written my 'goodness' paper. Fleming's Discourse should be compulsory reading for *all* undergraduates in electrical engineering, for it appears to have been written especially for those who feel oppressed by the formal lectures to which they are subjected in the 1980s and which suggest both that 'exciting research has all been done' and that 'electromagnetism is a dull subject anyway'.

Fleming declared of such students: "They are in a fertile field both for the investigator seeking to add to the sum total of existing knowledge, or to the inventor in search of applications in electrical technology for such acquired facts."

2.4 Boucherot, Zehden and Birkeland

The first decade of the twentieth century brought its own 'giants' in electric motor development, one of whom was the Frenchman, P. Boucherot (1869-1943), the kind of engineer who could design a turbo-alternator *this* month and the turbine to drive it in the next. A great protagonist of linear motors, he declared his frustration at the fact that with the maximum values of magnetic induction (B) and current loading (J) being what they were at that period, the normal magnetic pull on a motor surface $(B^2/2\mu_0)$ N/m^2 was of the order of 20-40 times the tangential thrust (BJ) N/m^2, yet only the latter was apparently usable in a machine with continuous rotary motion. Not to be entirely thwarted, he developed reciprocating machines in which he could utilise the magnetic (B^2) forces, and converted this into rotary motion by ratchet. His largest and most nearly commercially acceptable machine is shown in figure 2.4. It is interesting to note the re-appearance of an improved machine (with transverse flux) of this type in the 1970s (Hesmondhalgh *et al.*, 1973).

Figure 2.4 Boucherot's largest reciprocating 'magnetic' ratchet motor

Boucherot studied natural history in much the same way that several authors have done some 60 or more years later (Gamow and Harris, 1972; Laithwaite, 1973). Of a multiplicity of miniaturised magnetic machines he declared the possibility of making a conveyor system which operated "in the manner of a serpent" (its skin, that is).

Yet Boucherot is remembered in electrical machine development history mostly as the father of the double-cage rotor for induction machines, pioneering work which was later developed into the deep-bar rotor, the cast aluminium cage machines, in fact to the whole philosophy of direct, on-line-start induction motors. If for no other reason, the influence of this work on the design of the linear motors of the 1970s entitles him to a place in the history of these machines. His enthusiasm for the electromagnetic production of straight-line motion directly ensures him such a place.

While Boucherot was an engineer and philosopher and his ideas undoubtedly were over half a century ahead of their time, Zehden, whose work was published over roughly the same period as that of Boucherot saw the use for the linear motor as immediate. Like his contemporary, he was aware of the frustrating magnetic pull, some 30 times the useful thrust, going to waste, but whereas Boucherot had tried to *use* it, Zehden saw it as a fundamental limitation to linear motor development and invented means to *avoid* it. His reasons were sound. If a 50 ton vehicle was to be accelerated at $0.1\,g$, 5 tons of forward thrust must be associated with about 150 tons of downward magnetic pull—that is, the vehicle appears to 'weigh' 4 times its natural weight.

Half a century later, the subject of magnetic pull still carried a high significance in design offices, for rotary machines are never perfect, and especially in respect of cylindrical symmetry. In a perfectly centred rotor the radial pull should be completely balanced all round the machine. In a practical machine, that which is not is known as Unbalanced Magnetic Pull (UMP) and even in the 1960s the mystic figures which were included in design office procedure were not well understood, as exemplified by a paper which came into my hands for refereeing and contained the formulae of several firms for this particular quantity. Such was the diversity of opinion between different firms that for specific running conditions two particular formulae gave values for the UMP in the ratio ten to one! Since these figures had a bearing on the shaft diameter, their calculation could not be dismissed as a trivial one.

Zehden may or may not have gone through the same mental processes as I did in the 1950s, but certainly the end-product was the same. Figure 2.5 illustrates the stages in my argument. Initially I called my machines 'sheet-rotor motors' but since in high-speed traction the secondary was anything but a 'rotor', the term was soon replaced internationally by the title 'double-sided sandwich' motors.

Zehden was undoubtedly the 'father' of this topology (Zehden, 1905) and his first patent (applied for in 1902) was to be a milestone in the history of the subject, even though he, like Boucherot and many others before him, right back through Leonardo da Vinci to Moses, were only to see the Promised Land from afar!

Almost coincident with Zehden's patent on flat motors for rail traction came an even more ambitious series of five patents by Professor Kristian Birkeland (from September 16, 1901 to April 23, 1903), for they proposed to apply a d.c. linear reluctance motor as a noiseless gun. Today with our knowledge of the difference between electromagnetic machines and purely magnetic machines such as reluctance motors, it is easy to condemn Birkeland's work from its outset since it was an attempt to apply a magnetic machine to an application of the largest size (certainly in terms of power). Its inclusion here serves to emphasise further that the men of the 1900-1910 decade were certainly 'thinking big' in terms of linear motors. We may reflect on what a cornucopia they might have poured out had they had the advantages of modern materials and the results of changing 'fashions'.

2.5 Bachelet and World War I

Wars, however bad they are, have a habit of stimulating humans to heights that otherwise would apparently have to wait for decades. One need only contemplate the products of World War II in respect of jet engines, radar (the beginnings of computer technology) and medical advances for illustrations of this statement. It is also my own experience that man's compassion for his fellow man rises in at least as high a proportion as does his inhumanity to man during times of war. It is as if Good and Evil walked hand-in-hand.

At the start of World War I, a most ingenious engineer, Monsieur Bachelet, attempted to take the linear motor by the scruff of the neck and *make* the engineering world notice it. He set up companies to exploit, in particular, the properties

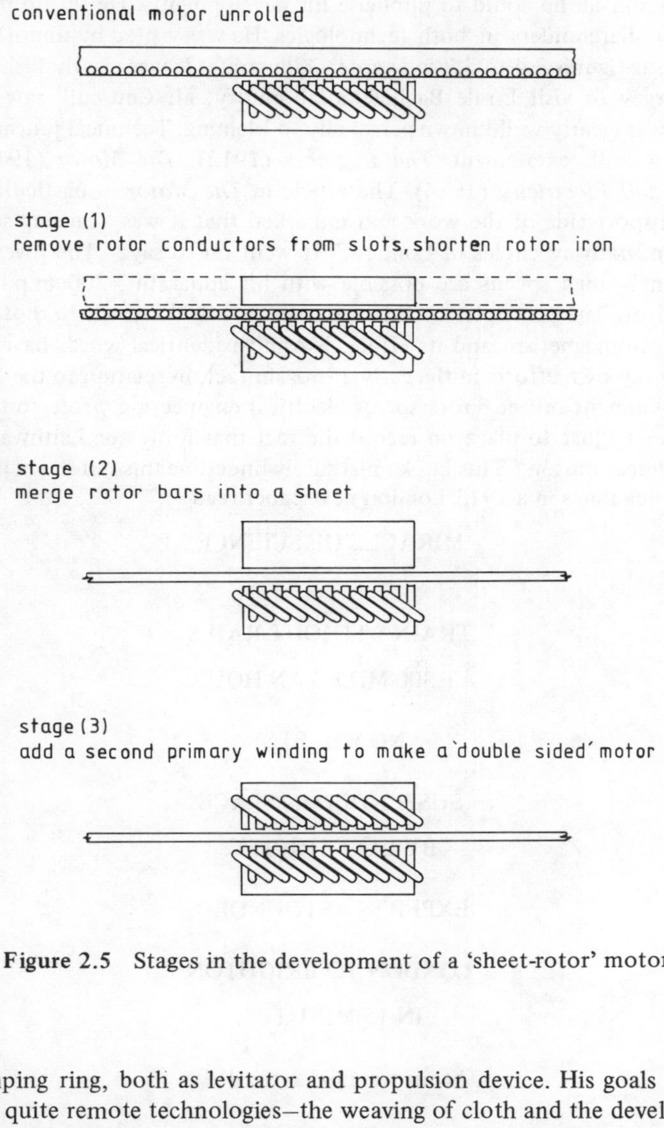

conventional motor unrolled

stage (1)
remove rotor conductors from slots, shorten rotor iron

stage (2)
merge rotor bars into a sheet

stage (3)
add a second primary winding to make a 'double sided' motor

Figure 2.5 Stages in the development of a 'sheet-rotor' motor

of the jumping ring, both as levitator and propulsion device. His goals lay in two apparently quite remote technologies—the weaving of cloth and the development of a levitated, high-speed railway. Readers who are familiar with my own efforts in these two fields may well ask whether my interest was indeed focussed by Bachelet's work, but the plain fact is that I was already concentrating on these two subjects before I ever heard his name. Can it be that the linear electric motor is as 'natural' a choice in these respects as apparently is the wheel to civilisation?—I think not, so far as shuttles are concerned, for it is still an uphill battle for those who would commercialise the linear motor for textiles, but high-speed ground transport is another matter.

Bachelet did all he could to publicise his developments. He set up public companies with shareholders in both technologies. He was visited by important people of the day, as figures 3.4 to 3.7 in chapter 3 illustrate. It was clearly fashionable for the aristocracy to visit Emile Bachelet's laboratory. Mr Churchill saw his experiments. He was clearly well-known personally to Fleming. Technical journals referred to his work with excitement—*The Engineer* (1912), *The Motor* (1914), *Junior Mechanics and Electricity* (1914). The article in *The Motor* in particular reported on the transport side of the work and remarked that it was "causing some stir in scientific and railway circles in London." It went on to say: "The inventor claims that extremely high speeds are possible with his apparatus, 300 m.p.h. being, in fact, talked of. The principles of this apparatus are well known to those who have studied electromagnetism and its effects." Almost identical words have been used to describe my own efforts in the early 1960s, in fact, in relation to the latter statement, one eminent retired Professor of electrical engineering wrote to the *London Evening News* "just to place on record the fact that Professor Laithwaite did not invent the linear motor." This book must surely underline this statement adequately!

Typical headlines in a 1913 London newspaper read

MIRACLE OF SCIENCE

TRAIN WITHOUT RAILS
AT 300 MILES AN HOUR.

NO WHEELS.

SUSPENDED IN SPACE
BY ELECTRICITY.

EXPERTS ASTOUNDED.

LONDON TO BRIGHTON
IN 15 MINUTES.

The air train has arrived! . . .

History shows us that, at least in the mind of the imaginative popular journalist, there has been no advance whatever in over 50 years, for compare the above with the *Daily Telegraph*, 1 May 1968:

300 mph HOVERTRAIN IN
BRITAIN BY THE 1970s
British Rail disinterest.

In respect of high-speed transport, Bachelet was too far ahead of his time to be successful. He was for high-speed ground transport what Charles Babbage was for computers. His work deserves more than a cursory glance and therefore is treated more fully in chapters 3 and 7 as appropriate. It will suffice here to say that although his ideas on transport may have been stimulated by the patents of Zehden, he first conceived the idea of an object in free flight—levitated, guided and propelled by the mystic forces of electromagnetism which Faraday had revealed almost a century earlier.

Whether it was the war, or purely the economics of the day which brought an end to his activities I have been unable to discover.

2.6 Japolsky

There was a shortage of linear motor enthusiasts between the World Wars, apart from the textile men (chapter 3), and the work of Dr N. Japolsky is therefore of special merit, for like so many of the 'greats' he was concerned for the future and his predictions included examples both of great foresight and of 'missing the big one'. Of his several papers, the classic is probably that of 1931 (Japolsky, 1931) in which he produced a number of new names to describe linear technology, none of which however, 'stuck'. For example, he called the linear motor itself a 'magnetofuge'. The secondary (moving) part, corresponding to the rotor of a conventional motor, he named the 'runner'.

The applications he dismissed as impractical included that to rail traction. Quoting Linev and Zelenay (but not Zehden or Bachelet) as having suggested the idea "at the beginning of the present century," he went on: "The magnetofuges had to be fixed underneath the trains: the rails were supposed to play the part of fixed runners. The commercial application was handicapped by the necessity of having a considerable air-gap for practical purposes, hence too great a reluctance in the magnetic circuit." But the big fish that really got away was when he considered the possibilities of high-speed transport. I can do no better than quote his paragraph on this topic verbatim.

"I suggested, in 1914, a traction on the same lines, but so arranged that the car should be suspended on the rail by the force of magnetic attraction without touching the rail. Special arrangements were suggested in order to stabilise this force. The speed might be very high, as there is no limitation due to the centrifugal force on the wheels. The calculation gave a quite favourable result, as to the possibility, and even the economic practicability, of such a traction. That, however, was before the development of aviation. Now, I think, the development of this idea has hardly any *raison d'être*, as air traffic solves the problem much more easily."

Alas, poor Japolsky, I knew him well! He was to die before oil began to influence world transport. 'The irony of fate' is an overworked expression I know, but how better is one to comment on the fact that if the system of the magnetically suspended vehicle using a feedback amplifier (Japolsky's 'special arrangements')—which was so vigorously developed in the 1970s in Japan, Germany and in the UK by British Rail—succeeds, it may be largely because of the *failure* of aircraft to meet the needs of the 21st century for short-haul internal flights. We have made nuclear submarines but we shall be a long time (if ever) before we make a nuclear aeroplane, for the submarine was forged as a weapon of war and aeroplanes are outdated as such. Aeroplanes however, have whetted our appetites for high-speed inter-city travel, and we will settle for nothing less in the way of inter-city speed when the last internal air flight is cancelled.

My suggestion above (in my Shakespearean mis-quote) that I knew Japolsky well is not quite true. But know him a little I did, for I was dining at the Savage Club in London as the guest of Fred Barwell when this small distinguished-looking man entered the dining room. Without explaining who he was, my host at once got up and brought over for introduction this elderly Russian with bright penetrating eyes. The year, I think, was 1964 and the topic of the moment was 'Sputniks'. He opened the conversation by saying how interested he was "to see a young man like yourself taking up a subject where I left off in my young days." Then his eyes brightened and pointing upwards he said quickly: "I think I know how they put their Sputniks in orbit." "Really?" I queried. He replied indirectly: "In my student days in Russia I had a friend named - - -*. We worked together on linear motors. I know that when I left Russia he continued the work all the time until now. Last year he was elected a Fellow of the Soviet Academy of Science for this work, so I think I know how they put their Sputniks in orbit." Considering that the first American space shot had yet to be made it was a plausible idea, for it was said that all the Russians' vehicles had gone into identical orbits because they were launched up a steep hill and not vertically. History records, I think, that Japolsky's speculation found no support.

In his earlier years he had quite a number of 'firsts' to his name. For example, in making a forging hammer he accelerated the field by imposing a frequency/time programme on the stator coils. His 8 cwt forging hammer is shown in figure 2.6. He pointed out that: "If you take a powder containing magnetic ingredients, the latter will be carried out by the magnetofugal field." This anticipated the 'rack and pinion' mechanism outlined in my paper of that title (Laithwaite and Hardy, 1970) and used, without being understood, in separating iron particles from slurry (McCarthy, 1922).

He was probably the first man to propel liquid metal (mercury) and the versatility of his mind can be seen by the agile leap from liquid to gas: "I have used this phenomenon (mercury pumping) for the purpose of demonstrating the properties of magnetofugal fields to students, but I think it may also be utilised for a practical purpose, viz. of ejecting air if we insert in that mercury stream suitably shaped ends of tubes connected with an evacuated space. . . . Now we shall consider the possibilities of application of the magnetofugal field to communicating

---

*I regret I cannot produce the name, for it was complicated and I remember at the time making no effort to try to remember it!

Figure 2.6 Japolsky's forging hammer

motion to electrons and ionised atoms, and even to neutral atoms. This field of application is quite new, and, I believe, very promising." So promising was it, in fact, that most of what followed in this direction is outside the scope of this book!

2.7 Liquid metal pumps

The first linear motors to achieve anything resembling commercial success were machines for 'Very Special Purposes', indeed in the 1930s it was a 'way-out' idea to use liquid metal for the rapid removal of heat from a locality. However the thermal capacity of a metal is high, and if its movement can be effected without physical contact—that is, without a conventional pump—a high degree of reliability can be obtained. Such reliability is required in highly contaminated atmospheres and especially that particular kind of contamination which at that time was a piece of 'things yet to come'—radioactive contamination. An early pioneer of those days was the late Mr R. D. ('Cannon') Ball, a great character who insisted on being addressed as 'Cannon' by masters and servants alike. He would proudly show you an enormous pocket on the inside of his jacket which he claimed was large enough to smuggle a blueprint out undetected! What he really used it for I never knew. It was always empty when I saw it. As for stealing other people's blueprints I would have thought it much more likely that other people would have wanted *his* sketches, for *he* was blazing the trail in linear motors at that time.

Liquid metal pumps were the first real linear motors to break the 'fashion' barriers of the time and become the first of the revival of the 'to hell with efficiency, can it be done at all?' slogan (see chapter 4). The liquid metal had to be contained in a thick walled tube of a non-corroding, non-alloying metal—that is, stainless steel. The magnetic flux had therefore to penetrate the tube walls before penetrating the metal itself, and despite their high resistivity the walls incurred heavy ohmic loss. (Stainless steel is not however as bad as its name implies, for some 'recipes' for it produce a more paramagnetic than ferromagnetic substance with resistivities many times that of copper.) A.c. liquid metal pumps had skin effect in the liquid itself because the tubes had such large bores, and the liquid suffered turbulence as a result of 'competition' between the linear motor which would have liked the liquid to flow fastest on the outside of the column, and the laws of hydraulics which wanted it fastest in the middle. When all its troubles were added up, few pumps achieved an efficiency much above 30 per cent and in at least one instance of which I have personal knowledge, they acted as a deterrent to adventure into Linear Motor Land, it being implied quite wrongly that all linear motors would have low efficiencies 'because they had big airgaps'.

The man who wrote the classic paper on liquid metal pumps was Dr L. R. Blake (1957). He classified them into three types, FLIP, SIP and ALIP, which stood for 'Flat Linear Induction Pump', 'Spiral Induction Pump' and 'Annular Linear Induction Pump'.

The first type is self-explanatory, the last is another name for tubular motors, but SIP requires further explanation. The liquid-carrying tube was constructed in the form of a helix, like a huge, closely wound tension-spring, and the whole inserted into the bore of a conventional induction motor stator. The fluid was thus constrained to move at a small angle to the travelling field direction, a feature later to become a stepping stone to the theory of short primary linear motors via a search for the brushless induction machine in the form of a sphere (see chapter 6). What a tangled and twisted story is that of a single piece of technology! I should add that, with FLIP in particular, not only true induction pumps were tried. D.c. equivalents in which the 'brushes' consisted of long copper bars in the wall sides (extending the full length of the d.c. magnet poles) were constructed. The liquid in such cases acted as its own commutator, having an infinite number of infinitesimal segments; surely the most perfect commutator ever made?

From liquid metal pumping to liquid metal stirring is a relatively short step. Before a liquid secondary had ever been tried there was reason to doubt whether it would work at all, for one can imagine the 'law of natural cussedness' working so as to break up the liquid into strata, rather like the process of lamination, in just such a way as to minimise current flow and force. For example, an annular tube containing mercury, when threaded on to a magnetic circuit carrying alternating flux, will be seen to contract at its narrowest point with increasing rapidity until the column actually breaks, producing a spark. Gravity then restores the connection, thereby reproducing the effect at some other point around the ring. This phenomenon is given the appropriate name—'pinch effect'. However, the day a linear motor first pumps is the day stirring and levitation of liquids also become possible (see chapter 5). Stirring liquid metal meant little more than fitting a set of polyphase coils under the vat of liquid and allowing the a.c. field to spread upwards in the liquid (increased resistivity due to high temperature coming to the rescue where skin effect might otherwise have put the scheme 'out of court'). More recent methods of liquid metal stirring are discussed in chapter 9.

2.8 World War II

Electrical machines were relatively in the doldrums so far as inventions were concerned, and academics, believing all the important machines to have now been invented, turned their attention to analysis, almost to a man. The two cornerstones of such philosophy were undoubtedly tensor analysis, which was literally 'preached' in the market place by the late Gabriel Kron (1942) and, to my mind, the more useful Generalised Machine Theory, first put in order by the American, R. H. Park (1929, 1933) and subsequently much developed by my friend and former colleague, the late Bernard Adkins (1957). Brilliant academic work though this was, it did tend to confirm that all profitable electric machines were now known and an electrical engineering student who said he was doing research on 'machines' meant 'computing machines'!

Of the only two linear motors of any significance to come out of the War years, one of them almost drove the last nail into the coffin of machine invention for it was short-lived and shown to have been unprofitable; another 10 years were to elapse before linear motor technology 'rose like a Phoenix from its ashes' (and the first Phoenix was not a very big one!). The project to which I refer was the aircraft launcher 'Electropult' built by Westinghouse (USA). A schematic diagram of the system is shown in figure 2.7.

The motor is at once seen to be a single-sided, straight equivalent of the conventional rotary motor, but with moving primary and elongated secondary, both necessary adaptations where 10 000 horsepower was dissipated in the active zone. Oddly enough the secondary carried insulated windings, which must have quadrupled the track cost compared with that of a sheet rotor motor—and there was plenty of track. The first to be built was a quarter of a mile long, the second, just over a mile. The performance was pretty fantastic. Acceleration of a 10 000 lb jet plane from rest to 115 m.p.h. in 4.2 seconds was achieved. Take-off speed was 225 m.p.h. after which the carriage that constituted the motor primary was brought

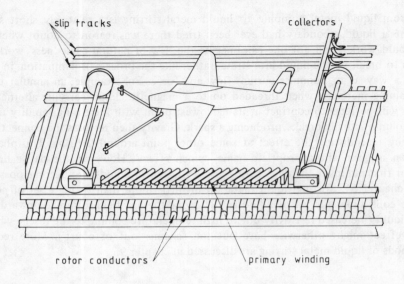

Figure 2.7 Schematic layout of the Westinghouse aircraft launcher

to rest by d.c. braking. It was certainly the fastest linear motor ever built up to that time and also, as one US journal title described it: "The Longest Induction Motor in the World."

Figure 2.8 shows the primary unit on its track. The herring-bone pattern corresponded to skewed slots in a conventional motor for the prevention of tooth-locking. The aircraft was attached to the primary by a tow cable. It was proposed to use the track for landing as well as take off, a sling from the incoming craft being caught up in a hook on the primary. The facts about the end of the Electropult were that it was superseded by the steam catapult as an aircraft carrier device, largely one presumes on the grounds of first cost and power supply requirements. Looking at today's developments in high-speed transport, it is fascinating to see how the Westinghouse engineers of 1946 stood as near to the transport 'super-system' of the 1980s as did Hippolyte Pixii in 1831 to the turbo-alternator of the 20th century. The reader is recommended to read the literature on the Electropult (Jones, 1946; *Westinghouse Engineer*, 1946) *after* having read chapter 7.

The other World War II development was not in fact completed until the early 1950s but it also had its origin in war planes and was, like the Electropult, a record breaker in its own right, although in this case the record still stands, for the terminal speed of the secondary exceeded 1300 m.p.h. at ground level (well on the way to Mach 2!). This project was different in almost every way from the modern transport-oriented linear motors. It was a d.c. machine with slip tracks and brushes—a fixed primary machine with a short secondary (only a few per cent of the energised primary was 'active zone' at any one time).

Figure 2.8 The Westinghouse 'Electropult' primary on its track (*courtesy of Westinghouse Engineer*)

Its purpose arose from a facet of wind-tunnel technique and scale-modelling which, like so many things, is not obvious until you are told of it. When an aircraft model is placed in a wind tunnel it is supposed to simulate the conditions in the open air as nearly as possible. Examination of the extent to which this is not met shows that if a real aircraft were to operate in identical conditions to those of the model, then it should fly inside a tube of only a few hundred feet bore with the tube walls moving along at the same speed as the aircraft!—a simple question of relativity after all. For supersonic tests where the tunnels were necessarily narrow, it was decided to attempt a better simulation of the real conditions by actually firing aerofoil sections at high speed.

Figure 2.9 Battery bank supplying the fastest-ever linear motor (1954) (*British Crown Copyright/RAE photograph*)

The d.c. supply consisted of accumulators, the extent of which can be judged from figure 2.9. The energy stored in the stator was very considerable and five air-blast circuit breakers were needed to switch it off. The sight of these breakers in action was most impressive and figure 2.10 is itself indicative of the mammoth task attempted (Shaw *et al.*, 1954). I find the most intriguing part of this experiment however, in the photograph of a missile emerging at supersonic speed (figure 2.11). The camera was almost fast enough to 'stop' the missile, but what of all the bright streaks that surround it? Presumably they were caused by particles travelling at speeds which made 1300 m.p.h. look like a snail's pace! So how did they receive their accelerating forces?—sub-harmonics in space or time harmonics? Back comes the immediate reply: 'You can't *have* harmonics in d.c. motors'—or can you?

The only personal touch I would add to this amazing experiment was that one of the three authors of the report on it was a fellow officer cadet at R.A.F. Cosford in 1944 and a contemporary at Farnborough in the later war years. He had had a share in the design and building of the fastest linear motor ever without knowing that I was at the same time developing linear motors in Manchester. I was ignorant of his work and he of mine. Communication is still one of our major problems!

THE EARLY INVENTORS AND THEIR PATENTS

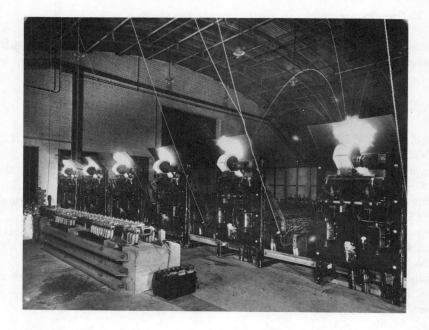

Figure 2.10 Circuit breakers switching off the field of the d.c. linear motor (*British Crown Copyright/RAE photograph*)

Figure 2.11 A missile emerges from the d.c. stator at supersonic speed. (Note the tracks of particles that must have reached a whole order of magnitude faster speed) (*British Crown Copyright/RAE photograph*)

2.9 References

Adkins, B. (1957), *The General Theory of Electrical Machines* (Chapman and Hall)
Birkeland, K. (1901), Norwegian Patent No. 11201
              (1901), Norwegian Patent No. 11342
              (1902), Norwegian Patent No. 11228
              (1903), Norwegian Patent No. 13035
              (1903), Norwegian Patent No. 13052
Blake, L. R. (1957), 'Conduction and induction pumps for liquid metals', *Proc. IEE*, Vol. 104A, No. 13, pp. 49-63
Fleming, J. A. (1891), 'Electro-magnetic repulsion', *The Electrician*, Vol. 26, No. 669, pp. 567-571 and 601-604
Gamow, R. J. and Harris, J. F. (1972), 'What engineers can learn from nature', *IEEE Spectrum*, Vol. 9, pp. 36-42
Hesmondhalgh, D. E., Tipping, D. and Sarson, A. (1973), 'High-torque low-speed motor using magnetic attraction to produce rotation', *Proc. IEE*, Vol. 120, No. 1, pp. 61-66
Japolsky, N. (1931), 'Moving magnetic fields in electrical engineering and physics', *Scientific Journal of the Royal College of Science*, Vol. I, pp. 105-126
Jones, M. F. (1946), 'Launching aircraft electrically', *Aviation*, Vol. 45, No. 10, pp. 62-65
Junior Mechanics and Electricity (1914), 'An experimental model electric railway', 1st June, pp. 246-247
Kron, G. (1942), 'The application of tensors to the analysis of rotating electrical machinery', *General Electric Review*
Laithwaite, E. R. (1965), 'The goodness of a machine', *Proc. IEE*, Vol. 112, No. 3, pp. 538-541
Laithwaite, E. R. (1973), 'Man made or God made?', *Electronics and Power*, Vol. 19, pp. 17-19
Laithwaite, E. R. and Hardy, M. T. (1970), 'Rack-and-pinion motors: hybrid of linear and rotary machines', *Proc. IEE*, Vol. 117, No. 6, pp. 1105-1112
McCarthy, J. (1922), U.S. Patent No. 1417189
Park, R. H. (1929), 'Two-reaction theory of synchronous machines–I', *Trans. AIEE*, Vol. 48, pp. 716-730
Park, R. H. (1933), 'Two-reaction theory of synchronous machines–II', *Trans. AIEE*, Vol. 52, pp. 352-354
Shaw, J. M., Sketch, H. J. H. and Logie, J. M. (1954), 'The theory, design, construction and testing of an electric launcher', *RAE Report AERO. 2523, E.L. 1484*
The Engineer (1912), 'Foucault and eddy currents put to service', 18 October, pp. 420-421
The Motor (1914), 'A high-speed electric repulsion railway', 12th May, p. 686
Tomlinson, C. (editor) (1852-54), *Cyclopaedia of Useful Arts, Mechanical and Chemical, Manufactures, Mining and Engineering*, Vol. I, p. 577 (George Virtue and Co., London)

Trombetta, P. (1922), 'The electric hammer', *Trans. AIEE*, Vol. XLI, pp. 233-241 (presented at the Spring Convention of AIEE, Chicago, 19-21 April 1922)
Westinghouse Engineer (1946), 'A wound rotor motor 1400 ft. long', Vol. 6, pp. 160-162
Zehden, A. (1905), U.S. Patent No. 732312

3 The contributions of the textile men

Chapter 2 describes inventions in linear drives in strictly chronological order. The applications of linear motors to weaving were deliberately omitted from that chapter for several reasons.

(i) There were far more patents filed between 1845 and 1950 on textile applications of linear motors than on all other applications of these machines taken together.
(ii) The textile men knew virtually nothing about electricity and could be said to be 'amateurs' by comparison with Bachelet, the designers of liquid metal pumps and those of the 'Electropult'.
(iii) The textile patents covered amazingly diverse *shapes* of linear motor, some of which did not re-appear for 20–30 years. Their ingenuity was of the highest quality but the economic fashions of the day (see chapter 4) were against them and no textile application of linear motors yet exists commercially.

The three foregoing statements nevertheless add up to the fact that the textile men who dabbled in linear motors made a real contribution to the subject and while they were probably unaware of each other's inventions, it seems probable that *some* of their work was known to later workers in other fields. If only some of the textile men had been aware of the potential for linear motors in those other fields, the 'Second Age of Topology' described in chapter 9 might well have started earlier, just as the invention of the induction machine might have occurred in the 1830s had not the inventors of that time been blinded by the demand to generate 'battery-like' current.

The invention list in this chapter is therefore to be interleaved chronologically with that of chapter 2, the two together covering the years 1846–1950 (and beyond in the case of textile applications only).

3.1 'Action at a distance' is the 'bait'

I have been told by different people on separate occasions that the first patent on linear motors was filed by the Mayor of Pittsburgh in 1890 and that it was an induction machine applied to loom shuttle propulsion. I have not however been able to trace the patent, but if it does not exist there is certainly a patent with the same objective in 1895. The popular concept of a 'flying shuttle' was of an object which literally flew through the air across a loom. In practice this did not occur, and a moment's thought suffices to show how complex the geometry of a loom would have to be to attain that (perhaps 'ideal') end.

Any object, given a horizontally directed blow and left to travel on its own will follow a parabolic path, so far as its vertical motion is concerned, and no-one is going to try to match such a path in the design of a loom. Nevertheless, the very name given to James Kay's shuttle of 1733 suggests movement without contact and, as with modern transport in which it is proposed to have ground vehicles 'hovering' clear of the ground, Tesla's invention promised immediate success if it could be applied in linear form. It is a fact, however, that just as at the time of writing (1985) no commercial shuttle is being propelled, so no fare-paying passenger was being transported by the electromagnetic mechanism invented in 1888 some 80 years later. It is also a fact, however, that sponsored research into electromagnetic shuttle propulsion was being actively conducted in 1972, while several countries are on the brink of achieving a successful high-speed ground transport system based on linear motor propulsion. The time interval of some 70–80 years during which progress in linear motors was extremely slow clearly needs an explanation. The latter is not simple and there are many contributing factors, not least that of the 'amateur' status of the textile inventors in the world of electrical engineers.

3.2 How a power loom weaves

Before embarking on the variety of linear motors invented by textile engineers it is necessary to appreciate how cloth is woven, so that the various constraints imposed by the application may be seen in the linear motor context. It is simplest and will suffice to describe 'plain weaving' in which each thread in the x-direction goes alternately under and over successive threads in the y-direction and vice versa.

The threads in the y-direction begin their passage through the loom by being pulled slowly from a 'beam' (a roller roughly as long as the width of the finished cloth with end cheeks which could be said to make the beam into an extremely wide pulley). These threads are all wrapped on to the beam, each thread being as long as is needed for the length of the finished cloth. These are the 'warp' threads. They are spaced apart by the correct amount after leaving the warp beam by passing between the teeth of a 'reed' (a metallic strip exactly like a comb but with the teeth closed at both ends). This reed or comb serves a second purpose as we shall see in a moment. Between the warp beam and the reed, each thread has been passed through an eyelet formed in a vertical wire which can be raised or lowered at will by attaching the top of the wire to another part of the loom mechanism. These wires and eyelets are called 'healds'. In a plain weave, heald numbers 1, 3, 5, 7 etc. are fastened to the same horizontal strip at the top of the loom. Heald numbers 2, 4, 6, 8 etc. are fastened to another so that either set can be raised or lowered. The two horizontal strips are so arranged that they rise and fall alternately. When one set of healds is raised, alternate threads form the top of a tunnel of threads called the 'shed', and in this position the shuttle carrying a bobbin of thread known as the 'weft' (in early days sometimes called the 'web') is passed horizontally through the shed from side to side of the loom, unwrapping weft thread as it goes. In making the journey across the loom the shuttle runs over a solid wooden beam known as the 'slay' (in early patents known also as the 'lay'). The weft thread thus inserted lies loosely in the warp shed and must be pushed forward tightly until it lies as close as possible, or as designed, to the previously inserted weft thread which is situated at the end of the cloth which has been woven so far (known as the 'beat-up point'). This beating-up action is performed by the whole of the slay which carries the reed being rocked until the teeth of the reed reach the beat-up point. The slay then retires to its original position, for the start of the next cycle of operations. The woven cloth is wrapped on to a second beam (the 'take-up beam'), whose slow rotation could be said to be responsible for the whole of the motion of the warp in the y-direction. One opening of the shed, passage of the shuttle and beating-up of the thread is known as a 'pick'. Lancashire looms would perform 60–150 picks per minute, depending on cloth width, and as the actual time of shuttle travel occupied perhaps one third of the total time of a pick, speeds of up to 60 m.p.h. are required by the shuttle. After each pick the healds change places and the shuttle returns through the new shed in the reverse direction. The lower set of warp threads rests on the slay so that the shuttle rides on the thread–slay combination rather than flying through space, indeed it generally has to force its way between the two sets of threads, since the shed may begin to close before the shuttle traverse is completed. Shuttles are therefore pointed at both ends.

The method of flinging the shuttle along the beam is to push it hard for a short distance (outside the limits of the cloth) by means of a piece of bone sliding on a rod. A recess in the bone houses the sharp end of the shuttle during this push, which accelerates it from rest to maximum speed just before it enters the shed, after which it is essentially 'on its own' until it enters the 'shuttle box' at the opposite end, where it is braked by rubbing alongside a strip of spring steel, thereby compressing the latter. It is finally arrested by striking the bone which is to repropel it on the next pick. To obtain such rapid acceleration, the bone is made the end of a whiplash of which the stock (known as the 'picking stick') is given a sharp twist by its mounting shaft, the latter carrying a cam which is struck by a projection on a shaft emanating from the main drive (see figure 3.1).

This system, invented perhaps 50 years ahead of its time, was used in Lancashire mills until it was 50 years *behind* its time. It was a clear case of mechanical engineering which was crude but effective. The pressures which finally forced it out of business included the demand for higher and higher weaving speeds and for greater and greater reliability. With the system described it was disastrous for the 'beat-up' to take place before the shuttle was clear of the shed. Thus the time required to stop the loom before the shuttle could be smashed through the warp threads, bending the reed in the process, tended to limit the top speed. In addition a worn piece of bone, picking stick or shuttle might cause an oblique drive which would

THE CONTRIBUTIONS OF THE TEXTILE MEN

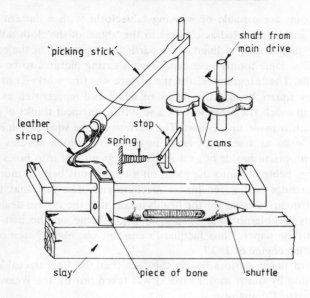

Figure 3.1 Mechanical shuttle propulsion by means of a picking stick with a whip-like action

cause the shuttle to pierce the roof of the shed and fly across the room, sometimes injuring other operatives. The likelihood of this occurrence was evidenced by the presence of a large diameter iron wheel at each side of the loom which, while allowing the operative to help the loom start by rotating this wheel by hand applied to its rim, also served as a guard against truly 'flying' shuttles with their lethal pointed ends.

3.3 The amazing Monsieur Jacquard

It is of interest, if only as a further link between the weaving and electrical industries, to explain that for weaving patterns more complicated than a plain weave there will be more than two horizontal strips to which the healds are attached. In the limit, provision can be made for each individual heald (and therefore each individual warp thread) to be lifted separately. In such an arrangement there need to be as many levers to lift the healds as there are warp threads. That such looms were made and used commercially with effect quite early in weaving history is quite remarkable, even by modern technological standards, but that the first of these looms was exhibited in 1801 is a tribute to the sheer genius of the French inventor, Joseph Marie Jacquard who, after finding himself a failure in business, lived to see his loom achieve such success that in 1806 the invention was acquired for the nation and Jacquard was awarded a royalty and a pension which he enjoyed for a further 28 years.

Jacquard looms are capable of weaving tablecloths with a pattern of flowers, leaves or even scenes, depicted as changes in the 'shine' of the cloth rather than by colours. These tablecloths, in linen, were particularly popular in the early part of this century. The same looms were capable of weaving pictures to be framed and hung upon walls. The 'programme' for the picture was first worked out on squared paper, each tiny square being either black or white and represented as an element of the picture in much the same way as a television engineer thinks of elements of his picture which is made up of lines. The pattern on the squared paper was transferred to the loom as a series of holes in cards into which needles dropped to operate the appropriate healds or as a series of pegs pushed into holes in strips of wood, the strips being fed into the loom on a continuous belt, in much the same way that a cartridge belt is fed into a machine gun. The fact that the weaving process is a 'lift/no lift' scheme for the healds means that the system deals essentially in binary digits—a hole/no hole or peg/no peg scheme on the belt—a black or white square on the paper. Thus Jacquard's concept was ahead, even of Babbage's famous computing engine of 1837.

The link with linear motors lies in the fact that the first traceable patent on shuttle propulsion by linear motor (1895) was taken out by the Weaver, Jacquard and Electric Shuttle Company.

3.4 Classifying shuttle drives

All the inventions proposing linear motor drive can be divided into one of two kinds.

(i) Stator units restricted in length to the shuttle boxes only.
(ii) Stator units mounted in the slay across the whole width of the loom plus shuttle boxes.

The first of these obviously demands much greater accelerations for shorter periods but has the advantage that it is not necessary for the shuttle boxes to move with the slay and reed during beat-up, although this puts a timing restriction on the parts of a pick. In mechanical looms the shuttle traverse, heald change and beat-up periods overlap. Even if the 'shuttle box only' linear motor is used, a more constant acceleration will be achieved than is possible with the whiplash mechanism. This is advantageous for it is possible that the peak acceleration may be large enough to break up the bobbin of weft carried in the shuttle, for the latter is wound on a core with no end cheeks. The reason for this is that only by pulling weft off a roll *endwise* can a sufficiently high speed of payout be achieved without breaking the yarn. The shuttle bobbin is hence mounted with its axis parallel to its direction of motion and is therefore highly susceptible to being broken by excessive acceleration. The 'all-across the loom' linear motor would, of course, have the shuttle under control throughout a traverse and could be arranged to be either accelerating or decelerating throughout this period, and the values of acceleration needed would be very low indeed compared with the peak values attained during a picking stick and whiplash action.

The first Jacquard patent of 1895 used an 'all-across the loom' motor. It was a reluctance motor in which an iron shuttle was handed on from magnet to magnet, as shown in figure 3.2. The magnets were switched on and off in sequence as required by means of a rotary switch driven in synchronism with the other mechanisms on the loom. It is easy to see that the disadvantages of this system were the downward magnetic pull on the shuttle, which must have been at least 10 times the horizontal driving force, the difficulty in the rapid switching of highly inductive circuits, this without the advantages of modern electronic switches (not that the latter solve the problem of inductive switching by any means), and the difficulty in accelerating rapidly from rest at each end. Crude as this patent may have been, it still represented a considerable advance in thought along a line which was subsequently to be explored and 're-invented' by many others. After all, the invention of the induction machine was but seven years earlier.

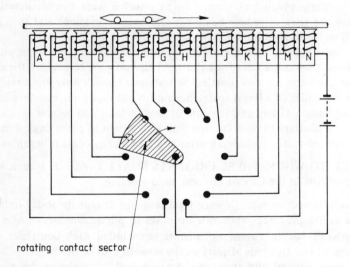

Figure 3.2 The first reluctance motor shuttle propulsion, 1895

3.5 A topological giant

Just as, in chapter 1, I first dismissed the caveman's hurling of a stone as the first game of cricket and then took a second look and decided that it might, from one point of view, have been just that, so there are simpler mechanisms involving magnetism which at first sight might be discounted as linear motors, in that mechanical power was required to produce mechanical motion, the magnet merely acting as a clutch. Yet it would be a pity to discount such systems entirely, for to do so would be to miss the patent of a most ingenious London man, J. Meeus, produced in 1844. For versatile, free-ranging thought in three-dimensional topology, this man can have few equals. He turned the conventional loom through a right angle, as it were, on to its side. Next he turned it through a right angle in the third axis—that is, front to back. Then he tried angles less than 90° in either axis alone or in combination. He tried circular looms in which the shuttle never stopped (and circular looms had a part to play in textile technology thereafter up to the present day). When he used linear looms in which the shuttle movement was anything other than horizontal (as it is in the conventional loom) he used gravity to propel or to assist in propelling the shuttle, so that drive in one direction only was sometimes possible. In one loom in which shuttle movement would have been vertical, he suggested dispensing with the shuttle and bobbin of weft entirely, each pick consisting of dropping a needle, carrying a single thread of weft through a very small shed and allowing the needles to drop off into a box at the bottom. This idea was to lie dormant for close on 100 years until Sulzers' so-called 'shuttleless loom' made a tremendous impact on the weaving industry in the late 1930s. Weaving speeds far in excess of anything which had been thought possible were then achieved by first miniaturising the shed to reduce heald lift and beat-up motions and then by firing across small weft-carrying 'bullets'.

Meeus's ideas did not stop there. He wrote "I propose to employ the pressure of the warp threads upon the tail of the shuttle; I also propose to employ the power of a magnet acting upon an iron shuttle." Wheatstone himself only just beat Meeus in at least the basic idea of a linear reluctance motor. At that time, courtesy came first in almost all human affairs and respect for one's 'elders and betters' was almost an art. It is a real pleasure to read the first page of a patent of those days, even though none of it related to the technical matter at all. In Meeus's case it began as follows:

"TO ALL TO WHOM THESE PRESENTS SHALL COME, I, Joseph Meeus of Ludgate Hill, in the City of London, send greeting."

This was followed by such flowery legal language as remains today only in such documents as are prepared when one proposes to purchase a house, or prosecute one's neighbour, but in Patent documents it included such courtesies as to be ". . . in praise of Her Glorious Majesty in the seventh year of her reign."

In the early part of the twentieth century and certainly in the nineteenth, Britain was the centre of the world textile industry, so it is fairly certain that inventions made overseas were covered by British patents as well as those of their own country. Accordingly, in the list of patents arranged in chronological order at the end of this chapter only British patents are given when the same invention is covered by patents in other countries. It is notable therefore, how Britain begins to be left out of the cover of other nations' ideas as the twentieth century progresses.

3.6 Classification using electrical considerations

Having said that electromagnetic shuttle propulsion systems can be broadly classified into 'shuttle boxes only' or 'all-across' systems, there are other forms of more detailed classification which are certainly more useful in a book about linear motors. For example, linear versions of d.c. machines, reluctance, hysteresis, synchronous and induction motors are all possible in theory, as also are repulsion, Schräge and the other forms of a.c. commutator motor, although none of the latter has ever been built to my knowledge.

Travelling magnetic fields can be generated in one of several ways, the simplest of which is by means of permanent magnets which themselves move, as was the case in Meeus's patent. The permanent magnets can be replaced by electromagnets. Then the physical movement of such magnets can be simulated by switching a series of stationary magnets on and off in sequence and finally a polyphase a.c. feed to a row of electromagnets gives the sinusoidal travelling field system of the true induction, synchronous and other rotary types of a.c. motor.

The patent list given as an appendix at the end of this chapter includes the following systems classified as outlined above.

(i) Moving permanent magnets pulling shuttles containing iron parts (reluctance machines).
(ii) Moving permanent magnets operating on shuttles themselves containing permanent magnets (a 'synthetic' kind of synchronous motor).
(iii) Switched electromagnets operating as a reluctance motor with iron-filled shuttles.
(iv) 'Jumping ring' type impulse drive from each end of the loom (a crude form of tubular induction motor).
(v) Induction motors proper which may be subdivided into single-sided, electrically, mechanically or both, 'sheet rotor' motors, tubular, etc.
(vi) Switched electromagnets pulling an energised shuttle with current collection (a form of d.c. linear motor).

In addition there are patents concerned with other aspects of textiles such as package winding. It is interesting that this list by no means exhausts the possible combinations. For example, there are no moving electromagnets operating as the primaries of reluctance motors, nor on permanent magnet shuttles, but one system has been particularly popular with inventors and equally unsuccessful in practice. This is the switched reluctance motor (group (iii) above) which began in 1895 and is still being patented, with various additions in detail as recently as 1971. More than twenty of the patents listed are of this type.

3.7 A curiosity

I now propose to describe a selection of the textile patents in more detail, those in fact which I regard as advances of some magnitude in linear motor thinking, but first there is one curiosity which needs explanation and this is concerned with the essential difference between a row of magnets which are switched, on and off, in sequence, the same row fed with polyphase a.c. and an equivalent row of moving permanent magnets, as shown in figure 3.3. The fundamental principles which separate the three may be stated thus: a permanent magnet is a source of m.m.f. and not one of flux. It is therefore to be seen as a *series* element in any equivalent circuit. If it is moved so that it induces current in a conductor, that secondary current will, so far as the goodness of the system allows, oppose the m.m.f. which was its creator and in the case of a permanent magnet it succeeds. The effective current source of the primary system is unable to react by supplying more current, so the resultant flux is reduced, as is the drag force, as the relative speed between magnet and conductor is increased.

On the other hand, a primary coil attached to a *voltage* source of a.c. and presented with secondary conductor in which to induce current, maintains, in the absence of imperfections such as primary resistance and leakage reactance, a constant flux at all times.

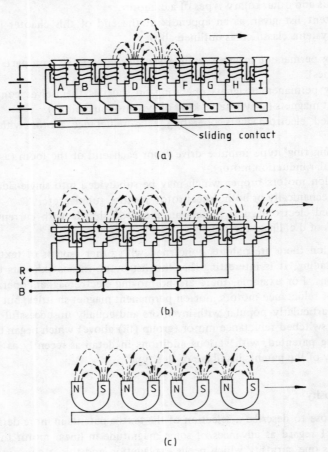

Figure 3.3 Three methods of producing travelling magnetic fields: (a) despite appearances to the contrary, the principal component of the switched field moves to the *left*; (b) the a.c. field is fixed by the impressed voltages and does not change substantially in the presence of secondary conductor; (c) the permanent magnet field behaves like a series-connected system and is largely neutralised by the presence of secondary conductor

The biggest surprise of all however is that, as mentioned in chapter 1, if the magnet system shown in figure 3.3(a) is switched so that two coils at a time are energised as shown and a conducting sheet is presented to the magnets as they are switched from left to right, D being switched off as F is switched on, the conducting sheet experiences a tangential force to the left! That this should be so can be demonstrated by deducing the instantaneous current flow in the sheet due to a coil being switched on or off, then using the lefthand rule to find the directions of the forces (Laithwaite, 1967). It suffices to point out here that if the switched coil system operates as a reluctance motor primary, the motion is to the right as our so-called 'common sense' would suggest. It is only as an induction drive that apparent reversal occurs.

3.8 Milestones revealed by the patent history

Following Meeus [1844] and Weaver, Jacquard and Co. [1895] the next patent proposed the use of an entirely different type of linear motor—a switched coil system primary operating on a shuttle containing a d.c.-fed magnet, the current being collected by brushes from slip tracks. This arrangement constitutes the first true synchronous motor, such as might have been suggested in the 1970s when, of course, the coil switching would have been effected by solid state electronic devices [Fairweather, 1897].

3.8.1 Concerning Monsieur Bachelet

Without doubt the next big step forward was made by Emile Bachelet whose electromagnetically levitated, reluctance-propelled vehicle was not only proposed as a transport drive as described in chapter 2, but as a flying shuttle mechanism. Indeed Bachelet formed a limited company to develop the textile side of his work. The Bachelet Shuttle and Loom Company had a capital of £100 000 divided into shares of £1 each and published an instructive booklet on the subject as a whole. A summary of the text on the first page reads as follows.

"The principle of Monsieur Emile Bachelet's invention is—

(a) TO LEVITATE, *i.e.*, hold suspended in the air by means of magnetic repulsion over the line of electro-magnetic coils, and
(b) TO TRANSMIT the shuttle, shell or other body by means of or through a series of solenoids or pulling magnets, quickly attaining a tremendous rate of speed which is regulated according to the amount of power used."

In the claims for his system he points out that: "It is as easy to weave goods from 20 to 25 feet in width as material only 3 or 4 feet wide," and that "There will be none of the awful din occasioned by the hammering of hundreds of shuttles." The booklet concludes with the words "It cannot be too clearly stated that there is either nothing in Monsieur Bachelet's invention, or his adaptation of what is for commercial purposes A NEW FORCE will have far reaching effects of a stupendous nature. There can be no middle course. Either the invention is of no commercial value whatever, or it will absolutely revolutionise all present methods of work in many directions, and bring corresponding profits to those persons who are far seeing enough to realise and have sufficient courage to invest money in the immense potentialities of the invention."

These last sentences might well have been written in 1986 for the question is not yet entirely resolved, but there is certainly proof that his ideas were not of 'no commercial value'. Certainly Bachelet made a big splash when launching his ideas, as photographs from the booklet show. Figures 3.4 to 3.7 carry captions which illustrate among other things that discussion of engineering projects was perhaps as popular for ladies of fashion of that time as were mathematical topics of the previous century. The caption on the Lord Mayor is either prophetic or written by a man who knew Bachelet personally.

3.8.2 The men who had it didn't know they had it

In 1923 the switched coil reluctance motor was applied to circular looms, again we must assume, unprofitably. In 1925, G. Gourdon patented a reciprocating reluctance motor (an iron armature attracted alternately by a pair of switched coils) and proposed to use it for driving various parts of a loom including beat-up, progression of warp and shuttle propulsion, the last being very like, if not identical to Bachelet's system. But if Bachelet failed to impress the financiers, what chance had they who came after? Perhaps the only mistake the subsequent textile inventors made was that they did not consider how many other industrial processes required linear and generally reciprocating devices.

Figure 3.4 The Lord Mayor, who remained nearly an hour, said, after a close inspection, "It is absolutely marvellous! This is something new, and what the possibilities are one can only imagine."—*The Standard, 9th May, 1914*

Figure 3.5 *The Times, 27th May, 1914* — The Duke and Duchess of Teck and their children visited M. BACHELET yesterday. They were accompanied by a large party, which included the Duchess of St Albans, Lady Moyra Cavendish, Lord Arthur Butler, the Hon. John and Lady Margaret Boscawen, the Hon. Claud Anson, and Sir Reginald and Lady Beatrice Pole-Carew

Figure 3.6 MR CHURCHILL, ever to the fore where progress is concerned, has seen it. "By George, it's great!" he said, "It is the most wonderful thing I have ever seen." — *Weekly Dispatch, 10th May, 1914*

Figure 3.7 Sir HIRAM MAXIM, the famous gun expert, visited the Laboratory and in a humorous remark said, "You have found a new kind of grease. It is greased lightning in actual fact."—*Illustrated Chronicle, 14th May, 1914*. [He is here seen with Professor FLEMING]

In 1932 a most comprehensive patent by Jasicek and Polnauer included the sort of linear topology which was to be re-investigated in the 1950s and harnessed for profit in the 1960s. These two inventors in particular held much of the topology of modern linear motor actuators in the palms of their hands without knowing the potential of what they held. Not only was this patent the first to propose an induction motor for linear shuttle propulsion, the inventors had many ideas which were 'firsts' in any field of linear motor activity. They used a new word, 'runner', to describe the secondary of a linear motor. They were aware of the problems imposed by magnetic pull on an iron-cored secondary, and of low efficiency caused by too large an airgap in the magnetic circuit. Like Bachelet, they were conscious, too, of that commodity which was often the subject of leading newspaper articles and television programmes of the 1970s—pollution by noise.

Their answer was to invent the 'sheet rotor' motor in which only a sheet of conductor moved between iron members which completed a good magnetic circuit. The variations on this theme which were listed included single-sided, electrical feed with double-sided magnetic circuit as in figure 3.8(b), double-sided, electric feed, as in 3.8(a), propulsion coils all across the loom or in the shuttle boxes only and perhaps best of all, they were more emphatic than Bachelet in pronouncing electromagnetic shuttle drive as "particularly advantageous for wide looms"—a feature which has not been commercially exploited up to the time of writing but one which I personally still maintain should be exploitable commercially.

Only one year later, Hopewell, perhaps without the knowledge of Jasicek and Polnauer's proposals, also designed a sheet rotor motor and included conducting

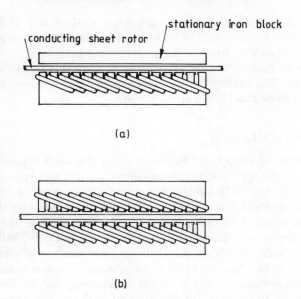

Figure 3.8 Some of Jasicek and Polnauer's shuttle-propelling arrangements, 1932

shuttle cross-sections as shown in figure 3.9. He described the travelling fields of an induction motor in a curious way by saying: "Two of said core elements are provided respectively at the opposed sides of the loom, each of the said core elements being adapted to be energised to produce a flux of variable density whose point of maximum density travels toward the centre of the loom." I can only assume that this refers to a back-to-back induction system such as I myself patented in 1956, having achieved switchless operation, and that Hopewell's moving 'point of maxi-

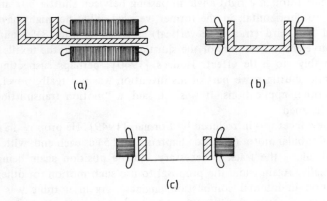

Figure 3.9 Cross-section through Hopewell's shuttles, 1933

mum density' is only the peak of a travelling sine wave of flux. However, he does describe also a switched d.c. form of his machine in which he may well have 'shaped' the flux wave. Hopewell was also conscious of the need to make the width of a sheet rotor motor greater than the width of the primary pole face so as to allow end currents to flow. The penalty for not doing this was calculated much later in linear motor history (Russell and Norsworthy, 1958).

3.9 Duplication of effort

After the patents of Jasicek and Polnauer and of Hopewell, what else was left to say? Apart from the application of linear motors to other reciprocating parts of the loom, apparently very little, although a lot of patents were to follow, most of which had perhaps just a little novelty of their own but spent a long time explaining already-known facets of the subject and it is doubtful whether there would have been many of the original claims left if the patents had ever been challenged in open court. But of course they were *not* challenged because no-one made profit from them. Whether or not this was because the developers were amateurs in electrical machine design, although expert in textile engineering (and no doubt in business also) we are about to discover, for development work towards commercialising the linear motor-driven shuttle was certainly being continued in the 1970s by people expert in machine design, who could draw on the knowledge of linear motor development which had occurred in other fields such as high-speed transport, liquid metal pumping and stirring, linear actuators, etc.

Eisselt [1936] after re-describing a reluctance motor system proposed a supply frequency higher than 50 Hz to enable pole pitch to be reduced; in the light of the rule about the size of a magnetic machine, clearly this was a progressive step. Grondhal [1937] described a reluctance system also but proposed putting the primary coils on top of the shed to use magnetic pull to relieve the weight of the shuttle from the lower warp threads (only to re-introduce the problem on the upper warp, I would have thought). In the USA, Wilson [1938] turned the plane of the magnetic circuit through a right angle in passing between shuttle box and shed; in the latter it was horizontal, in the former, vertical. What he might have achieved had he tried the third (transverse vertical) axis (see chapter 8)! Young [1939] brought photo-electric detectors of the shuttle position on to the textile scene, but again presumably, to little effect. Hodges [1940], perhaps suspecting that the electromagnetic shuttle drive part of his invention was not really novel, proposed using it for other applications. It was, he said, a "motion transmitting means." Still nobody listened.

Another new facet was introduced by Forman [1949]. He proposed a horizontal 'jumping ring' tubular motor (see also chapters 2 and 5) at each end, with additional switched coils along the track if necessary, shuttle position again being detected photo-electrically. Again came the proposal to use such motion for other purposes (valve operation in internal combustion engines). Again nothing was done, not until 1968—that is, when the jumping ring principle was applied successfully to high-speed circuit breakers (Freeman, 1968).

The shuttle propellers continued their search for the shape which would best use the forces of magnetism or of electromagnetism. As the patents illustrate, they were caught up between whether to choose permanent magnets/electromagnets as reluctance devices, or induction/synchronous motors. We can see why experiments left them still unable to decide between magnetism and electromagnetism for a shuttle is a 'small' thing in some senses, but like a reasonably large induction motor in that its speed is high enough for its propeller to be 'good enough' (see pages 76-78).

3.10 Ingenuity continues

Johnson and Johnson [1951] produced new thinking in magnetics by putting rollers or wheels on both sides of the lower warp threads as shown schematically in figure 3.10. This idea was developed further by Linier and Latieule [1954] so that either both wheels of one pair were themselves the magnets or transmitted magnetism through smaller magnets than had previously been possible (figures 3.11 and 3.12).

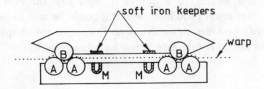

Figure 3.10 Entirely new concept of shuttle drive by Johnson and Johnson, 1951

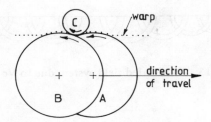

Figure 3.11 Development of Johnson and Johnson patent by Linier and Latieule, 1954

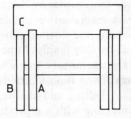

Figure 3.12 Orthogonal view of the system shown in figure 3.11

Meanwhile in France, von Vago [1952] had produced a new electromagnetic induction system in which the pole pitch of the stator coils was progressively increased towards the middle of the slay, as shown in figure 3.13. On the same theme of controlling shuttle speed during its entire run, rather than leaving it free to run between shuttle boxes, Filter [1954] proposed d.c. braking magnets at each end of the run to act as clamps during the beat-up and take-up period. Haberhauer [1959], in Germany, made the shuttle itself switch on the next coil ahead in a reluctance device, which therefore also exhibited an accelerated field system. His original cam-switched coils were later replaced with electronic devices—Haberhauer [1959]. When 'rapier' looms were introduced (a rapier loom replaces the bobbin-carrying shuttle by threading the weft through a very small shed, a separate thread at each pick, by means of a very long needle or rapier), Weil [1971] applied a tubular induction motor to the task of pushing and withdrawing the rapier.

One thing perhaps emerges above all else as a reaction to reading the published work on shuttle propulsion by magnetic or electric motors. No-one, Bachelet apart, wrote about it, boasted about it or tried to sell it. The reason is plain. No-one was succeeding! Despite a wealth of topological ingenuity the cost of linear motor shuttle propulsion was always too high, and this could be said to be due to the even greater ingenuity of the loom designers, who fashioned the shape of a Lancashire loom, so that it was not only effective in use but cheap to buy.

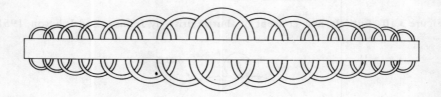

Figure 3.13   Accelerated field system due to Von Vago, 1952

3.11 An amateur textile engineer tries his hand

At the risk of being accused of making this book a part autobiography as well as a history I should like to conclude this chapter by describing the events which led up to my interest and later intensive specialisation in linear machines for I too, can be called a 'textile man'. Born among the cotton mills of Lancashire and my best schoolfriend a member of a mill-owning family, I knew more about weaving by the time I was 14 than I did about electric motors. In my schooldays the average cotton mill machinery in Lancashire was driven almost entirely from a central steam engine and flywheel, with a network of belting and shafting throughout the factory. A weaving mill which was experimenting with a separate electric motor to drive each loom was considered 'way out in front' so far as textile technology was concerned. Let it also be recorded that other European countries (notably Switzerland and Sweden) were at that time thinking about and working on entirely new types of loom and concepts of weaving. This work was soon to contribute to the virtual downfall of the Lancashire cotton industry. There is surely a lesson to be learned here for the whole of industry. The Lancashire loom with its 'flying shuttle' was, when first conceived in the nineteenth century, 50 years ahead of its time. It must surely rank (along with the umbrella perhaps) as one of the best 'value for money' machines that technology has ever produced. Even in the 1950s a Lancashire loom which would weave 48 inch cloth could be bought for around £500. The irony is that this fact also contributed to the downfall of that self-same loom, for users were slow to buy a Swiss loom costing four times as much which would weave only twice as fast. The era when the engineering 'fashion' was reliability had not yet dawned (see chapter 4). Besides, there had grown up a kind of 'engineer' in Lancashire (I use engineer in the strictest sense of the French word—*ingénieur*—an ingenious one) who has perhaps been without parallel in the history of technology—the Lancashire 'tackler'. Such a man was solely responsible for the repair, maintenance and smooth running of a batch of looms entrusted to his charge.

The dynamics of a flying shuttle were not understood formally. The slay, on which the wooden-cased shuttle ran, was also of wood—not the best material for reducing wear. As the slay changed shape the tackler would re-adjust the shuttle flight by unscrewing one or two nuts and inserting bits of cigarette packet as washers here and there, indeed it was said that a tackler who did not smoke was at a considerable disadvantage! The Swiss engineers meanwhile dabbled in this 'new fangled electricity' and tried new types of shuttle drive. In particular one system was developed in which the shuttle was blown from side to side by compressed air.

My best friend's father's mill bought one of these looms in the late 1940s (this company was undoubtedly in the forefront of progress in the country) and I was shown the new loom on the day of its arrival. I asked what controlled the opening and closing of the air jets and was told that it was done by an electrical solenoid operated from a 12-volt car battery. Knowing of the existence of the Electropult, I remember making the almost casual remark that "if you were going to bring electricity to a loom you might as well go the whole way and fire it electromagnetically." "Could that be done?" my friend inquired. I promised him I would make a model and show it to him.

I was at this time an undergraduate at Manchester University and it was customary to take a job in industry for two months during long vacations as an almost necessary qualification for the degree. I found myself working within an organisation such that I had access to a coil winder. Missing my lunch for 18 days, I wound one coil secretly during each lunch hour and threw it over the barbed wire enclosure (a common precaution during the World War II era). In the evenings I would cycle down a country lane and retrieve the coil of the day from the long grass. Thus I built my first linear motor, shown in figure 3.14, which is now housed in the Science Museum, South Kensington.

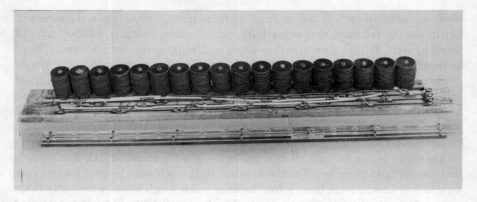

Figure 3.14 The first linear motor stator built by the author, 1948

3.12 A naturally self-oscillating motor

I decided that the main difficulty in applying linear motors to shuttle driving would be the rapid reversals of current flow necessitated by changes in the direction of the driving magnetic field at a frequency greater than 1 reversal per second (the thyristor not yet being on the market). I realised that not the whole of the primary track coils need be switched. There could be, I argued, an unswitched portion at each end and a reversible portion in the centre, as shown in figure 3.15. This would allow time for the switches to operate while the shuttle was in the fixed direction field zone. I tried to find how small this central portion could be so as to be adequate to supply frictional losses.

While thus occupied I learned that it was possible to suspend conducting objects on an alternating magnetic field system. I was told by someone (and 20 years later, when one undertakes to write history, one wishes one had enquired at the time as to the source of the information and made a note of it!) that at a "recent exhibition" (at a place unspecified) he "had seen an egg fried in an aluminium pan which was floating in space and there was no apparent source of heat under the pan." This, I thought, would be a part solution to my shuttle problem—if only I could make it float! I would have produced the first truly 'flying' shuttle, supported and propelled perhaps by one and the same set of coils. Assisted by a new member of staff, George Nix, I embarked on a contributory line of research—electromagnetic levitation. The subsequent interplay between this subject and the further researches into new and better linear motor drives became increasingly advantageous and is very much still in evidence at the present time as a result primarily of the new fashion of high-speed transport discussed in chapter 7.

It took a year for me to realise that, even without a floating shuttle and a quite sizable frictional force to fight, I could reduce the length of the switched centre-section to zero! Invention is seldom the cool, premeditated, clever process usually assumed by those who have never invented. It is much more often the result of

Figure 3.15 Back-to-back motor with a reversible centre piece (lapsed patent of 1948)

finally observing something which should have been obvious months ago! So it was with the self-oscillating induction motor.

Almost every student of electrical engineering at that time was taught that the torque/speed relationship of a rotary induction motor (and therefore, by inference, of a *linear* induction motor) was of the form shown in figure 3.16, the different curves corresponding to different ratios of R_2/X_2.

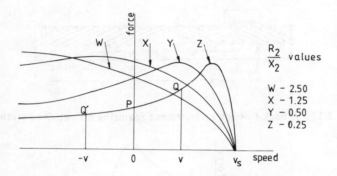

Figure 3.16 Force/speed curves of linear motors for different ratios of secondary resistance to reactance

Consider now a linear motor of the form shown in figure 3.15 from which the central, switched section has been removed entirely. The resulting layout is shown schematically in figure 3.17. It will be assumed in the first instance that the R_2/X_2 ratio gives a force/speed relationship of the form shown at Z in figure 3.16. Suppose a short rotor is initially released from rest at a point A distant x_1 from the centre of the track (figure 3.17). It will at once be accelerated towards the centre of the track by the force OP (figure 3.16). As its speed increases, so does the accelerating force (a condition seldom met in 'natural' force phenomena which usually display a damped simple harmonic motion). Let us suppose that by the time the rotor reaches the centre it has attained a velocity v corresponding to an accelerating force vQ. Beyond this point the rotor travels to the right under the influence of a magnetic field which is travelling to the left. So far as the primary field system is concerned the rotor is now travelling *backwards* and its effective value of slip is greater than unity, and in fact its operating point on curve Z, figure 3.16, as it crosses the track centre moves from Q to Q'. The rotor is subsequently brought to rest under the influence of retarding forces represented by the ordinates from Q' to P, which are seen to be everywhere *less* than those from P to Q, which accelerated the rotor from A to the track centre. In the absence of friction therefore, the rotor reaches a point B, distant x_2 from the centre, before coming to rest, where $x_2 > x_1$. The next run towards the centre now begins and having a greater distance to go before meeting reversed field, the rotor reaches a greater speed v at the centre. Each subsequent distance x increases so long as the accelerating effects are greater than those when decelerating. If we imagine the graph of figure 3.16 to have been folded about the y-axis as in figure 3.18, an easier appreciation of the action is possible for now we see that the amplitude of oscillation must continue to increase until the point at which the two curves cross over has been passed and the rotor settles down into a steady cycle. When the rotor is resisted by a load force, such as Coulomb friction load of the form shown in figure 3.19, the effect modifies the cycle to that shown.

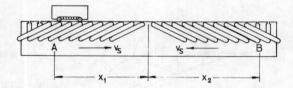

Figure 3.17 A back-to-back arrangement requiring no switched section, 1956

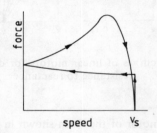

Figure 3.18 One of the force/speed curves of figure 3.16 folded about the vertical axis to simplify explanation of amplitude build-up

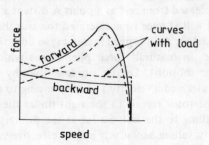

Figure 3.19 A similar diagram to figure 3.18 taking into account Coulomb friction

It follows from the foregoing arguments that unless the speed/force characteristic is *rising* at zero speed—that is, the graph has a peak to the *right* of the origin—no self-sustaining oscillation can exist.

The first back-to-back motor to be tested (now also in the Science Museum, South Kensington) is shown in figure 3.20, having each half three feet long and wound so as to give the field a synchronous speed of 25 feet/second. The rotor consisted of a 9 inch long laminated steel core with a ladder-type copper bar and end-connection grid let into its face. All primary coils of the same phase were in parallel. The machine failed to oscillate continuously and its degree of failure was obviously considerable. The rotor conductors were sunk deeper into the slots in an attempt to increase X_2. Various desperate inventions were tried, in one of which steel rings were fitted around the end bars of the rotor grid. Surely these must increase X_2 enough to make $R_2/X_2 < 1$?—for this is the condition needed for a speed/force peak to occur to the right of the origin. We pressed on harder with our levitation experiments in the belief that excessive friction, aided and abetted by that arch enemy magnetic pull, was responsible for the failure to oscillate. Impatient for new levitation apparatus we increased the voltage on the back-to-back machine in the hope that with roller bearings on the rotor we could improve the drive force/friction ratio. We increased the applied voltage until the motor insulation finally

Figure 3.20 The first self-oscillating linear motor

broke down and the windings were totally destroyed. We were not to know for several weeks that our Guardian Spirit and not a Demon had apparently afflicted us thus, for I remained convinced that switchless operation should be possible. I asked the workshop for a rewind and a technician, Arthur Hill, asked if he might wind the new coils in thick wire and put all coils in one phase in series. "It would be easier to wind," he said. I agreed and as soon as the new winding was complete, *any* of the rotors we had tried previously oscillated magnificently, building up from incredibly small initial amplitudes.

It had been suggested much earlier that if the back-to-back system had failed to oscillate, an alternative system could be used whereby the much-discussed central portion of a track of the form shown in figure 3.15 need not be switched at all for it could consist of single-phase windings only. A single-phase motor was, after all, known to be capable of accelerating in *either* direction, the direction of the accelerating force being governed only by the direction in which it was already moving. It was even pointed out that the whole track could be wound single-phase, reversal being effected by a mechanical spring at each end of the track. The only trouble was that none of our linear rotors would run on a single-phase track! My former teacher and guide Professor F. C. Williams often used a phrase "In the bright light of hindsight, all things are obvious." When you cannot understand a phenomenon, the problem may appear as an unscalable mountain, yet in an instant the whole mountain can disintegrate. If it were not so, few people might be engaged in fundamental research, for there is no substitute for the witnessing of mountains of ignorance crumbling away in a few seconds. When I knew far less about induction motors than I do now, the failure to make a single-phase linear motor operate suggested that there was something fundamental in the linearisation process which prevented it from behaving as a rotational machine was known to behave. It was a fundamental which, in a way, was too simple for me to notice. It took just over a year before the mystery was revealed. It was during this time that I had burnt out the back-to-back oscillator and had it rewound. Of course, once a device works, it is much easier to change parameters, dimensions, materials to see what makes it *worse*. In the days *before* it works it is a whole degree of magnitude more difficult to know what will make it better.

In the case of the burnt-out winding however, we had made it better by *accident* —and there were no 'shades of grey between the black and the white'. It was virtually an all-or-nothing exercise—series or parallel. I must confess that the idea of all the coils in one phase group in series and the phase groups in parallel did not occur to me at that time. However, the tantalising results of experiments during the year of ignorance showed that any rotor which would oscillate continuously in the back-to-back connection would run when the stator was connected for single-phase operation only. If it would not oscillate, it would not run single-phase. The converse was also true. All rotors which would run single-phase would oscillate back-to-back and vice versa. The preciseness of this relationship led me to realise that a single-phase stator–rotor arrangement was essentially the same as a back-to-back oscillator in which the designed natural amplitude was, say, 50 feet but whose track was limited to 5 feet and terminated by end-springs, for the 'journeys' along

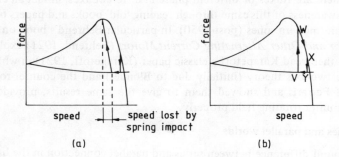

Figure 3.21 Self-oscillation cycle when end-springs are incorporated: (a) imperfect springs, (b) perfect springs

the speed/force relationship with imperfect end-springs were as shown in figure 3.21(a). If the end-springs were perfect, the behaviour would be as in figure 3.21(b). This diagram is virtually identical to those appearing in elementary textbooks to 'explain', in the simple terms which Ferraris had coined over 50 years ago, the action of the single-phase induction motor.

There exists, so the teachers and books of my student days insisted, in the air-gap of a rotary single-phase induction motor, a pair of counter-rotating fields which are the precise equivalent of the pulsating field of a single coil carrying a.c. (see figure 3.22). One may draw the speed/torque characteristics of each field separately and add them to discover the total torque (or force in the linear case). Once again this can be done by folding the diagrams about the vertical axis, when the net torques are seen to be the vertical intercepts between the curves, and therefore, when running on load, the same characteristics, the same criterion $R_2/X_2 < 1$, obtain for both back-to-back and single-phase operation.

What most (although happily not all) textbooks failed to point out was that a single coil carrying alternating current does not represent the single-phase induction

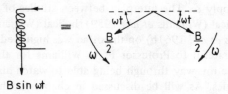

Figure 3.22 A single pulsating field is equivalent to a pair of counter-rotating fields

motor windings except at standstill, for in the presence of rotating secondary conductors there are fluxes of different phase in different axes, in fact an elliptic field. My own awareness of this came through reading 'old' books and papers on machines rather than 'modern' ones (post-1950). In particular Behrend's book on *The Induction Motor and Other Alternating Current Motors* (Behrend, 1921) explained all to me and I then read Karapetoff's classic paper (Karapetoff, 1921) in which he compared the two-axis theory (initially due to Blondel) and the counter-rotating field theory of Ferraris and showed them to give the same results, provided that one applied counter-rotating field properly.

3.13 Series and parallel worlds

The profound difference between series and parallel connection in the back-to-back oscillator was that having a very much shorter secondary than primary, well over 90 per cent of the applied voltage (in the case of series connection) was dropped in the non-active coils, so the speed of the rotor made little difference to the overall impedance; that is, the machine worked at virtually constant current and it was a hundred times easier to make $X_m/R_2 > 1$ than to make $X_2/R_2 > 1$, the conductivity of copper and permeability of free space being the values they are.

I recorded my more generalised findings in a paper in 1965 (Laithwaite, 1965a). They were that in any machine not having cylindrical symmetry, series connection was essentially different from parallel connection and not merely an exchange of amps for volts. Linear motors had now, perhaps for the first time, begun to reflect knowledge back into the commercial rotating machine world, for the above pronouncement affects vitally the unbalanced magnetic pull experienced in a rotary motor as the result of a departure from axial symmetry, whether it be due to an offset rotor, inconsistent rotor end-ring joints, a non-circular cross-section of rotor or stator, a locating slot in either rotor or stator core, or a preferred grain orientation in the core steel and teeth.

3.14 The 'DNA' molecule of machine engineering

What was for me a much more important concept was that I had discovered for myself the very heart of an electrical machine—the DNA molecule of electromagnetism, if you like—a factor which had appeared again and again in my calculations and which in the researches into spherical motors about 1953-1956 (see chapter 6) we had represented as a time constant τ multiplied by the angular frequency of the supply ω. The interplay between studies of linear motors on the one hand and spherical (Williams *et al.*, 1959), helical (Williams *et al.*, 1961a) and log motors (Williams *et al.*, 1961b) on the other was highly educational for me and I shall be forever grateful to Professor F. C. Williams for the opportunities and guidance which came my way through being able to watch him work on this latter collection of machines. As will be discussed in chapter 6, the spherical motor in particular was a foretaste of 'the shape of things to come'.

During the years 1957-1963 I had become more and more aware of the necessity of being able to express in physical terms (and therefore to 'understand' in the only way we know when dealing with four-dimensional phenomena) the overall 'quality' (for want of a better word) of an electromagnetic device. Through a series of arguments about flux cutting and flux linking explanations (which were later to rage in print in letters which appeared in the *Journal of the Institution of Electrical Engineers* between October 1962 and October 1963), and a knowledge of the relativistic explanations of both a.c. and d.c. phenomena, I finally decided that neither a magnetic flux nor an electric current had any meaning in the absence of a closed circuit and that the interlinking of magnetic and electric circuits was all that should be needed to specify the quality of a machine. About the same time I remember noting that while the electric circuits of machines had undergone a fantastic number of 'shape' permutations in the form of interpoles, compensating windings, Schräge motor and N-S motor windings, etc. during perhaps a hundred years of progress, the magnetic circuit had stayed virtually unchanged throughout this time, being a single-turn, short, fat loop.

Faraday, however, had placed considerable emphasis on the magnetic circuit, declaring that the design of such could be likened to the problem of designing electric circuits using bare copper wire in a bath of salt water. This undoubtedly is one of the main reasons for the non-development of magnetic circuits from about 1890 to the 1950s.

At first I despaired of ever being able to write an algebraic expression which represented the quality of an electromagnetic machine, whatever its form. It seemed, I thought, like asking someone to define in mathematical symbols how the quality of an oil painting might be judged. It was surely a matter of taste on the part of the user or viewer. It was no help to be able to measure the reflection coefficient of each minute part of the oil painting nor to know the frequencies of the light radiation emitted therefrom.

But with motors it was different. The quality of an electric machine did seem to be capable of being measured as some function of the electric and magnetic circuits alone. So I set down equations for the two circuits. Since the tangential force developed by any electromagnetic machine could always be calculated per unit of equivalent pole area as B × the current loading J, and since force × velocity = power (the quantity which usually interests us most), the generation of the maximum values of B and J, combined with some function expressing the surface velocity, must be, I argued, of vital importance.

The final step occurred when I took the view (more for fun than as a serious exercise) that

$$\text{goodness} = \frac{1}{\text{badness}}$$

and that badness was the product of resistance (in the electric circuit) and reluctance (in the magnetic circuit) (Laithwaite, 1965b).

Later (Laithwaite, 1968) I wrote the equation for goodness in terms of the physical parameters of the electric and magnetic circuits of any electromagnetic machine. This equation was, I now realise, to become the bedrock on which all my subsequent work was to be based. It was a great comfort to me to learn later that the same quantity was being used by Russian engineers under the name 'Magnetic Reynolds Number'. It is, after all, as important a quantity in electromagnetic dynamics as the Reynolds Number is in fluid dynamics.

This 'textile man' had at last made a positive contribution to the subject of linear motors, only to find that it embraced the whole of electromagnetic machinery. The engineer is only concerned with sifting the profitable from the unprofitable, whether it be in hardware or in his concepts, and for the academic at least, the concepts are more important than the hardware. Nor need he claim that his concepts are 'better' than someone else's. What is a most useful 'explanation' for one man is confusion for another.

In the period 1950-1970, hitherto flourishing textile firms came to grief. We should never forget that in the broadest sense, in technology as in Nature, 'progress' generally means 'death' for certain 'species' of the population.

3.15 References

Behrend, B. A. (1921), *The Induction Motor and Other Alternating Current Motors* (McGraw-Hill, New York)

Freeman, A. T. (1968), 'Latched ultra-high-speed d.c. circuit-breaker', *The Engineer*, Vol. 226, pp. 212-215

Karapetoff, V. (1921), 'On the equivalence of the two theories of the single phase induction motor', *Journal of the American Institute of Electrical Engineers*, Vol. 40, pp. 640-641

Laithwaite, E. R. (1965a), 'Differences between series and parallel connection in machines with asymmetric magnetic circuits', *Proc. IEE*, Vol. 112, No. 11, pp. 2361-2375

Laithwaite, E. R. (1965b), 'The goodness of a machine', *Proc. IEE*, Vol. 112, No. 3, pp. 538-541

Laithwaite, E. R. (1967), 'The moving window motor', *Electrical Review*, Vol. 181, pp. 126-128

Laithwaite, E. R. (1968), 'Some aspects of electrical machines with open magnetic circuits', *Proc. IEE*, Vol. 115, No. 9, pp. 1275-1283

Russell, R. L. and Norsworthy, K. H. (1958), 'Eddy currents and wall losses in screened-rotor induction motors', *Proc. IEE*, Vol. 105A, No. 20, pp. 163-175

Williams, F. C., Laithwaite, E. R. and Eastham, J. F. (1959), 'Development and design of spherical induction motors', *Proc. IEE*, Vol. 106A, No. 30, pp. 471-484

Williams, F. C., Laithwaite, E. R., Eastham, J. F. and Farrer, W. (1961a), 'Brushless variable-speed induction motors using phase-shift control', *Proc. IEE*, Vol. 108A, No. 38, pp. 100-108

Williams, F. C., Laithwaite, E. R., Eastham, J. F. and Piggott, L. S. (1961b), 'The logmotor—a cylindrical brushless variable-speed induction motor', *Proc. IEE*, Vol. 108A, No. 38, 91-99

Appendix: Chronological order of patents on electromagnetic textile devices

1844   Meeus, J.   British   10 196
Reluctance device in which a moving permanent magnet pulled along a shuttle containing a piece of iron.

1895   Weaver, Jacquard and Electric Shuttle Co.   British   12 354
Switched reluctance drive all across the slay.

1897   Fairweather, W.   British   19 745
Switched stator coils, electromagnet in shuttle with current collection. A linear synchronous motor in shuttle boxes only.

1911   Bachelet, M.   British   9573
Switched single coil for shuttle at each end, levitation with mechanical guidance all across the slay.

1923   Electromagnetic Loom Corporation (New York)   British   215 733
Switched reluctance shuttle drive for circular looms.

1925   Gourdon, G.   British   242 701
Beat-up and progression of warp both achieved by reciprocating motion of an iron armature between a pair of single switched coils. A reluctance impulse device also for shuttle propulsion, very similar to that of Bachelet [1911].

1932   Jasicek, A. and Polnauer, F.   British   374 741
Sheet rotor shuttle in induction motor system. Comprehensive claims covering single-sided and double-sided stator systems either in shuttle boxes only or all across the slay, or in circular looms. Reversal by contacts on a rotating cylinder. Many shapes of sheet rotor shuttle conductor. Systems in which stator is sectionalised and switched by the shuttle itself. Some sections short-pitched for regenerative braking.

1933   Hopewell, F. B.   British   398 917
Double-sided sheet rotors or linear cage rotors as shuttle. Stator a.c.-fed when driving, d.c. applied for deceleration. Shuttle box drives only.

1936   Eisselt, G.   British   483 816
Linear reluctance motor shuttle drive all across the slay. Conscious of stator end-winding bulk, suggested higher-frequency supply than mains and smaller pole pitches. Cage-type rotors.

| 1937 | Grondhal, K. T.
Switched reluctance shuttle drive with stator magnets on top of shed to compensate for shuttle weight. | British | 507 429 |

| 1938 | Wilson, P. B.
Single-sided linear motor shuttle propulsion with stator below shed, with double-sided stators in shuttle boxes driving horizontal flux through the shuttle. | USA | 2 135 373 |

| 1938 | Bowles, E. L., Farnes, W. and Smith, G. H. B.
Almost identical material to Hopewell [1933]. | USA | 2 112 264 |

| 1939 | Young, Y. L., Junior
Single-pole stator, switched synchronous motor in shuttle boxes only. Shuttle electromagnet fed from brushes. Switching point detected by photo-electric cells. | USA | 2 146 611 |

| 1940 | Hodges, P.
Moving permanent magnet primary system for reluctance shuttle drive. Patent applies to devices other than looms also; described as 'motion transmitting means'. | British | 522 046 |

| 1949 | Forman, J.
Impulse shuttle propulsion by 'jumping ring' principle at each end. Additional coils at intervals along slay if necessary. Coils switched by photo-electric sensors. Quotes patent 605 747 for use of device in valve operation in internal combustion engines. | British | 626 719 |

| 1950 | Dalle, O. and Servillat, G.
Reluctance device including tubular stators. | French | 967 269 |

| 1951 | Johnson and Johnson Ltd
Moving permanent magnet primary, reluctance shuttle drive all across slay with triple wheel combination to prevent magnetic attraction from wearing warp threads. | British | 696 457 |

| 1951 | Ockermann, G. and Mulet, P.
Contact-switched reluctance shuttle drive. Double-sided stator all across slay. | French | 989 730 |

| 1952 | von Vago, P. I.
Induction motor shuttle drive. Cage-type secondary. Stator track all across slay and of variable pole pitch for improved acceleration and regenerative braking. Final deceleration by reversed plugging as result of switching two phases. | French | 1 009 908 |

1953	Purdy, B. B. Fixed permanent magnet track all across slay for shuttle guidance only.	USA	2 647 542
1953	Olivier, L. and Dehors, R. Reluctance device in shuttle boxes only but with elaborate timing circuitry.	French	1 033 191
1954	Filter, W. Single-phase, double-sided reluctance shuttle device all across the slay with d.c. magnets at each end to act as braking clamps.	British	709 299
1954	Linier, C. and Latieule, Y. A wheeled shuttle drive similar to that of Johnson and Johnson [1951], but in which the wheels themselves are magnetised.	British	715 494
1954	Linier, C. and Latieule, Y. Extension of previous patent to include systems in which wheels carried only induced magnetism supplied by non-rotating permanent magnets separated from the wheels by a small airgap.	British	720 027
1954	Linier, C. and Latieule, Y. Minor modifications to earlier two patents only.	British	720 028
1956	Laithwaite, E. R. and West, J. C. Self-oscillating back-to-back linear induction motor.	British	763 362
1956	Josephy's Erban G. Double-sided linear motor shuttle drive with double-sided cage-type secondary.	French	1 120 559
1959	Haberhauer, K. Self-switched reluctance shuttle device all across slay and therefore an accelerated field system. Shuttle guided through pairs of fork prongs between which warp threads fall.	German	1 066 958
1959	Haberhauer, K. Cam-switching of earlier patent replaced by electronic switching with sensing devices to determine shuttle position.	German	1 072 569
1959	Chevallier, M. Reluctance shuttle device, tubular stator.	French	1 201 684

1959	Sulzer Brothers Ltd Guidance device only for shuttle using permanent magnets. Similar to Purdy [1953].	British	924 902
1960	Electronica Textil S.A. Switching circuits for pulsing a coil at each end of the slay for reluctance shuttle propulsion. Very similar to Bachelet [1911].	French	1 227 885
1961	Laithwaite, E. R. Self-oscillating motor, double-sided, all across the slay shuttle propulsion with sheet rotor. Similar to Hopewell [1933], and with d.c. magnets for retardation and pause at each end of a traverse.	British	866 772
1961	Williams, F. C. and Laithwaite, E. R. Self-oscillating motors with end-springs added.	British	876 795
1961	Laithwaite, E. R. Non-tubular, axial flux motors with wound secondary. Also optimum connections at centre of a back-to-back system.	British	883 837
1961	Grandi, G. A. Moving horseshoe electromagnet primary for reluctance shuttle drive.	Dutch	264 671
1962	de Fliedner, A., Pollak, I. and Parekh, S. Rows of horseshoe stator electromagnets all across the slay, switched for reluctance shuttle operation with upper set also providing lift.	Swiss	362 992
1962	de Fliedner, A., Pollak, I. and Parekh, S. Concerned only with improvements in a switched stator coil system using light beams, photocells and relays.	Swiss	363 623
1963	Bergerdorfer Eisenwerke-Astra Reluctance motor shuttle propulsion, not particularly novel.	Belgian	636 811
1965	Electronica Textil S.A. Phased stator coils operating on reluctance-type shuttle with variable stator coil pitches to provide acceleration and deceleration.	Swiss	395 892
1970	Sigrist, E. Propulsion of trolley around circular rail for feeding different spools to a shuttle mechanism—that is, not itself concerned with shuttle propulsion.	Swiss	525 309

1971 Weil, A. French 2 060 258
Tubular induction motor applied to the propulsion of
the weft-carrying device (a long thin rod) in 'rapier
looms'.

1971 Nogoya City French 2 082 105
Impulse-reluctance shuttle system with a primary coil
at each end of the run and one coil in the centre. Very
similar to Bachelet [1911].

1971 Basart, J. B. USA 3 590 879
Tubular reluctance shuttle gun with electronic switching
and d.c. available for accurate positioning of the shuttle.

1971 Linka, A. USA 3 618 640
Moving primary permanent magnet system with horseshoe permanent magnets on both primary and
secondary.

4 'Fashions' in engineering

4.1 The task of the engineer

An engineer is first and foremost a scientist. He is undoubtedly an *applied* scientist and one whose ultimate objective is the profitable manufacture of articles for himself or for the organisation which employs him. Academic engineers may argue that they are as concerned with profitable *concepts* as they are with hardware and that the concept is more important than the machine. To this extent they run alongside the pure scientist, seeking the advancement of knowledge for its own sake, yet with at least half an eye on the profits and with problems many orders of magnitude greater in complexity than any with which the pure scientist is concerned. An engineer may be faced with a problem whose solution lies in a set of 19 simultaneous differential equations which contain perhaps 82 variables known to the engineer and possibly a further 63 which are relevant to the problem but unrevealed to him. He must of necessity fix the values of all but perhaps 21 of the unknowns and then seek the use of a computer to find the best solution formally. But no-one can tell him which 21 to choose, nor even whether his 19 equations have even interpreted the problem correctly.

In such a no-man's land he is hand-in-hand with his medical colleague, who faced with a malignant disease must let the patient die or try *something*. It is not strange that the engineer fails to produce a unique solution, that his product is seen to be the result of 'art' more than science. As such, it can be criticised, more by some than by others. Another engineer would have done it differently. The product becomes a matter of opinion and because it *is* subject to opinion, it joins the ranks of many other products such as literature, painting and sculpture, and of course, clothing. It has, in fact, its own history of Fashion.

4.2 The importance of 'fringe subjects'

Fashion is inextricably interwoven with almost all aspects of human life, including religion. A history of any facet of engineering must therefore be concerned, in part, with what might be called 'the fringe subjects', which have been initiated perhaps by no more than a single word in public or a single thought in private conversation which led the originator, his colleagues or even his enemies to branch out into new fields of study.

The Industrial Revolution clearly had a great deal to do with textiles and, as indicated in the previous chapter, they were the lifeline which perhaps held on to a basic need which found its outlet in linear motor development. If so, then I could argue selfishly that there was a period when that lifeline was but a single thread, which was I myself between the years 1946-1953. Of course, no lifeline need exist for a subject to be re-opened at any time by a mind with imagination, and speculation of what the present state of the linear motor art might have been without my own contribution is both unprofitable and unscientific.

4.3 Ignorance really is bliss!

What is fact is that I applied the principle of the Electropult to the shuttle without the knowledge that many others before me had tried and failed. Such ignorance is almost vital in research and equally so in invention. Had we known, in 1954 in Manchester, of the papers of Shturman (1946) and Shturman and Aronov (1947) years earlier, we would have almost certainly abandoned our work on edge effects in short primary machines, for the analysis of the Russian papers was extremely complex, but accurate, and showed the heavy extra losses incurred by those who dared to put a discontinuity in a rotating electrical machine. As it was, a theory emerged at Manchester based on simpler concepts, sweeping aside even such important factors as leakage reactance, but showing how to design *against* edge losses, rather than merely calculating their values for any given configuration. Such is the power of ignorance. The would-be inventor must struggle daily to empty his mind of all earlier rules which he has been taught or has taught himself. Put all this material into a kind of half-tone background, to be drawn upon when required but not to influence original thinking, and you have the makings of a fascinating life, for often you can dip twice into the pool of background knowledge only to find that the new idea is black on one piece of sacred 'evidence' of experience but white on another!

4.4 Digression on a theoretical dilemma

Such was the case of the angled-field principle which we tried to exploit in a brushless variable-speed motor in Manchester in 1953-1959. This example is such a classic case of the above that it is probably worth digressing here for a moment to elaborate upon it. Figure 4.1 shows a linear motor represented as a row of alternate magnetic poles which are travelling at speed v_s in the direction shown. They influence a conducting sheet of non-magnetic material, A, by induction, but the conductor is constrained in such a way that it may only move along a path making an angle θ with the direction v_s. What is the terminal speed in the absence of friction?

Figures 4.2 and 4.3 provide two alternative answers. In figure 4.2 we plotted a vector diagram of velocities and showed that once the conductor had reached a

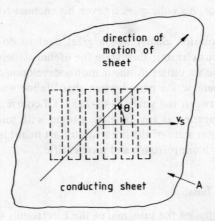

Figure 4.1 The angled field principle

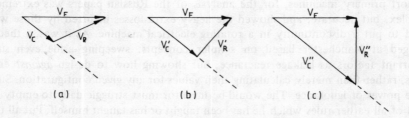

Figure 4.2 In each vector diagram the dotted line indicates the direction in which the conductor is constrained to move

V_F = field velocity

V_C, V_C' and V_C'' are conductor velocities

V_R, V_R' and V_R'' are corresponding relative velocities

It is difficult to see how further driving force can be obtained when the value of V_C rises above that shown in (b), for, as shown in (c) the relative velocity V_R'' has a *backward* component along V_C''

velocity value of $v_s \cos \theta$, its velocity *relative* to that of the field was perpendicular to the motion of that field and hence no more thrust could be derived from that source. Thus the terminal speed

$$(v_0)_1 = v_s \cos \theta \quad \text{that is, always smaller than } v_s$$

Figure 4.3 shows the same polar array moving over a glass tube containing a single steel ball and set at angle θ to v_s. This represents a *synchronous* motor as

'FASHIONS' IN ENGINEERING 87

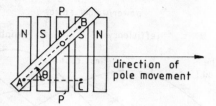

Figure 4.3 The ball-in-slot analogue

opposed to the *induction* motor of figure 4.2, but the effective synchronous speed is, in each case, the target for the argument. In the synchronous case, the ball must at all times remain in line with the pole which it is shown as 'occupying' so that as the field moves from A to C, the ball must, of necessity, reach B and thus its velocity, for unaccelerated motion of the field must be

$$(v_0)_2 = v_s/\cos\theta \quad \text{that is, always } greater \text{ than } v_s$$

The reader should refer also to chapter 6 where this topic is further discussed in relation to spherical motors.

4.5 The bad old days

Returning to the subject of fashion, in the 'bad old days' there were real 'working classes' who are distinguishable from any class of society today in at least two ways.

(i) They were generally exploited by the management, the wealthy, ruling classes.
(ii) They worked at maximum capacity, driven by the ever-present threat of total poverty, misery and starvation.

In those days to replace a man by a machine was generally too expensive, even if the machine was *very* good, for human labour cost hardly anything. When a machine *was* accepted it had to be seen to be highly 'efficient', even though this word was used both in its strict scientific sense (that is, power output/power input) and loosely to describe such concepts as an 'efficient typist'.

4.6 A 'fashiongraph'

In those days, the man whose motor was 85 per cent efficient triumphed over the man whose motor was 80 per cent efficient, even though the latter cost only half as much to build. But 'first cost' soon caught up with efficiency as a fashionable commodity, only itself to be subsequently overtaken by other quantities. I have attempted to draw a 'fashiongraph' in figure 4.4 to show the rise and fall of the popularity of each commodity as a criterion of 'goodness' with the passage of time.

In the beginning it was a question of 'Can it be done at all?' This is the happy hunting ground of the inventor whose products may be the subject of conversation

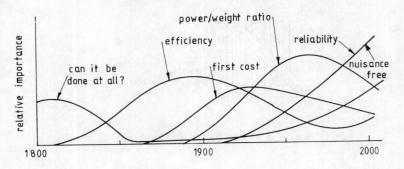

Figure 4.4 A 'fashiongraph'

of a nation, for short periods. Then the devices are put to the test and profit motive appears, first in the form of efficiency in its strictest scientific and therefore simplest sense. 'First cost' is increasing rapidly at the turn of the twentieth century, to be followed by 'power/weight' ratio, a criterion undoubtedly influenced by the developments in transport (both in quantity and in speed). Heavier objects cost more to transport (especially in aircraft), it takes a heavier crane to load and unload them, a larger concrete block for a heavier machine to stand upon, and so on. Size, and more especially weight, made their presence felt in almost every aspect of the economics of a job. Efficiency was very much on the decline. Power/weight and first cost became almost a part of each other, although this was not always true in the electrical machines field. An aircraft generator for 80 kW might weigh only 80 lbs. A conventional alternator for 50 Hz a.c. and the same output might weigh 10 times as much, but the aircraft alternator cost more than the 800 lb machine. So power/weight dominates even first cost in the aircraft era. (It was estimated that in 1952 it cost £1 per year to lift each extra 1 lb mass in an aircraft's equipment. Today, the monetary figure must be considerably higher.)

4.7 Extrapolation of the fashiongraph

Can we stand back and see ourselves in perspective with what has gone before and assess properly the relative fashions of today? I think we can do more in that we can use a crystal ball and continue the 'fashiongraph' into the twenty-first century.

With the release of atomic energy in the form of a bomb came the immediate question as to whether such an energy source could be tamed, and the 'Can it be done at all?' fashion is at once revived. The realisation that the fossil fuels must come to an end, as surely as must the life of a living creature, made atomic energy a 'must'. The beginning of the space age strengthened this renewed fashion a great deal. Ask not what it costs to send a man to the moon—ask what it costs to get him back!

4.8 The great pollution bandwagon

In the 1970s it was impossible to listen to all radio or watch all television programmes for one week without hearing the word 'pollution'. In its broadest sense it could mean anything from pollution of the atmosphere or of water by chemicals to pollution of the air by noise. It was a subject which became the source of the total income of 'experts' who studied and advised on it, wrote 'prophet of doom'-type articles for the popular press, etc.—and yet in the USA perhaps the most complex computer program in the world was used to assess the kind of life we are unconsciously 'planning' for our grandchildren in the twenty-first century. The results were alarming—an almost inevitable human catastrophe within that generation, but never, under any set of constraints, arising from pollution. Yet 'nuisance-freeness' must now have a place in the fashiongraph and is certainly going to rise, at least until the end of this century.

So far as linear motors are concerned the possibilities of the controlled use of atomic energy alone were sufficient to produce a most ingenious linear synchronous reluctance motor, working on the Vernier principle (Lavelle and Orpen, 1964) as shown in figure 4.5, for pushing the control rods of nuclear reactors in and out. These motors have virtually zero efficiency and an appallingly low power/weight ratio but can do the job, and do it perhaps for centuries without attention. So can liquid metal pumps perform their task in nuclear reactors, and so simple are they in construction that should a winding burn out, it can be replaced by remote control entirely. For this reason, the double-sided 'FLIP' was preferred to the permanently linked 'ALIP' (Blake, 1957).

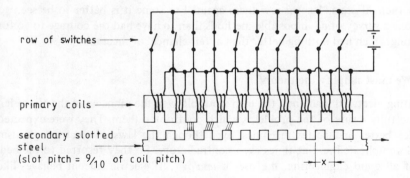

Figure 4.5 A linear vernier reluctance motor. As the lower member is moved a distance x to the right, a region of high flux density travels the full length of the lower member, and vice versa when used as a motor

4.9 The danger of breeding a race of robots

The growth of a new industry automatically attracts the best brains in related fields of technology. Thus did the electronic valve and the transistor deplete the ranks of the heavy electrical engineering men. The computer added to this trend as leading universities in the world began to scrap their experimental machine laboratories. This last exercise represents of course, the death of invention in that subject. It reminds me of seeing masses of tennis players, all coached in the 'proper' way to play the game, being beaten at Wimbledon by a man who never had a lesson in his life and was most 'unorthodox'. In other sports the 'proof of the pudding' is more obvious. High jumping is a religion until Fosbury goes over flat on his back, breaks the world record and a new fashion is born in that instant.

4.10 Contrasts in industrial outlook

In linear motors as in many other facets of human endeavour, only success succeeds. So it was that the first linear motors to sell commercially were small machines, the cost of development of which was probably less than £1000. Only when these were seen to work did anyone try a bigger machine as a commercial proposition. I have it on good authority that the world's first commercial linear actuators were developed with the following philosophy: 'We had the opportunity to spend a certain amount of money on producing a linear motor for commercial uses. This meant we could either do a design study on them, or simply make a thousand of them without trying to get the design right. We made a thousand and we've sold five hundred of them already'.

Contrast this with a small team of Government employees who were investigating methods of improving a piece of test equipment and came to me to ask whether a linear motor would do the job. On being told that it was ideal, they were visibly shaken, asked me to put it in writing, which I did, and returned to report the fact to their Masters. I never saw them again! For some it is better to be seen to have done a survey of all impossible methods than to have had the courage to go for something which had nothing better than a 'fair chance' of success.

4.11 'We must simplify the theory'

In rotating electrical machines, the men who followed the pioneers of the Victorian era were fully aware of the problems which confronted them. They were expected to make better and better machines, even though the laws of electromagnetism seemed complicated and at times even contradictory. So they resorted to the bedrock of all good engineering, the use of *analogy*. 'If electric current behaves like water flowing in a pipe, I shall hope that it will do so in an entirely untried situation. I shall go on using it until it fails me. All analogies fail ultimately. My skill will be judged by how far I *know* I can go'. Such might have been the religion of an engineer of the 1900-1920 era. So Heyland invented the circle diagram for the induction motor, Ferraris the counter-rotating field theory for single-phase machines, a theory subsequently misinterpreted by generations of teachers who used 'field' to mean 'flux' when it should have meant 'm.m.f.'. These theories served generations of machine designers virtually until the advent of 'generalised machine theory' begun by R. H. Park (1929 and 1933) and developed by Adkins (1957), as quoted in chapter 2. I well remember Bernard Adkins visiting me in Manchester in the early 1950s and looking a little disturbed by some of my linear motors. I guessed that he must have been conscious of the difficulty of applying generalised theory to travelling fields which had a 'start' and a 'finish'. 'Is generalised theory really *general*?'—he must have been thinking. There is really no conflict here. The old theory can often be applied to the new machine, albeit somewhat laboriously and in the end generally unprofitably, and to this extent the academic is justified perhaps in declaring concepts to be more important than hardware.

4.12 Useful theory or formal discipline?

My own feelings on the generalised theory were that the subject of electrical machines was in danger of becoming a 'discipline'—to be taught in the manner of Latin, formal logic or even mathematics, simply as a means of training the mind. Computers, we were told later, would be designing all electric motors by the 1980s. For me the linear motor was the means by which this danger was averted, for we now know that all rotary machines follow a strictly cylindrical and therefore uni-dimensional topology. In two or three dimensions, what remains to be invented far exceeds that which has been done (see chapter 9).

4.13 Unification for its own sake

A second limitation on inventiveness is often imposed by a desire to 'unify', or at least to reduce all things to three or four very 'fundamental' things and generally to forget that what is regarded as fundamental is also subject to changes of fashion. The alchemists of old would have it that all things were made from earth, fire, air and water. Maybe they were right but I doubt it, although one has to be most careful in this modern age of engineering, for almost anything is possible *at a price*. After all the alchemists, whose mention would raise a smile at any discussion in the early part of this century, had dreamed of changing one element into another and to a limited extent this we can now do—at a price.

By the time I was at school the physicists had got themselves nicely sorted out again. There were protons, neutrons and electrons, and that was all. Looking back we are inclined to regard them as Little Jack Horner in the nursery rhyme! Soon there were positrons, then mesons, then others until I lost count of the number of 'fundamental' particles which we were told existed. Who can doubt that the fashion in physics in the 1950s was 'get yourself a new particle'?

4.14 Industrial 'pros and cons'

In that same period I was struggling for recognition if not even for survival in academic life. I was the more baffled by the fact that many industrialists who visited me would demonstrate to me conclusively that linear motors were just 'not on' commercially. It was pointed out that the greatest tangential force obtainable from a linear motor was of the order of 10 lbs/sq. inch of pole surface (and this only on short-term rating or with a very elaborate cooling system). Compared with the tens of thousands of lbs per square inch available by hydraulic means, the linear motor came a very poor second. Yet others encouraged me with such statements as 'You will never get an oil leak', and 'Linear motors are more convenient'. How does one measure 'convenience' in physical terms—or for that matter noise, nuisance and reliability? I was told that a large post office parcel sorter in the USA had gone over to linear motors because of oil leaks in their standard system.

So it appears that fashions in other topics had a very important effect in linear motor developments. Without nuclear power there would have been no liquid metal pumps. Liquid metal pump designers never achieved a greater efficiency than 50 per cent and had come to accept such figures as profitable, within their own specialised needs. But at least one industrialist carried his knowledge of the inherently low efficiency into regions beyond liquid metal pumps and implied that *all* linear motors must, of necessity, be of low power factor and efficiency (Armstrong, 1967). I am indeed grateful for this public display of antagonism towards commercial linear motors for it pointed out to me earlier than I might otherwise have noticed that if dimensionless quantities such as efficiency and power factor could be determined by the measurement of a single length (the length of the airgap), this fact might be unique in the scientific world. Be that last statement right or wrong, it set me out on a quest for the other linear dimension which must couple with the airgap length to provide the 'magic ratio'. It is now a fact of history that I found that dimension in 1964, that it was the wavelength of the travelling wave of magnetic field and that 'good' machines could be designed with large airgaps, and this discovery, coming at that time, made it perhaps my biggest contribution to the subject, except for the discovery of the transverse flux topology (chapters 7 and 9).

4.15 Automation

If there is one outstanding fashion in a 'fringe' subject which is dominant today it is that of 'automation'. There is probably more effort being directed towards the *control* of linear motors than towards their improvement as machines. Automatic conveyor belts, controlled accelerators, sliding doors, travelling cranes are all demanding better and cheaper controls, for linear motors themselves are inherently cheap to produce, being even easier to wind than are their rotary counterparts.

One other aspect of fashion is worthy of inclusion in this chapter. Fashion *implies* change, whether it be for better or for worse—and in many subjects who shall say which? In clothes, a French couturier can take a lady's skirt hemline from the ground to the thigh overnight, but in engineering the change has to take place, of necessity, much more slowly. The time taken to change from British Units to metric is measured in decades, whereas change to decimal coinage takes only months.

4.16 Design or evolution?

When fashion changes slowly there is a possibility that 'design' may be largely replaced by 'evolution', where each 'new' machine is based on the previous one—and this is evolution proper. In my 1971 Chairman's Address to the Power Board of the Institution of Electrical Engineers (Laithwaite, 1972) I declared that the first teapot ever to be *designed* had just appeared on the market! I exaggerated, of course, but merely to point out that most teapots were very ill-designed for the purpose intended and this largely because the shape of a teapot was 'traditional'. How much better are we in electrical machine design? Already we fight the danger of the old dogmas dominating what should be essentially new thinking in linear motors. The double-sided sandwich motor was the first shape to be considered for high-speed transport, since it made the track as cheap as possible and eliminated the dreaded 'magnetic pull' in a single stroke. But the first of these qualities is now in question and fear of the second was destroyed in a simple short letter in the *IEE Proceedings* entitled 'Unbalanced magnetic push' (Freeman and Laithwaite, 1968).

Evolution is almost unavoidable in industrial 'design' offices (a contradiction in terms surely?) for no manufacturer can afford to design every machine *ab initio*, starting with Maxwell's equations and using Conformal Transformations at the end of every tooth. So the designers use well-tested formulas which lead directly to equivalent circuits and quick answers. But now it is simply a question of where you stop—if the simple formulas are acceptable on the grounds that they are 'well tested' it is a short step to building the customer a machine which is very little different from a well-tested machine that has been available 'off the shelf' for years, for this also is well tested. This attitude is much encouraged by the profit motive.

4.17 Accountants become fashionable as dictators

It has been fashionable for quite some years now for the leaders of industry to have had their basic training at a School of Economics rather than an Engineering Department. This is nevertheless a fashion which never troubled the pioneers of the nineteenth century. Never for them was an accountant looking over their shoulder daily and asking 'Will it make a profit next year?' There is however the famous story of a lady who is supposed to have asked, 'But Mr Faraday, what *use* is it?' and received the sharp rebuff, 'Madam, what is the use of a new-born baby?'

Fashions move on and although modern economics can never afford science to be the luxury it was in Faraday's time, at least one engineer is encouraged by the Space Program and hopes that engineers as a community can keep in mind the slogan which is engraved around the Lamme Medal of the IEEE: "The engineer views hopefully the hitherto unattainable." I hope that this is a fashion which never changes.

4.18 References

Adkins, B. (1957), *The General Theory of Electrical Machines* (Chapman and Hall)

Armstrong, D. S. (1967), 'Application of the linear motor to transport', *Railway Gazette*, Vol. 123, pp. 145-150

Blake, L. R. (1957), 'Conduction and induction pumps for liquid metals', *Proc. IEE*, Vol. 104A, No. 13, pp. 49-63

Freeman, E. M. and Laithwaite, E. R. (1968), 'Unbalanced magnetic push', *Proc. IEE*, Vol. 115, No. 4, p. 538

Laithwaite, E. R. (1972), 'The shape of things to come', *Proc. IEE*, Vol. 119, No. 1, pp. 61-68

Lavelle, P. M. and Orpen, V. C. (1964), *Sperry Report no. 153* (AERE, Harwell)

Park, R. H. (1929), 'Two-reaction theory of synchronous machines–I', *Trans. AIEE*, Vol. 48, pp. 716-730

Park, R. H. (1933), 'Two-reaction theory of synchronous machines–II', *Trans. AIEE*, Vol. 52, pp. 352-354

Shturman, G. I. (1946), 'Induction machines with open magnetic circuits', *Elektrichestvo*, No. 10, pp. 43-50

Shturman, G. I. and Aronov, R. L. (1947), ' "Edge effect" in induction machines with open magnetic circuits', *Elektrichestvo*, No. 2, pp. 54-59

5 Electromagnetic levitation

5.1 Different worlds

A great deal of literature and much pontification have emerged from the last two decades on the subject of specialisation—or rather on its opposite, the 'broadening' of education. Since 1960 I have watched with interest, and sometimes with horror, the inroads which the educationalists, many of whom never did any research in science *per se*, have made into the educational institutions and their traditions, and this activity is by no means confined to Britain. Perhaps it had its origins in the 'Science makes War' movement which followed when the full impact of Hiroshima and Nagasaki had been felt by the man in the street who then, inadvertently or otherwise, brought his opinions to bear on his teenage children who then marched annually to Aldermaston or to any place rumoured to be dabbling in germ warfare and the like. But I think not. The broaderners consisted, at least in part, of those who felt a need to compete with their University colleagues who were more gifted in the art of research. Others were genuine crusaders with a deep sense of responsibility for the Destiny of Man. What is surely a fact is that the broadening process overgrew itself like a neglected greenhouse plant and in the last decade alone I have seen the 'broadening' (usually said to the accompaniment of a gesture in which the arms are thrown apart) of primary school teaching, secondary school curricula, sixth form courses, undergraduate teaching, the Master's degree and the Ph.D., in rapid succession. Not all establishments by any means joined this bandwagon, but nevertheless I feel that I can now look forward to the broadening of the D.Sc., the F.R.S. and perhaps ultimately the Nobel Prize!

But of course, it all depends on one's definition of 'broadening'. If the science side of it consists of writing new books in which Heat, Light, Sound, Magnetism and Electricity appear together in one volume of a series of books, one of which deals with Waves, Forces and Action at a Distance, another with Symmetry and Analogue, this is fine to my way of thinking, but any attempt to mingle Sociology and Atomic Physics will spell *disaster* for those who participate and for the organisations whose members have been so taught. Now this is only one man's opinion and I am merely exercising what I think has always been the right of a historian, namely to write his own 'slant' into his train of facts.

The vision of science as a great jigsaw puzzle covering the whole earth, in which millions were trying to fit pieces together on different 'fronts', quite unaware that the pieces which they needed badly were being examined on the other side of the world, has been with me for many years. It is only recently that I have come to know the immensity of the puzzle and the microscopic proportions which have been made to fit. But if I can stake my own claim to have put any pieces together, then I think I would like it said by others that I have *cemented* together the work begun by Bachelet in 1914 and have extended the knowledge of induction by that very process of *combining* electromagnetic levitation and the theory of electrical machines.

I think, for example, that it was obvious from the outset that levitation systems involved electric and magnetic fields in a three-dimensional space. What was less obvious was the restriction of elementary rotating machine theory to a one-dimensional world in which the student was concerned only with magnetic flux which entered and emerged from primary and secondary pole surfaces at right angles to those surfaces. In other words, one could get near enough to the prediction of a machine's performance by consideration of the radial dimension only. Perhaps the principal consideration responsible for this limited approach was the necessity to impart information to the student by means of books—essentially collections of flat pieces of paper on which the ability to be able to represent the so-called 'essential' parts of a machine on a single cross-sectional drawing was a 'godsend' indeed. These diagrams are, after all, only an elaboration of that popular seaside confection, also manufactured traditionally in cylindrical form, in which one of the principal attractions is the fact that wherever it is cut the words 'Blackpool Rock' can be seen.

An interesting point about the manufacture of seaside rock, for those who have never seen it made, is the fact that it starts as a cylinder of large diameter (about a foot) and is rolled out thinner and thinner until it is the designed selling thickness. During the rolling, the outer layers yield throughout the material, so that the coloured lettering first appears to be a completely random affair, as illustrated in figure 5.1.

More detailed design of machines for industry gave the impression that the machines were actually being given the full three-dimensional treatment by calculation of slot leakage, zig-zag leakage, etc., to take account of dimension no. 2, and of end-winding leakage and shaft voltages to bring in dimension no. 3, but it was only like having a bite at a cream cake in which the confectioner had smeared a thin layer of cream on top of a plain spongecake, whereas the full delight of a confection was only to be revealed when the entire inside of a cake was as interesting to the palate as the top decoration was to the eye! As in so many facets of machine theory however, one must seek an 'enlightened' viewpoint from the past rather than from the present, and the most elaborate books were produced nearly a century ago in which stiff, cut-out pages were laid layer upon layer to show the machine in every detail, just as if you were literally taking it apart with a spanner. The pages were coloured on both sides, giving views of both back and front, inside and outside and leaving no part unillustrated.

ELECTROMAGNETIC LEVITATION

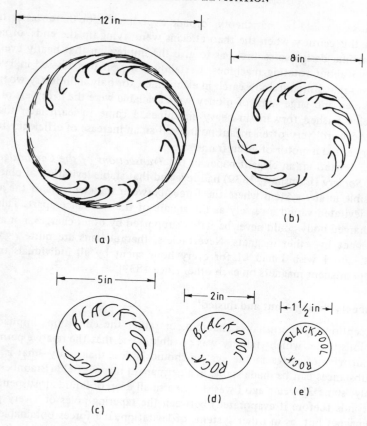

Figure 5.1 Stages in the rolling of Blackpool Rock

As with many technological developments, it seems as if we had to get worse before we could get better!

5.2 Obsession with cylindrical geometry

Historically one must acknowledge Fleming as the first 'levitator', as already dealt with in chapter 2, even though the 'jumping ring' needed mechanical guidance laterally. Bachelet was undoubtedly the first man to recognise that induction forces could be used in two dimensions on the same piece of secondary conductor (he still used mechanical support laterally) but his ideas fell on barren ground at the time. Whether our obsession for cylindrical geometry stemmed principally from the need to print books or with the deeper-rooted pride of Man, that he had outwitted Nature by his exploitation of the wheel, is of little consequence perhaps, but not for another 50 years was there an attempt (outside the textile industry) to combine forces in two dimensions in a single machine.

Of course it could be that theory forbade experiment, for there was certainly a period in this century when the theoreticians were 'tying up the ends' of conventional machine design so securely as to give the impression that nearly everything was known about rotating machines. Certainly I myself was advised in 1949 (on graduation) that I should do research in electronics, for only in industry where large teams of men and large sums of money were available were the 'frontiers of knowledge' being pushed forward in heavy machines. I came to learn later that such 'pushing forward' very often meant no more than an increase of efficiency from 92 to 93 per cent in a motor of given frame size.

As for levitation, an elegant paper in the *Transactions of the Cambridge Philosophical Society* (Earnshaw, 1839) had revealed that stable levitation of objects was not possible in any system where the forces between the elements of the levitator and the levitated varied inversely as the square of their distance apart. This meant that a charged body could never be stably supported by other charges, nor a permanent magnet by other magnets. Nevertheless, theoreticians are quite rightly disbelieved, and I wish I had £1 for every hour spent by all individuals trying to balance permanent magnets on each other since 1839!

5.3 A theory 'brought out and dusted'

The exceptions to Earnshaw's theorem, or rather the conditions implied by his calculations under which stability was possible, were that the relative permeability or permittivity effective in the system should be less than unity—that is, diamagnetic substances can be made stable in a permanent magnet field (Braunbek, 1939). Certainly, some 35 years ago I was shown a small drop of liquid air suspended for a few seconds (before it evaporated) between the tapering poles of a very powerful electromagnet but, as in other systems of levitation, the forces obtainable were of no more than 'museum curiosity' value.

In an interesting review of the whole subject of levitation (or 'anti-gravity' as some would call it) it was pointed out (Boerdijk, 1956) that there were *four* methods by which it could be achieved; namely

(1) reaction forces, as in rocket technology
(2) radiation pressure (very small forces at very great expense)
(3) diamagnetics in stationary magnetic fields
(4) quasi-stationary magnetic fields.

It is interesting to note that the last of these embraced *both* of the two systems which are now being hotly contested in the high-speed transport game (see chapter 7); namely

(4a) induction at low frequency into secondary conductor
(4b) feedback systems using magnetic attraction limited by detector and feedback amplifier.

Two more diametrically opposed systems it would hardly be possible to find, as will be illustrated also in chapter 7.

My work in this chapter has been simplified enormously by a publication (Geary, 1964) in which almost every paper of note, even vaguely connected with electromagnetic levitation, was not only referenced but given a drastic condensation which varied from a page in length to zero, depending on its importance. The fact that this author included my own paper on 'Linear induction motors' in the *Proceedings of the IEE* for 1957 fills me with misgivings as to the adequacy of this book, for I can only afford to look at papers whose title suggests linear motors or levitation almost directly. Geary must have scanned virtually all papers and patents on induction in order to provide such a comprehensive book.

Returning to the subject of exceptions to Earnshaw's theorem, the search for an equivalent to *ferromagnetism* on the 'diamagnetic side' of μ_r = 1.0 might be compared to the Gordian Knot of legend. This famous knot is reputed to have been tied by the Phrygian King Gordius to bind the pole to the sacred wagon in the Acropolis of Gordium. An oracle proclaimed that whoever untied the knot should conquer Asia. Many tried and an equal number failed, for we are told that no ends of the cord were showing. Then came Alexander the Great who cut the knot in half with his sword. 'Cheat!' might exclaim all those with any sense of justice, but Alexander went on to conquer not only Asia but all of the then-known world. (A friend of mine would declare at once that Alexander was to be complimented for his 'lateral thinking'.)

The electromagnetic analogy is apparent when one considers the superconductor. Niobium-tin, one of the alloys which are best suited for this purpose, has a true value of μ_r very little removed from unity (like *untying* the Gordian Knot), but when used in an electric circuit it *behaves* as if $\mu_r \approx 0$ (*cutting* the Knot, surely!). Geary credits a Russian (Arkadiev, 1945) with the first publication of superconducting levitation in which a bar magnet 15 mm long was supported over a superconducting lead plate at the Institute for Physical Problems of the Academy of Science of the USSR. A letter to *Nature* from the same author followed in 1947, including a photograph of a 4 × 4 × 10 mm bar in suspension.

Geary, pointing out that only Maxwell (1881) and Jeans (1925) had commented on Earnshaw's theorem in the 100 years following its publication, referred delightfully to the papers of Holmes (1937) and Tonks (1940) in this context by saying "Earnshaw's theorem, ... was brought out and dusted." He also effectively passed comment on the inefficacy of the Patent system by the following passage:

'... at least six patents were granted for, or included descriptions of, arrays of permanent magnets intended or seemingly intended to be capable of producing levitation ... One can only conclude that these inventors either did not try out their schemes before patenting them, or else assumed that it would be possible to obtain a position of stable equilibrium after further experiment. The first of these patents, that of Snell (1940), has been cited by patent examiners against more than a few later patents granted for magnetic suspensions: one wonders, therefore, whether the existence of these six patents has brought about the rejection of any applications made for patent cover for physically realisable magnetic suspensions."
—and fair comment too!

Patents which were granted and undoubtedly worked were the results of the German Kemper using method 4(b) in the list described earlier on page 98 (Kemper, 1934). Using an electromagnet operating via a good magnetic circuit into the underside of a steel girder and a sensing device to detect proximity to the girder (the sensing device feeding in turn an amplifier which then controlled the magnet strength) Kemper was without doubt the 'father' of the so-called 'Maglev' systems of the late 1960s in connection with high-speed transport—described in full in chapter 7. By 1938 Kemper had suspended 210 kg. His basic arrangement of magnetic circuit is shown in figure 5.2.

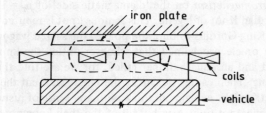

Figure 5.2 Basic arrangement of Kemper's patent, 1938

Of the electromagnetic self-stabilising methods (listed as 4a), the first of the 'moderns' and possibly of all time was exhibited at the New York World's fair in 1939 (Bedford *et al.*, 1939). Their system is shown in cross-section in figure 5.3. The input of 40 kVA at 60 Hz floated a $19\frac{1}{2}$ inch diameter disc at a height of $3\frac{1}{8}$ inch, a remarkable achievement as a mere 'starter' in the subject. Loaded with 13 lbs of non-conductive material, the suspended height was reduced to $1\frac{3}{4}$ inch. Peer (1940) went on to develop more elaborate systems of 3 or 4 coils placed at the corners of triangles or squares respectively. Geary comments on interesting 'spin-off' topics, recording such facts as that "the disc will return to its central position after being displaced 4 inch off centre and may be thrown into the levitating field from 10 ft. away," and "At certain critical temperatures a disc 3/32 inch was found

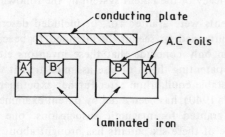

Figure 5.3 Bedford, Peer and Tonks' induction levitator, 1939, shown in cross-section

to emit a loud note, apparently due to resonance with the frequency of the levitating field at critical resistances of the disc material." Finally, this interesting passage: "The use of a shading coil to produce rotation of the disc is described. Use of the device for advertising displays is envisaged: its possible application as a centrifuge is mentioned."

This last application, like many others in linear motor techniques proper also had a 'pupation' period, in this case some 34 years, for only in 1973 did the medical centrifuge become the subject of further research. The combined levitation and rotation can now be seen as the very first 'electromagnetic river' (see chapter 7, section 7.10) albeit in this case a 'whirlpool' rather than a 'river'. Regarding the acoustic resonance described, this is a clear practical indication (if one is ever needed) to demonstrate the fact that a vertical axis levitation is but a curious form of single-phase, linear induction motor. In addition to the steady component of tangential force produced by shading ring, split-phase coils or rotation alone, all single-phase motors produce a pulsating force at double supply frequency.

The only difference between the Bedford, Peer and Tonks' levitator and those which we developed at Manchester in the late 1950s was the extra outer magnetic ring, shown in the cross-sectional view of figure 5.4 (compare with figure 5.3). Only

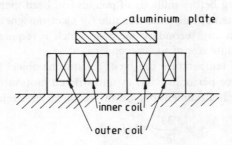

Figure 5.4 Cross-section of a levitator of 'modern' times

after we had discovered accidentally for ourselves that a small piece of aluminium held just above the centre of such a double-coil system would experience a downward attractive force, did I read of the attempts of Lovell and others to utilise a similar force in a Bedford, Peer and Tonks' levitator to remove non-ferrous particles from human eyes (Lovell, 1955). Lovell's experiments had two quite distinct effects on my own research in the early 1960s. First, they suggested to me that all stable electromagnetic levitation systems in which the primary coils were entirely *below* the suspended body also produced, *of necessity*, a 'cone of attraction' as shown in figure 5.5 which must cut the suspended object for the latter to be stable. This untrue assumption on my part was to inhibit my initial thinking on the possibilities of large levitation systems (such as those with possible application to passenger vehicles), since it appeared that a levitator supplied at 50 Hz had a maximum

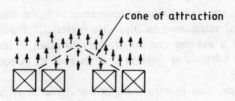

Figure 5.5 The 'cone of attraction' above a levitator. Conducting pieces within the cone are attracted, outside it they are repelled

suspension height, by the very *nature* of electromagnetism (in the same way that there is a maximum size of bird that can fly or a minimum size of shunt generator which can ever self-excite).

Secondly, as if to compensate for this negative thinking, I learned from Lovell's failure to remove tiny particles of metal from the eye that his failure *was fundamental*, and in the process of pursuing what was later to become my 'goodness factor' (Laithwaite, 1965), I first became aware of the 'size effect' which I was to develop subsequently to cover the whole range of magnetic and electromagnetic devices and which enabled me to see clearly which high-speed transport proposals were 'non-starters' long before millions of pounds had been spent on their development. Expressed mathematically, the size rule for electromagnetism relates the rate of temperature rise in any secondary member which is required to be moved by induction, to the absolute size of such secondary.

Thus if the rate of temperature rise be $\partial T/\partial t$ and the object to be moved has an acceleration F/M (force per unit mass) in unrestricted motion, the absolute linear dimension (L) to which all dimensions of any secondary are related is governed by

$$\frac{\partial T}{\partial t} = k(L)^{-n} \qquad (5.1)$$

where k is a constant of proportionality and n is greater than unity. Stated in practical terms, this means that if acceleration of a mass is attempted by induction, other things (including shape, cooling method, etc.) being equal, the conducting mass will ultimately melt before it moves, as it is made smaller and smaller.

5.4 Levitation melting

While some experimenters were bemoaning this fact, others were exploiting it, and the desirability of melting small specimens of metal *in vacuo*, without contamination by physical contact, inspired many workers to produce successful solutions to the problem by induction in which the secondary currents both suspended and melted the metal. In such a system, primary coils could be placed in any position, including vertically above the specimen to be lifted, and such a position was desirable when the melted metal was finally to be dropped into a receptacle. As Geary points out, although the first patent on this subject was filed in 1923 (Muck, 1923) the practical techniques did not appear until some 28 years later when there followed a flood of papers, mostly by American authors, between 1951 and 1958. Be that as it may, Muck's patent certainly held the key by declaring the advantages of coils above the specimen, later to be known as 'attractors'.

The major difference between the apparatus for levitation melting and that of larger levitators for transport, however, lies in the frequency of the supply. Where levitation at the lowest cost is the target, the system should clearly be as large as possible, as indicated by equation (5.1). If small enough, a levitator will melt the secondary whether the designer wishes it to be so or not, but in the 'no-man's land' between these extremes, the designer must seek what is in effect artificial deterioration of performance—that is, a larger secondary loss per unit of force applied. In the several important papers on the purely metallurgical applications the designers were, in the main, physicists who were unlikely to come to an engineer's viewpoint of the problem, which is that the appropriate analogy in these problems is to regard the forces which oppose those due to gravity as being supplied by *moving* magnetic fields. For each system the slip is unity, so that the power dissipated in the secondary increases with increase in synchronous speed. Since 'effective pole pitch' is set by the dimensions of the specimen, the only variable is frequency.

The experimenters of the 1950s are not to be regarded as 'slow developers'. Remember: 'In the bright light of hindsight all things are obvious!' Nor were the high-frequency systems apparently the result of massive calculations. At least I have failed to find any formal design technique for the shape of levitator which became popular as a metallurgical melting system, as shown in figure 5.6, other than that the best way to hold liquid is in a basin with a concave surface facing upwards, so they built a basin-shaped coil and found that it worked! The other design feature which was forced upon the experimenters was that the primary losses were also high and needed to be literally washed away by using tubular conductors for the primary coils through which water could be pumped continuously.

A great deal of discussion went on concerning the question of whether a liquid metal which was free to move in any direction at a microscopic level could ever

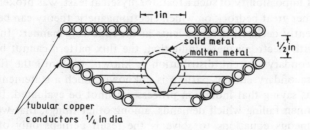

Figure 5.6 Cross-section through a water-cooled primary system for suspending liquid metal

truly be contained by induced current alone, for a ring of mercury in a trough surrounding a 'jumping ring' core is known to be in constant agitation as a result of the narrowest sections 'pinching' and parting, only to be continually rejoined by the effects of gravity. Surface tension, of course, plays a part, but the ultimate conclusion of most authors appeared to be that oxide film forming on the surface, as the result of what little air remained in the melting vessel, was a vital contribution to the stability of shape of the molten mass (generally of a 'pear' shape, as shown in figure 5.6). Without oxide film, many metals dripped from the bottom as soon as they become molten.

Another facet of levitation is its application to attempts to produce higher and higher rotational speeds of objects *in vacuo*. Once bearing contact can be eliminated entirely, speeds a whole order of magnitude greater than otherwise obtainable become possible. Without doubt the authority on this subject is Jesse W. Beams, no fewer than 40 of his papers being listed by Geary. In 1969 I met one of the authors of the *Guinness Book of Records*, the late Ross McWhirter, while on holiday and pointed out to him that the highest rotational speed produced by man was not 'in the Book'. I quoted a paper by Beams recording one million revolutions per second. Not long afterwards, Ross McWhirter wrote to me to say that he had communicated with Beams whose world record now stood at 1.5 million revolutions per second (90 000 000 r.p.m.)! At such speed the periphery of a needle of 0.001 inch diameter spinning on its own axis would need to sustain a radial acceleration of over a hundred million times the acceleration due to gravity, without doubt one of the most amazing feats of technology of modern times. Some of the vacua necessary for speeds of this order were such that when the rotating magnetic field drive was removed the rotor reduced in speed by 1 per cent over a period of two hours. But then, men have made a 'tilt meter' that can detect lunar-produced 'tides' in a cup of tea!

5.5 Levitation with a degree of freedom

Another experimenter in levitation, E. C. Okress (Okress *et al.*, 1952), and myself both found it possible (independently) to levitate a metallic object with a single coil. The apparent impossibility of such a feat, for myself at least, was broken only when I realised another great bedrock on which electromagnetic theory can be erected, and I have frequently declared it to students in the following manner: 'In any problem concerned with electromagnetic induction, the flux pattern cannot be determined until the secondary current distribution is known, yet until the flux pattern is known, the secondary current pattern is unknown'. Such a statement is not to be interpreted as saying that induction processes cannot be evaluated. It states, however, that human failing which demands, among other things, that we shall always have simultaneous equations to solve is the result perhaps only of our lengthy process of 'explaining' electromagnetism in terms of classical physics. As in life, if you tell one lie, you usually have to tell another to cover it up!

The lesson to be learned from the statement on purely physical considerations, however, is that the apparent lack of 'ingredients' required to produce a desired effect on an object will often be solved by the existence of the object itself, often giving therefore apparently impossible achievements. For example, the primary coils of a single-phase induction motor are apparently incapable of producing any kind of rotating field, 'and therefore', doubtless cried the pessimists of 1880 'it can never run'. There is a piece of prose by Minnie Louise Haskins, made more famous by its quotation by King George VI in speaking to his people at Christmas 1939 when all seemed uncertain in World War II, that could well still be used to put heart into those who fear the limitations of electromagnetism: 'And I said to the man who stood at the gate of the year: "Give me a light that I may tread safely into the unknown" and he replied: "Go out into the darkness and put your hand into the hand of God. That shall be to you better than light and safer than a known way." ' So might a prophet of 1880 have said to Tesla: 'Ask not how the rotating field shall be produced. Only place a rotor in the path of the pulsating flux and give it a push.'

Levitation by induction was achieved, I am convinced, by just such acts of faith, for permanent magnets came before electromagnets and Earnshaw's theorem came up like barbed wire each time stable levitation was considered. For me, stability was achieved initially only through the mechanism of setting up inward-travelling fields by feeding inner and outer coils of the levitator shown in figure 5.4 with currents of appropriate phase difference. Later I discovered that the inner coil could be disconnected from the mains and simply short-circuited on itself to act as a 'shading ring' with the lagging current which this action implies. It was several years before it was realised that the levitated object *itself* might provide the lagging current and therefore the inward-travelling field needed for stability. I know not by what thought processes Okress arrived at the system shown in figure 5.7, which is only two-dimensional to the extent that it represents a cross-section of a long tube supported by parallel wires which ultimately were joined at the ends of the 'track' to form a very long narrow single coil. Dr Nix and I arrived at the same system some time later without the knowledge of Okress's work.

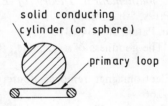

Figure 5.7 A single coil can support a conducting ball or cylinder

Such systems were to reveal a further 'secret' of electromagnetic induction. The tubular secondary of figure 5.7 need not extend the full length of the primary. When a short length of tube is suspended over a long coil, *no change* in operating conditions occurs as the result of movement along the coil—that is, movement perpendicular to the plane of the diagram in figure 5.7. This fact should have made its full impact far earlier than it did, for in my Manchester days (prior to 1964) we demonstrated a similar system, using twin primary coils and a flat plate, in which the latter can likewise move *with zero resistance* in a horizontal plane. Our only satisfaction in those days was that we had produced a 'spirit level' (without spirit!) and that in this respect the device was *perfect*, in that when at rest, any force greater than zero would start it moving. We did not know that we stood at the gates of a much bigger city within which electromagnetism was to be practised in *three* dimensions. But the story of levitation must properly be halted temporarily here, for historically it awaited the spur of the application to high-speed transport before it moved on significantly.

5.6 References

Arkadiev, V. (1945), 'Hovering of a magnet over a supraconductor', *Journal of Physics (Moscow)*, Vol. 9, No. 2, p. 148

Bedford, B. D., Peer, L. H. B. and Tonks, L. (1939), 'The electromagnetic levitator', *General Electric Review*, Vol. 42, No. 6, pp. 246-247

Boerdijk, A. H. (1956), 'Levitation by static magnetic fields', *Philips Technical Review*, Vol. 18, pp. 125-127

Braunbek, W. (1939), 'Free suspension of diamagnetic bodies in magnetic fields', *Zeitschrift für Physik*, Vol. 112, pp. 764-769 (in German)

Earnshaw, S. (1839), 'On the nature of the molecular forces which regulate the constitution of the luminiferous ether', *Trans. Cambridge Philosophical Society*, Vol. 7 (published 1942), pp. 97-112 (Paper read 18 March 1939)

Geary, P. J. (1964), *Magnetic and Electric Suspensions* (British Scientific Instrument Research Association)

Holmes, F. T. (1937), 'Axial magnetic suspensions', *Review of Scientific Instruments*, Vol. 8, No. 11, pp. 444-447

Jeans, J. H. (1925), *The Mathematical Theory of Electricity and Magnetism* (Cambridge University Press, 5th edition)

Kemper, H. (1934), German Patent No. 643 316

Laithwaite, E. R. (1965), 'The goodness of a machine', *Proc. IEE*, Vol. 112, No. 3, pp. 538-541

Lovell, W. V. (1955), 'Electromagnet removes nonferrous metals', *Electronics*, Vol. 28, No. 9, pp. 164-166

Maxwell, J. C. (1881), *A Treatise on Electricity and Magnetism* (Oxford University Press, 2nd edition)

Muck, O. (1923), German Patent No. 422 004

Okress, E. C. *et al.* (1952), 'Electromagnetic levitation of solid and molten metals', *J. Appl. Phys.*, Vol. 23, pp. 545-552

Peer, L. H. B. (1940), British Patent No. 535 871

Snell, M. S. (1940), British Patent No. 539 409

Tonks, L. (1940), 'Note on Earnshaw's theorem', *Electrical Engineering (New York)*, Vol. 59, No. 3, pp. 118-119

6 Academics and industrialists

6.1 Background

There are so many facets to almost any subject worthy and old enough to deserve a historical record that to tell the whole in its proper time sequence would be to lose the reader in a sea of facts and details, some related, others not at all. Electric motors and generators 'came of age' over almost the same period that engineering was becoming clean and respectable as a profession. Although technology (inextricably entangled with engineering so that it can never, in my view, be unravelled) preceded science, indeed paved the way for it, scientists were regarded for centuries as belonging to the upper class, the intelligentsia, so closely related to philosophers as to allow overlap. In such a world, technology was not recognised as a subject and engineers, by definition, did not appear until there were 'engines' for them to look after. Even then an engineer looked after his engines much as a shepherd did his sheep.

Even in the early part of the twentieth century, science as a whole was almost a 'middle class' occupation compared with studies of the classics. I myself attended a Grammar School where the headmaster was a scholar of Latin and Greek and a Minister of the Church (without Parish). He regarded all his pupils who elected to pursue science in the sixth form as slightly deranged mentally and saw little point in studying physics except as a discipline, and therefore those who pursued it could only be faced with the possibility of teaching the stuff to the next generation in order to perpetuate it. (In fact his outlook on physics coincided almost exactly with my outlook, at that time, on ancient Greek!)

I am sure that World War II brought home to many the tremendous intellectual content of applied science, together with, in my case, the realisation that, even if my own school was unaware of it, there existed recognition of the extensions of the disciplines of pure science into more than a score of engineering subjects, all of which could be read at universities. After that the question: 'Where does pure science end and applied science begin?' could never again be answered, but for the first forty years of the twentieth century there was a similar question which had an obvious answer: 'Where does an engineering academic end and an industrialist begin?' The universities and the factories were as far apart as the gymnasium and the monastery. University graduates who dared to set foot in a factory were inevitably greeted by the manager 'who'd worked 'is way up from 't shop floor' with a sentence such as: "Ah suppose tha' thinks tha' knows it aw!" (in my home county of Lancashire, of course). In a small way perhaps, this watershed inhibited linear motor development for the industrialist would make a linear machine, basing his designs on conventional rotary machine practice, find it to have an efficiency of 20 per cent and a power factor of 0.1, and abandon it for the rest of his career. The reason for the low values of these, in part, still fashionable quantities, was not only the lack of theoretical ability but the low speed and small size of applications for which the linear motor was suggested. (In the 1950s I had a standard form of reply letter for correspondents who had 'invented' the application of linear motors for car windscreen wipers.)

So the academic tended to dislike the industrialist and the industrialist both distrusted and feared the academic—distrusted because 'theory never works in practice' and feared because the managing director might reveal some chink in his 'armour of experience' when confronted by the academic in the presence of some of his own staff. Had not the 'long-haired Professor' long been a music-hall joke and his caricature the subject of comedy films?

6.2 Portrait of an academic

Perhaps it was World War II which came to the rescue again when the ridiculous Professor became almost indistinguishable from the 'Back Room Boy'—a war hero if ever one existed! It reminded me of a young lady who was quite accurately described as 'long and lanky' until she inherited half a million pounds and overnight became 'tall and stately'. The image of a Professor 'stumbling across ideas' was transformed into the Scientist making 'inspired guesses'. 'Men ahead of their time' became a common compliment to those whose ideas were so abstract that they could not be understood. It was true of some, not of others. The image of a scientist as a hermit, neglecting himself and his bodily needs in the cause of science also found support and the genius of Oliver Heaviside had long been recognised by the 1950s, supported by a special volume of the *IEE Proceedings* to celebrate in 1950 the century since his birth. If further example be needed then the Russian-born Dr N. Japolsky was surely everyman's 'boffin'. Moreover, he is my link with the past as I was privileged to meet, just once (see chapter 2, pages 41-43), this genius of the 'old school' who became the legendary academic, regarded perhaps by today's undergraduates as part legend, part fact.

It was not until I had begun to write this book that I realised that Professor E. M. Freeman, my present colleague and friend at Imperial College, had known Japolsky well and I am greatly indebted to Professor Freeman for allowing me to recount some of his personal recollections of this amazing man, for me a true inhabitant of the 'first age of topology'. Professor Freeman recalls Japolsky's philosophical attitude towards the 'unbelievers' of one of his theories, a much more patient and tolerant attitude than was my own at the time of the closure of the

Tracked Hovercraft project (see chapter 7). "Aristotle had to wait 2000 years," reflected Japolsky, "so why should I worry?" This book records that he did not have to wait more than 50 years to see the application of his ideas. Of his forging hammer he claimed such accuracy that: "With this magnetofuge hammer I can crack nuts without breaking the kernels."

Professor Freeman described for me a visit he made to Japolsky's lodging, a description from whose value I should only detract by any attempt to modify:

"At his invitation I visited him at his flat in Highgate. The flat was in a chaotic state with books and papers everywhere. Poor Dr. Japolsky was apparently very worried about the possibility of being evicted because he was using one bedroom as a laboratory. We went into this room and he proudly pointed out a set of cupboards of unusual shape. He opened the doors to reveal a bedstead. He was, he claimed, therefore still using the room as a bedroom. Although this was within the letter of the law (according to Japolsky) the whole ruse was negated by a gigantic assembly standing in the middle of the floor. This object was the reason for my visit. Japolsky required an extra pair of hands in order to conduct his experiments. The reward would be great 'cudos' [sic]. The apparatus terrified me. It looked like the command module of an Apollo in the launch position. Japolsky had obviously not changed his equipment suppliers since the day he declared that all experimental equipment could be made up from articles bought at Woolworths. The object was constructed from copper wire, bits of wood and an enormous quantity of paper—any paper—used as an insulator. Paper was everywhere, giving the impression of being ankle deep. It was a potential home for mice! He described the apparatus as his 'magnetofugal umbrella'. It was not awfully clear just how it worked. A rotating field (electric?) would cause air to be rotated and thrown out of the base and the whole assembly would rise in the air. Electrically it looked extremely dangerous.

"I tried to persuade Japolsky that he would be better employed writing up his life's work rather than trying to conduct further experiments. As I expected, he was not interested in writing, he only wanted to go on experimenting. I had to decline the invitation to work with him on the grounds that I was already committed to work one day a week at Imperial College, apart from my daily tasks at King's College.

"I believe his experiments were not the important things. They were a means to some other end. Any newcomer to the staff at King's during Japolsky's stay was soon tracked down and invited to his room for a chat. He was extremely slow in speech and he still retained a very heavy Russian accent. He would be extremely persuasive and it was only with the utmost difficulty that you could get away without promising him some assistance.

"He had absolutely charming manners and was a great favourite with wives of staff at social gatherings. He would always appear at these functions in formal attire. He did everything slowly, *very* slowly. Putting on his glasses or getting a book out of a cupboard seemed to take minutes. Some months after he had left King's I saw him in the Strand during the evening rush hour. Everyone was in a hurry except Japolsky. He was carrying unprotected several 5-foot long glass rods before him like a spear. I had a word with him and asked if he was going far with the glass. He was 'going home on the Northern line'—I doubt if the rods survived!"

He once embarked on an extraordinary project to make a six-strand, flat cable. People told him that it was available commercially but he was intent on making his own. His criterion of excellence was still: 'as bought at Woolworths'. The winding machine took months to build. It was made from a breadboard, four television table legs, two silver-covered wedding cake bases, two rubber heels, razor blades, Sellotape, knitting needles and door finger plates. The cable that came out of it had to be seen to be believed.

In cupboards around the electrical department Japolsky kept bits of equipment which were regarded by many as the products of an eccentric mind. But for his sternest critics there was always the feeling that he was close to something. (There was a story that he once designed an electromagnetic cannon for the Czar.)

At his memorial service on 29 June 1972, the address was given by Professor James Greig who was Head of the Department in which both Freeman and Japolsky worked, and perhaps the best description of the latter comes from this address:
"No one could be long in Japolsky's company without recognising an intellect of exceptional power and a mind which was a storehouse of knowledge both scientific and cultural. He had a fund of reminiscence, particularly of his early days in Russia. He would recall how, shortly after the Revolution he exchanged a few words with Pavlov when they stood together in a queue.

"He could be a delightful companion, yet one had to approach his companionship with a little caution, for as he was once described to me, he was a man with no sense of time. . . .

"In a way, he exemplified one of the great Christian virtues, courage in the face of rejection and defeat.

"In the words of the Apostle Paul he was 'able to withstand in the evil day and, having done all, to stand.' "

6.3 The importance of a great industrialist

The Japolskys of academic life were no strangers to me during my Assistant Lecturer days in Manchester. The slogan 'publish or perish' was just a joke, until in 1952 I was told that I would not be re-appointed at the end of my three-year appointment as I had not published any work. It was not a joke after all! As a result only of the efforts of my head of department, Professor F. C. Williams, a 'stay of execution' of one year was granted to see if I really had the makings of an academic. It was during this period of great uncertainty that I observed a certain lecturer in botany who was spoken of in hushed tones by his students as 'the only man in the world who could identify, *with certainty*, one particular species of moss'. I was duly impressed, and even more so when the lecturer got himself a Chair of Botany. I enquired further of the species of moss and was told that it was only to be found in the Chilean Andes above 10 000 feet, so who cared? Somebody cared—he got a Chair!

I determined to get me one of these mosses—an 'electromagnetic moss' of course, but like the real moss, a great rarity. I would get the largest orange box I could find and stand on it at the Hyde Park Corner of the academic world and declare as loudly as possible: 'I am the world's leading authority on ? ?'. It only took me an hour or two to fill in the blank space at the end with 'Linear Motors', for the Westinghouse aircraft launcher Electropult had fascinated me for some seven years. I was not to know until the industrialist who mattered most entered my world, that my 'moss' grew in everybody's garden—how lucky can you be?

By 1957 I had won myself a doctorate for my work on linear motors and had not only been appointed to age of retirement but had been promoted senior lecturer.

So when the industrialists began to enquire of the linear motors which I claimed I had made my own, they were all warned off by such statements as, 'They are much too low in efficiency because of the large airgap', 'There is too much magnetic pull' and all the phrases with which my own enthusiasm had been blunted prior to my Ph.D. (and, oddly enough, are *still* being used by people in high places when such phrases lead to the desired answer!)—all, that is, but one.

Gilbert Vaughan Sadler was a member of the North Western Centre Committee of the IEE, as was I, in 1958. Gilbert, I knew, worked for Vaughan Cranes (in Vaughan Street) in Manchester (although I did not recognise the possible connection of his middle initial for many years). At committee meetings he would ask me about linear motors and whether he could come and see my experiments. Once in the lab, he would argue that low efficiency was of hardly any consequence in moving the gantry of a travelling crane. I was wrong, he said, to condemn linear motors as I was now doing. I was right *before* I got my Ph.D. There was a great future for linear motors. In the end he won the argument and I agreed to help him design the first motor for a crane traverse. It was also the first time I had made money out of linear motor studies!

Throughout the trials and setbacks of my days with the British Transport Commission in the early 1960s, Gilbert never lost faith and we became friends as well as business partners, both aspects of which grew increasingly until Gilbert's untimely death in 1973. He was a man full of vigour and of good business sense, just the sort of man we needed to bridge that gulf between the academic in his ivory tower and the industrialist who would never admit to ignorance about any aspect of his business. He experimented with products as I did with concepts of magnetic circuits. His thinking was never 'out of date'. His conversation could stimulate me as could that of several of my academic colleagues. Casting aside the glamour of high-speed transport he had his sights set on such a multiplicity of uses for linear induction motors that he could have hoped, as did the painter Deakins when painting a 2 ft × 1 ft canvas with a palette knife in 20 minutes, to have a piece of his work in every home in Britain. Like Moses, however, he was only allowed to see the Promised Land from afar and he died on his way to read his IEE paper on the applications of linear motors (Sadler and Davey, 1971) to the IEE local centre in Hull.

6.4 The new generation

But in many other places the division between industrialists and academics was being bridged. Consulting for industry by university lecturers was encouraged by allowing the latter to keep their fees either wholly or in part. It became fashionable in the 1960s for Professors to become Company Directors also. From the opposite direction, prominent industrialists became part-time Industrial Professors of Universities. Research workers outside universities were allowed to register for higher degrees, including the Ph.D. As the decade neared its end the future of linear motors for slow speeds was assured, even though there was no official fanfare of trumpets to announce the fact. Such a fanfare was reserved for the high-speed transport applications which captured the imagination of newspapermen and public alike.

There was a new generation of young engineers springing up who had put aside fears of world catastrophe by nuclear bomb and had a genuine concern for such new topics as social science, while famine and pollution were to replace atomic war as the Great Black Shadow over mankind.

The definition of pollution was often broad enough to cover pollution by sound and by atmospheric disturbance, as well as actual chemical substances poisoning air and sea alike, and in such a climate all-electric propulsion for high-speed travel was a 'natural'. As will be seen from chapter 7, however, some of the fanfares were blown too loudly and some were out of tune.

My 'moss' grew in private gardens rather than in fields, and certainly not yet on the railway tracks of the world.

6.5 John Lowe

I find it quite impossible to write a history of linear motors without reserving a corner of it for a most remarkable man. I met him originally in connection with high-speed transport but, since his production of hardware has always been severely limited, I choose to describe his activities here on the grounds that, so far as linear motors are concerned, he has never been for me, either industrialist or academic. More than that, I would describe him as almost a 'species' apart and if not *Homo sapiens*, then surely *H. sapiens sapiens* would be more fitting for, on the subject of industrialists, wisdom poured from him like a torrent. Academically I could never fault him on induction motor theory and I will always recall in this connection his first visit to me in Manchester in 1958.

When he slipped in modestly the fact that he had had no formal technological education, having obtained a first degree in Mediaeval Languages at Oxford, my reaction was one of plain disbelief. He spoke of his 'consortium' of companies but I was never to know more than that about his personal connections. What I do know is that he obtained a pit-head winding pulley (some 15 ft in diameter) and had it delivered by road from the Potteries to Trafford Park, Manchester, with police escort. There he had the use of premises of a civil contractor and set up the wheel as a base for a circular aluminium track around which he proposed to run a 400 Hz motor which George Nix and I built for him, and Hugh Bolton (now a member of my staff) helped him put together with a 400 Hz generator and prime mover.

His interest in high-speed transport could be traced back further than my own for in August 1958, together with Sir Alfred Bossom, a report was prepared for the Rt Hon. Harold Watkinson M.P., the then Minister of Transport. Among his other business associates was Sir Frederick Snow and I remember a champagne party which was held at the latter's offices in Southwark Street, London SE1, to introduce the Gentlemen of the Press to 'Magnarail'. Previous to that, he had set up a lunch date in London with myself and representatives of Alcan Industries who, as a result, offered me a mile of aluminium plate free of charge on which to test a high-speed motor. The 'climate' at British Rail at that time was most unfavourable and neither that track, nor John Lowe's 15-ft wheel track ever materialised, for he never found sufficient funds for a big enough prime mover for his 400 Hz supply, and eventually the civil contractors in Trafford Park forced him to dismantle the system.

Readers may like to refer back to this section after having read chapter 7 and reflect that

(i) John Lowe proposed in 1958 precisely the same form of magnetically levitated and propelled vehicle as the most advanced form of feedback amplifier system proposed in 1974.
(ii) the building of either the wheel track or the mile of aluminium at that time would have changed the whole of transport history.

I like to think that I may still count John Lowe among my friends although there have been times when I have given him just as much 'cold shoulder' as was ever meted out to me in the same context. He has taught me much about the ways of industrialists. He would refer to large industrial empires (both nationalised and otherwise) as 'dinosaurs' and told me a long time ago that I must learn to live among dinosaurs. "All you have to do," he said, "is to learn how to move faster than they do, which is easy." Another quote which I shall always remember is: "The only place to be is in the Driving Seat."—I try, John, I keep trying!

6.6 The 'clean-up' men

In many social systems, whether they be of men or of insects, there are those with a flair for exploration, a natural desire to be 'first', and they are invariably followed by a larger group which then 'saturate' the newly won territory, exploring every corner of it, classifying it, making it fit existing knowledge. The army calls it 'mopping up'. Engineers may call it 'analysis'.

There is no suggestion intended in this concept that the 'clean-up men' are in any way inferior or of less moral fibre than the 'front men'. Those skilled in mathematical manipulation may embark on analyses which are in no way forced upon them and they are, to this extent, front men also. But it has always been my experience that when an entirely new invention has been made, its originator saw it as a *physical* concept first and then set down its equations, even though his first paper on the subject may be written 'maths first'. Yet the interplay between industrialists and academics is such that analysis carried out by the man with a pistol at his head (the industrially employed analyst) may unconsciously be solving the problem of the device yet to be invented.

So, in a way, it was with linear motors. The textile men did their experiments almost entirely without analysis. Had they enjoyed commercial success, the theory of edge effects in linear motors might well have been advanced by 50 years or more. Even as it was, the theory of a discontinuous pole array in an a.c.-produced magnetic system was first attempted for purely commercial reasons. Russian engineers in the 1940s had tried to capitalise on that same effect that had frustrated Boucherot 40 years earlier—the utilisation of magnetic pull. If the top half only of a conventional stator were used to drive a complete rotor (figure 6.1(a)), then the magnetic upward pull on the latter might compensate, possibly completely, for the force of gravity, thus relieving the bearings of their load. A second glance at figure 6.1(a), however, suffices to show that the driving thrust will also require bearing reaction, being no longer symmetrical cylindrically. The ideal 'arch motor', as such machines were called, is therefore that shown in figure 6.1(b) where electromagnetic thrust and magnetic lift are combined to produce a net thrust of zero on the rotor bearings.

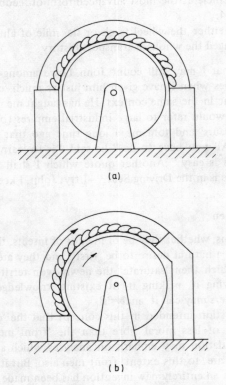

Figure 6.1 (a) An arch motor whose magnetic pull can lift the rotor in its bearings. (b) An offset arch to utilise both drive and radial forces for rotor lift

This condition can clearly obtain at only one speed and load, but good design, preceded by a knowledge of the dependence of flux and current loadings on the slip (arch motors were all induction machines) could clearly minimise rotor bearing pressures under all required running conditions, and a knowledge of the effect of edge transients was therefore vital.

The problem was tackled by Shturman and Aronov who set out the problem in the form of third-order partial differential equations which they proceeded to solve. Their mathematical model was a good one and it revealed many of the effects due to the entry of unmagnetised rotor teeth into the 'active zone' of the stator and their subsequent extraction at the exit edge. They were able to take secondary leakage reactance into account and they set out their conclusions in a positive manner, indicating clearly the new effects which the edge transients introduced. For example, they indicated clearly that the rotor bars in the active zone would carry heavy currents at zero slip and there would therefore be rotor ohmic loss on no load. They predicted extra ohmic loss at other values of slip and a complete redistribution of airgap flux density, depending on slip and the other parameters of the machine.

For the purpose of predicting the performance of a piece of machine which had already been built, this approach was excellent. The one thing it lacked was the ability to tell designers how to improve an existing design, for the equations were so complex that only a process of changing one parameter and repeating the complete solution would give any indication of whether the change had even been in the right direction. Couple with this the fact that their analysis came six years before the world's first large-scale commercial computing machine became available, and the limitations of such analysis become obvious.

This example of the failure of analysis to promote better machines is especially important in the computer era, for it does not follow that just because a machine exists that can solve equations ten thousand times faster than can a man with a desk calculator, it has the time to optimise the design of a device for which 19 equations relating the possible parameters of design can be set down but for which 81 parameters exist. Even if only 10 possible values of each parameter are admitted, there are 10^{62} sets of equations to be solved and despite 'hill climbing' techniques there are problems for which an analytical attack is a pretty unrewarding exercise.

But the sudden availability of these calculating wonders had an almost immediate effect on the academic. Problems on conventional motors could now be attacked in greater detail. Their basic equations could now be set down to include several more variables and their solution would enable prediction of performance to have a greater accuracy than ever before. A new 'rat-race' began among engineers. The phrase 'paralysis by analysis' was coined in university circles. 'Predict the performance of an existing machine to 1 per cent greater accuracy' was perhaps only rivalled by the 'get yourself a new fundamental particle' fashion in the academic hierarchy of physics. But none can deny that the products of both were beneficial. In particular, the academic who had this new-found power of more accurate prediction was much wanted by the industrialist. So there was a steady flow of graduates, now even those with Ph.D.s, into industry as employees; there was an increase in the number of academics used as consultants to industry, until of course industry bought its own computing machines as well as its high-level academics.

So far as teaching was concerned, formal mathematics had certainly taken a knock. Number theory replaced equation-solving at a quite early stage in school curricula, and my own daughter was taught binary digits at the age of 10. The world-famous Massachusetts Institute of Technology scrapped its electrical machine laboratories in favour of desk calculators and access to large-scale computers and, I was told, lived to regret it. As a teacher myself I tend to divide the teaching of applied science into *inductive* and *deductive* teaching and as an engineer, to condemn the latter in even more scathing terms than 'paralysis by analysis'.

6.7 The blending of electronics and 'machines'

In deductive teaching you set down the most general set of equations covering the problem. Solve them (if you can) and then the applications are simply a question of putting $a = c = f = k = p = z = \beta = \theta = 0$, $d = g = \alpha = \pi/2$ ... etc. (You tend to run out of symbols in the end!) In inductive teaching, on the other hand, you do a very simple case first, just to get the hang of it. Then you introduce additional variables one at a time, assessing the difficulty in solving at each stage, and stop when it hurts! It is fairly obvious that the inductive method belongs to the man who 'sees' the problem physically and knows the solution instinctively before he ever sets out the first equation. Such a man am I. It is also clear I think that the inductive method is much more likely to enable the investigator to spot the 'arch-enemies' in a system and to see how he might improve a design fundamentally without ever solving an equation.

With linear motor theory, the second theoretical wave was undoubtedly begun by F. C. Williams. Working as a postgraduate student and later as a staff member alongside such a man was like rubbing shoulders with a colleague wearing a suit of gold. Some of it was almost bound to rub off on to you and you would be the richer for it. I was honoured indeed that the penetrating brain which had produced dozens of patents on electronic circuit techniques and scores of papers in learned societies should even notice my linear motor work, let alone join in and allow me to study the mechanism of invention.

From an almost chance remark by 'F.C.' that a linear motor placed across a disc, as shown in figure 6.2, could produce a brushless variable-speed motor simply by changing its effective radius he was within days of appreciating the fact that merely 'aiming' a linear motor at a different radius, as in figure 6.3, was enough to change the effective synchronous speed. He had replaced *translation* of the linear motor by the much more space-profitable *rotation* of it. Having established that it worked in principle he rapidly changed to spherical topology and was soon testing quite sophisticated spherical motors as large as 14-inch diameter rotors (Williams *et al.*, 1959) (figure 6.4).

It was in this connection that the edge effects of linear motors reared their ugly heads and, unaware of the Russian papers, we set about understanding and predicting performance of arch motors, which were perhaps better described by Shturman

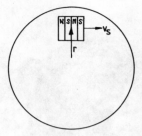

Figure 6.2 The 'angled field' principle had its beginnings in this simple concept of a linear motor on a disc

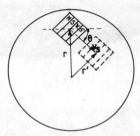

Figure 6.3 The linear motor can be 'aimed' at a smaller radius without actually being put there

himself as "Induction Motors with Open Magnetic Circuit." As I said earlier, it is possible that had I known in 1953 of the papers of Shturman and Aronov I would have abandoned work on linear motors, but knowing 'F.C.' he would, I think, have been spurred on all the more to find a physical explanation for the strange phenomena predicted by analysis alone. As it was, I was able to appreciate the physical nature of edge effects as they were unfolded before me day by day by the Grand Master of electronic techniques.

He saw the entry edge of an arch motor as a thing in itself. Sweeping aside all secondary trivialities such as ohmic resistance of conductors, leakage of magnetic circuits and reluctance of airgap (!), he produced his first 'model' of a 'short-stator' machine (as we then called arch motors). This is shown in figure 6.5. Each secondary tooth, A, resisted a change of flux on entry completely since it was surrounded by a loop of zero resistance (figure 6.6(a)). Hence there could never be any flux under the active zone—not by any means a useless deduction, for it was not radically

(a)

(b)

Figure 6.4 (a) Spherical motor of F. C. Williams, 1958. (b) Advanced form of spherical motor, later built commercially in a 5-ton prototype

changed in the vicinity of the entry edge when the first relaxation towards a real machine was made by considering the secondary to have a finite conductivity only. In this second stage the ohmic resistive volt drop in the secondary must be supplied by a rate of change of flux, which must therefore be constant, giving the result that the flux density in the airgap rises from zero at entry to a maximum at exit—the basic flux shift towards the back edge was established and understood in an extremely simple model (figure 6.6(b)).

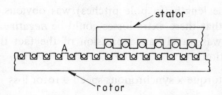

Figure 6.5 Theoretical model to illustrate edge effects

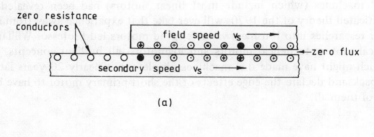

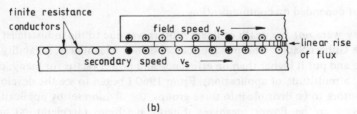

Figure 6.6 Using the model of figure 6.5: (a) any change of flux is resisted successfully by a perfect short circuit; (b) in the presence of secondary resistance a steady $d\phi/dt$ is needed to sustain the ohmic drop

The most difficult step was undoubtedly the next (slip $\neq 0$). F. C. Williams saw the entry edge as the generator of two waves. In anti-phase at the entry edge, one wave travelled at primary (synchronous) speed and the other at rotor (secondary) speed. These waves produced a 'beats' effect as they moved in and out of phase, setting up both 'in-phase' (B_p) and quadrature (B_q) components of airgap flux. The extent of the beats depended upon the fractional slip, s, and on the length of the block (np, where n is the number of poles and p the pole pitch). Comparison with rotary machines at once revealed the extra secondary loss predicted by Shturman, but now its origin was understood and its fundamental relationship to slip and to active zone length (in pole pitches) was obvious. The new and quite surprising effect was that these extra losses could be *negative*. This did not imply perpetual motion. It was merely an expression of the fact that the fundamental induction motor equation

$$\text{torque} \times \text{synchronous speed} = \text{rotor loss}$$

could be broken under sustained conditions of transient injection.

The second important fact was that the model of figure 6.6(b) had been retained without change for the case of finite slip, but retracting the ability of the gap to demand magnetising current and of course leakage flux was neglected in all cases. Thus $\int_0^{np} B_q \, dx$ was always zero in a similar treatment of a rotary motor, and another new concept was born—the production of reactive volt-amps not attributable either to magnetising current or to leakage flux.

The model was 'completed' by the addition of magnetising current, but never so far as secondary leakage was concerned. Nevertheless, the mysteries of 'short stator' machines (which include most linear motors) had been revealed and no sophisticated theory of the 1970s will ever take that experience away from me.

Our researches into brushless variable-speed motors led Professor Williams from spherical motors to helical motors and logmotors—all brilliant concepts, any one of which might have made d.c. machines obsolete, but nearly 20 years later I can look back and declare the edge effects of the short-primary motor to have been the death of them all.

6.8 A lot depended on communication

That they were *not* the death of linear motors was due to the industrialist alone, to men like Gilbert Sadler who put a price on silence, cleanliness, reliability and convenience and put it higher than on efficiency and power factor for many, one might even say a multitude of applications. From 1960 I began to see the development of linear motors to be divisible into three groups, the division set by application alone. There were to be Power machines, Energy machines (accelerators) and Force machines. Energy/weight ratio and force/weight ratio were respectively the counterparts of power/weight ratio in the continuously rated machine. Likewise energy/cost and force/cost corresponded to power/cost. In some applications the 'can it be done at all?' fashion returned.

It was not long before Fred Eastham and I were asked by the industrialists to put our exit edge theory to the test, as it emerged, one of the most severe tests to which it *could* be put. The Motor Industry Research Association (MIRA) wanted to smash road vehicles into a concrete block at a pre-determined speed, on the understanding that all acceleration must have ceased immediately before impact. The space for acceleration was, of course, limited. The desired speed was $32\frac{1}{4}$ m.p.h. (+ 1/4 m.p.h.; −0). This meant less than a 1 per cent tolerance on terminal running speed using a primary block with a small number of poles. I think we used a combination of numerical flux plotting and the known fact that the equivalent 'airgap' of an open-sided linear motor was $(1/\pi)$ times the pole pitch. We achieved an accuracy of better than 1 per cent and, whatever other historians may write, this was a milestone for us from which we began to design with confidence.

The importance which I place on this result is perhaps emphasised by subsequent events abroad. In the USA, in particular, the Department of Transportation in Washington D.C. had placed a study contract with AiResearch, a subsidiary of the Garrett Corporation. The resulting report contained a computer program for the prediction of short-primary machine performance. This took account of secondary leakage reactance which none of the 36 papers with which I had then been associated (and all of which were quoted in the report) had been able to do. Their predictions of performance at high values of slip were far better than were ours at Manchester, but all the Garrett speed-thrust curves ultimately passed through the synchronous speed point, which we knew to be untrue in practice, especially at low pole numbers. 'A prophet is not without honour, save in his own country' as we know, and far more use was made of the Garrett program in Britain than was of our Manchester programs. This is not a case of 'sour grapes', it was a plain fact.

If there was a 'no-man's land' between academics and industrialists, it was without doubt due to a bad system of communication. It would be easy to put the blame on *either* the academics *or* the industrialists for the lack of communication between them. But while it was certain that failure to communicate was, and still is, one of man's greatest failings, it was not restricted to the academic/industrialist interface, for academic seldom met academic and industrialists were, by definition, in competition with each other. It is therefore not surprising that to an outside observer various groups appeared to be tackling the *same* problem in a variety of ways, even publishing their results in various Proceedings of learned societies, without ever having time to know the relevance of the other fellow's work to their own.

A statistician of this time estimated that when a paper was published in the Proceedings of a 'learned society' (as opposed to a trade journal) it would, on average, be read from start to finish by 1.6 people in the whole of time—if it were on physics, chemistry or engineering, that is. If it was a mathematical paper the figure was 0.2 people in the whole of time!

A single example will serve to illustrate the breakdown of the communication system, for which blame could be apportioned to

(a) the printing explosion

(b) personal jealousy and 'competitiveness'
(c) apathy (and there was plenty of that!)
(d) the ascendancy of the accountant as overall dictator
(e) the academic's legendary 'ivory tower'

depending on which viewpoint you took. Those who put the emphasis on (a) might with advantage refer to the Bible to discover that it was ever thus: 'Of making many books there is no end; and much learning is a weariness of the flesh' (Ecclesiastes 12, xii).

But as to the example, linear induction motors with liquid rotors have obviously something in common with induction stirrers of liquid metal. Liquid metal pumps are only one stage removed from the ordinary double-sided sandwich motor, yet the designers of stirrers clearly knew nothing of the properties of the simplest of solid-secondary linear motors.

From about 1966 I had been aware that a conducting cylinder placed on a single-sided linear primary with the cylinder axis parallel to the primary slots would roll from end to end of the motor in the *opposite* direction to that of the travelling field. In 1970 I published a paper (Laithwaite and Hardy, 1970) that included a simple physical explanation of the phenomenon, including the fact that it applied to steel cylinders as well as those of non-ferrous metal and was effective on all sizes of the latter down to microscopic ferrous dust. But my paper was itself only a simplification of the more complete treatment by Dr Brown of Sheffield University a year earlier (Brown, 1969). In his quantitative treatment Brown showed that the inherent rotating flux component over an open-sided motor was equal to the purely linearly travelling field that was usually the designer's target. This meant that in a liquid stirring system in which a circular coil arrangement was to produce a swirling effect in the liquid metal above it, the rotating flux component was of much more use in *mixing* the liquid than the designer's so-called 'rotating field', for the latter rotated about a vertical axis and under such influence the liquid could remain in layers. The rotating component described by Brown operated, so to speak, 'top to bottom'. Yet a paper on stirrers published in the same year as Brown's (Sundberg, 1969) took no account of the horizontal axis rotations whatsoever—a clear case of the 'Little Jack Horner' disease from which most scientific papers suffer, at least in part.

6.9 Series connection

Another fact which should be recorded here, if only to save future historians the job of searching through mountains of paper work, is that F. C. Williams' experiments and theoretical arguments of the early 1950s left a mark on linear motor design which remains with us still, namely, all our short primary linear motors have all the coils of each phase connected in series. If this had been otherwise, an entirely different theory would have been needed. Being a 'pentode man' (F. C. Williams could make more use of a single pentode than most electronic engineers of his day could of a whole boxful of vacuum tubes) he was self-educated in the 'constant-current' concept. This approach often landed us in a lot of cross-talk when our variable-speed motor papers were read at local centres of the IEE. The industrialists would protest loudly that large electrical machines 'were never run at constant current', while we would insist that our theory was easily convertible into constant voltage predictions, for what we were really calculating was dynamic impedance not expressible in terms of a simple equivalent circuit, as was any conventional rotating machine. It took some time, I must confess, before it hit me that 'constant current' in this context was to be translated as 'series connection'. It was not until 1965 that I understood the fundamental difference between series and parallel connection (Laithwaite, 1965) and not for a further two years did I appreciate that every electric equivalent circuit could be converted into a magnetic equivalent circuit, thereby 'inventing' the concepts of magnetic inductance and capacitance (which I called 'transferances'), and magnetic impedance which I called 'concedance' (Laithwaite, 1967). Magnetic equivalent circuits are as 'natural' to a parallel magnetic circuit, and therefore to a series electric circuit into which it transforms, as electric equivalent circuits had been for nearly a century for the conventional, but nevertheless man-made, parallel world of constant voltage.

Throughout the 1960s and early 1970s engineers throughout the world devised linear motors in a wide variety of shapes and sizes, but I know of only one which, to date, was not series-connected. This was a *short-secondary* machine developed for impact extrusion of metal at Manchester College of Technology (now UMIST) and whose energy efficiency was raised from 3 to 28 per cent by changing from series to parallel connection. At least as much work, both theoretical and experimental, remains to be done on parallel connection, or what is probably more fruitful, judicious use of series/parallel hybrid connections, as has been completed so far on the series-connected machine (especially if the pole pitches are non-uniform). However original we may at times think ourselves to be, we prefer in the main to 'follow my leader' along at least reasonably well-trodden paths. This could be, indeed *has* been, termed the process of 'brain-washing', although not in anything like so extreme a form as political brain-washing and generally, unlike the latter, not intentional. But there is no brain-washing of the technical form quite so effective as self brain-washing, which makes the process of inventing largely one of brain-emptying rather than one of filling!

As the industrialists began spending more money on linear motor development, the academic clean-up men really went into business and whole books on edge effects alone are now on sale. Happily a blending of industry and university activity has occurred as part of a general pattern of development, and it is unlikely that the technologist will ever again re-build his ivory tower or the industrialist think that theories are 'just for the birds'.

6.10 References

Brown, R. (1969), 'Travelling magnetic waves in electrical machines described by rotating vectors', *Proc. IEE*, Vol. 116, No. 6, pp. 1011-1013

Laithwaite, E. R. (1965), 'Differences between series and parallel connection in machines with asymmetric magnetic circuits', *Proc. IEE*, Vol. 112, No. 11, pp. 2074-2082

Laithwaite, E. R. (1967), 'Magnetic equivalent circuits for electrical machines', *Proc. IEE*, Vol. 114, No. 11, pp. 1805-1809

Laithwaite, E. R. and Hardy, M. T. (1970), 'Rack-and-pinion motors: hybrid of linear and rotary machines', *Proc. IEE*, Vol. 117, No. 6, pp. 1105-1112

Sadler, G. V. and Davey, A. W. (1971), 'Applications of linear induction motors in industry', *Proc. IEE*, Vol. 118, No. 6, pp. 1044-1056

Sundberg, Y. (1969), 'Magnetic travelling fields for metallurgical processes', *Spectrum*, Vol. 6, No. 5, pp. 79-88

Williams, F. C., Laithwaite, E. R. and Eastham, J. F. (1959), 'Development and design of spherical induction motors', *Proc. IEE*, Vol. 106A, No. 30, pp. 471-484

7 The high-speed transport game

7.1 What to leave out, not what to put in!

The previous chapter has hopefully set a background against which the story of *high-speed* linear motors can be told, for nowhere else in the subject, so far, has there been emotional involvement, public awareness, international competition, industrial espionage and political intrigue!

Where to begin is obvious—with Michael Faraday, of course. But we must proceed rapidly, jumping 70 to 80 years to Zehden (1902) and to Bachelet, then on to Kemper (1934) (surely the 'father' of Maglev), on again to Bedford, Peer and Tonks (1939) for induction levitation and finally to the Westinghouse 'Electropult' of 1946, the first high-speed linear motor ever to be built. While I myself see the Electropult as the proper beginning of high-speed motors, there was a gap of 13 years (with one exception*) until the world really began to stir itself about the possibilities of the contactless drive that linear electric motors had to offer. Maybe we had to see a Hovercraft in action or to contemplate space travel seriously before the 'magic' of electromagnetism was brought home to us. Certain it is that I can guarantee to interest an audience for an hour, using only an open-sided linear induction motor and a few bits of metal, whereas similar efforts with a 'rolled-up' conventional motor with iron-cored rotor would risk yawns, clock-watching and all the other general discomfort which an audience can convey to a none-too-interesting lecturer. If the 'end product' is public transport, then the interest is so much the greater, for people of all ages like listening to a story-teller for they may identify themselves with the characters or the action, in this case may imagine what it would be like to be carried over the ground at 300 m.p.h.

7.2 The railway experiments

Once the newspapermen get these thoughts the flood gates are opened and every lift of an eyebrow of those developing the technology is seen as meaningful. So it

---

*The d.c. machine described in chapter 2.

was with the first efforts of Fred Barwell and myself to try out the feasibility of linear motor drives for railways. In 1960 we built an 80-foot track in the laboratories of Manchester University (figure 7.1). Having put a seat on this vehicle and given rides to daily newspaper reporters (acceleration 0.5 g), we had all the publicity we needed for a time. Readers should note that in those days we put a seat on every vehicle that was big enough to take it. During the taking of evidence by the Select Committee of the House of Commons following the closure of the Tracked Hovercraft project in 1973, the Minister of Transport said that it was obvious that the Advanced Passenger Train was more developed than the Tracked Hovercraft because it had seats in it. At the time these were seen as words of wisdom for a vehicle with seats can give rides to journalists who will make enthusiastic reports and the project escalates from there. On this argument, however, the Manchester machine of 1960 was more developed than the full-scale vehicles of Germany and the USA in 1975! Its pole area measured 24 inches × 2 inches and it

Figure 7.1 The 80-foot track in the laboratory at Manchester University, 1960

produced over 200 lbs of thrust, accelerating and decelerating a passenger at over 0.5 g. (For comparison, the wheels of a sport car start to spin on the road at 0.3 g.)

A picture of the experiment was published in the American magazine *Mechanix Illustrated* in August 1961.

The Press therefore attached great importance to our move to the Gorton Loco Works late in 1962 to set up a 100-yard track with a motor of some four times the thrust of the laboratory machine. That we *did* fit a seat—and headlamps (!)—is evidenced in figure 7.2 which shows the author making a point to Mr C. C. Inglis (Director of Research, British Transport Commission) and Dr F. T. Barwell, then Director of Electrical Research.

I remember asking Fred Barwell just why we were building the Gorton track since we knew a linear motor designed for 34 m.p.h. (sync.) would work well. His

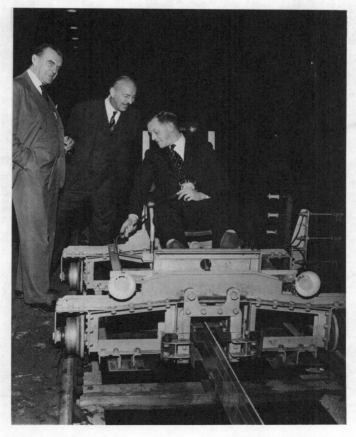

Figure 7.2 Seats come first—then headlamps! The author demonstrates to British Transport Commission Directors C.C. Inglis (left) and F. T. Barwell, 1963 (*courtesy of the Liverpool Daily Post and Echo Ltd*)

Figure 7.3 H.R.H. the Duke of Edinburgh talks to Dr Barwell as they inspect the Gorton vehicle after its transfer to B. R. Labs, Derby. Lord Beeching is on the Duke's immediate left (*courtesy of British Railways Board*)

reply reassured me that we had a common purpose in mind: "To get some money to build a bigger one." It was then that I first knew that the *seat* was the most important part of the whole equipment. The Gorton machine was transferred to the British Rail Research Laboratories at Derby and in due course shown to the Duke of Edinburgh (figure 7.3). It will be noted that for this demonstration the passenger seat had been removed!

Facts to note about the Gorton machine were that with a speed of only 30 m.p.h. and a goodness factor of 4, it developed 90 h.p. at an efficiency of over 70 per cent, for 14 years later the 'low efficiency' of linear motors was still 'seen' as a stumbling block.

It was while riding on the Gorton machine, which was a double-sided sandwich motor, that the problem of track joints first occurred to me. The vehicle itself ran on conventional railway track but the linear motor was connected to it in such a way as to allow relative lateral movement so that the motor could be guided separately from the track aluminium by guide wheels. Side pressure from these wheels could cause the plate to bend and, if a joint between any two adjacent lengths of track was insecure the side-wheel pressure might shear the fishplate connecting the two pieces so that the front edge of the primary would make a head-on collision with the oncoming sheet end (see figure 7.4). Discussing this with American engineers in 1973 I discovered that the same kind of thought had occurred to them also: 'What happens to all the aluminium sheet in that event?' The American test vehicle had three seats abreast. They told me that the engineers always sat on the seats at either side. The centre seat was for distinguished guests only—the 'Senator's Seat' they called it!

The other important point about track joints that we discovered at Gorton was that, even when using a 34 m.p.h. vehicle with a pole pitch of only 6 inches and a goodness factor no better than 4, a recording of input current against time gave the

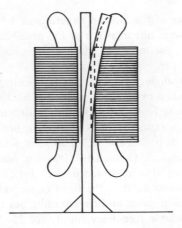

Figure 7.4   End view of a split reaction plate to an oncoming double-sided primary

form of trace shown in figure 7.5. The joints are clearly 'visible'. The motor obviously lost thrust when it passed over a joint for the reason illustrated in figure 7.6. The problem would be worse at larger pole pitches. The spacings between the transients on the current trace were sufficiently accurate for us to use them as measurements of acceleration.

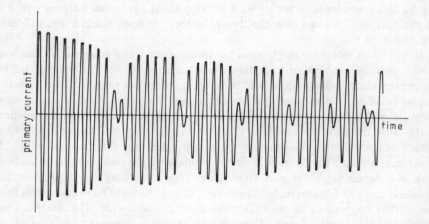

Figure 7.5 Variation of input current with time locates the track joints easily

7.3 The 'Aérotrain'

Even during the existence of the Gorton experiments I became alerted to the fact that overseas engineers were not unaware of the possibilities of non-contact transport—'low level flying' is the term I like to use, for it has many aspects in common with those of aircraft proper.

In France, Monsieur Jean Bertin had been experimenting with a vehicle suspended by air pressure and propelled by an airscrew since 1961. He named it the 'Aérotrain' and was proposing speeds so high that rocket propulsion might ultimately supersede airscrews. There was talk (and patents) in 1962 in both France and the UK of speeds of 250 m.p.h. British Rail's inability to fund a test track prompted me to apply to the Science Research Council in 1966 for funds to build a disc rig on which aluminium, steel or aluminium-faced steel discs, 7 ft in diameter, could be rotated using a 'linear' motor built in the form of an annular sector so as to avoid the problem of different slip values at different radii, but to retain all entry and exit edge effects. Given a variable frequency supply for 25–250 Hz a large amount of almost invaluable data could quickly be accumulated at real speeds up to 300 m.p.h.

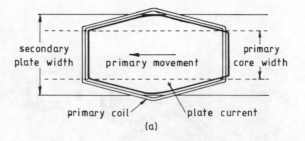

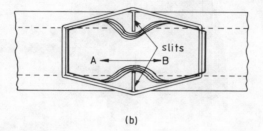

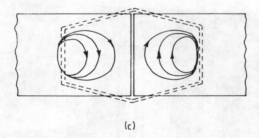

Figure 7.6 The discontinuity in current flow occasioned by a track joint: (a) in a continuous sheet, primary and secondary currents mirror each other; (b) the effect of slits is seen to expose unbalanced portions that produce only leakage-reactance-creating flux; (c) a total break leaves the primary coil facing virtually an open-circuit

(the maximum peripheral speed). Practical measurements, I argued, were sure to be the one thing anyone would need who wanted to enter the high-speed transport game. No sooner had my equipment been assembled (figure 7.7) than I learned of a similar rig in Japan, and another in Italy—other countries, too, were economising to the extent of preferring a disc rig to a long track, initially.

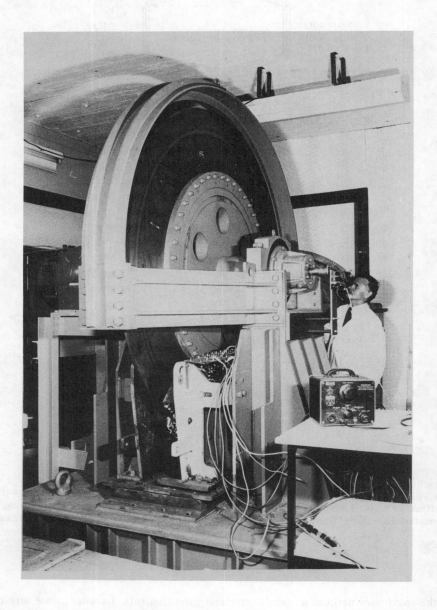

Figure 7.7 7-foot diameter disc rig installed in the labs at Imperial College, 1969. Maximum peripheral speed 300 m.p.h.

As early as March 1961, I had received a small deputation from the National Research Development Corporation's subsidiary company, Hovercraft Development Limited, a company set up to exploit Christopher Cockerell's invention. They were considering the possibility of forming a further separate company, to be called 'Tracked Hovercraft Limited' (THL), to utilise more effectively the principles of the hovercraft. It was obvious that a vehicle hovering 5/8 inch above a smooth, concrete beam would waste less air and therefore make less noise and less of an atmospheric mess (turbulence) than would a land/sea vehicle with a large clearance. The NRDC people had come to study the possibilities of linear motor drive, for at that time the proposal had been to drive by airscrew, as had Monsieur Bertin's. The choice of a noisy, gale-making airscrew seemed a curious one to me, until I remembered that the men who had been to see me were aerodynamic engineers and I recalled a legend of an Englishman who was a pillar of the RSPCA and was appalled at the cruelty to animals in Spain. He decided to set up a branch of the Society in Spain. On arrival he found many sympathisers but they warned him: 'Señor, you needa tha money. You do nothing withouta tha money!' 'True,' said the Englishman, 'what's the best way of raising money here?'—'A bullfight' came the unanimous reply. For such reasons, perhaps, did NRDC first propose airscrews and did Bertin retain his airscrew drive longer than did all others. (His first linear induction motor drive did not appear until 1968.) What was surprising to me, as a novice in the industrial scene, was the speed with which news was passed to interested parties thousands of miles apart. In 1962 I received a letter from the National Aeronautics and Space Administration asking for linear motor design information to help them build acceleration test equipment.

Another piece of writing of the early sixties, dismissed without a second thought no doubt in Whitehall, appeared in *New Scientist* from the pen of Timothy Johnson, and what an accurate prophet he turned out to be. Under the heading 'Science and the Paymasters', he wrote on the subject of lifts as follows:

"The drawback is that a linear motor lift costs about £40 extra. This may not seem much on equipment that can cost £10 000, but the biggest customers for lifts are local councils, and they are bound to take the lowest satisfactory tender. 'I've known a £4 million contract lost for less than £1000' remarks Dunlop ruefully. But he remains confident that 'as people discover it's an indestructible, rugged component', the linear motor will come into favour.

"What the linear motor needs to develop rapidly is someone big enough to carry the cost and effort of designing a modular range of motors and controls for a mass-market application, setting them at a price which is realistic from the user's point of view, and then waiting for rates to grow to a volume which justifies the price. . . . One is left with the uneasy feeling that, despite our current world leadership in linear motor design, British industry is too tired and conservative to try anything so adventurous."

I wish everyone had believed in our 'world leadership' at that time. By 1976 I had wished it even more.

7.4 The formation of Tracked Hovercraft Limited

Four months were to elapse, following the initial discussions with the NRDC representatives, before I received a request from a Board member of that organisation to visit me in Manchester to hear the linear motor story again first hand. The meeting took place on the 28 July of that year (1961). I spent a whole day with Dennis Hennessey who was later to become Chairman of the new company.

It is necessary to specify 'of that year' in connection with the start of the project for although I expected some rapid action following the visit of the 'head man', it was in fact 1962 before I next met the same people who had first discussed the project with me nine months previously. "This time," they assured me, "we have got the green light." 'Amber' might have been a better description, for not until 1967 was the Tracked Hovercraft company actually formed.

It was at this point that I had some misgivings about NRDC's handling of the project. Six years to reach that decision was a long time and I recalled my assistant lecturer days when I was Secretary to the Inaugural Conference of the world's first Universal Digital Computer (the Ferranti Mark I). By the time the conference was held, F. C. Williams, Tom Kilburn and their research team had completed the prototype of Mark II, later to become 'Mercury'. These were the days when many hours were spent in trying to name projects in such a way that their initial letters spelt out an interesting word.

News hounds at the Computer Conference wanted a name for the machine and someone came up with the 'MADAM' (Manchester Automatic Digital Adding Machine). It was agreed among the academics that this was a good name, for at times the machine did appear to have a mind of its own! 'What shall we call the Mark II?' was an obvious question. One morning one of the research team bounced into the lab. with 'Got it!—BISON'. How did he make that one? 'Binary Indicating Summator of Numbers.' he said. 'That's not clever,' said another. 'I know,' added the inventor, with a glee that showed there was more to come, 'but those of us who have worked on it will always remember that it stands for 'Built In Spite Of NRDC.'

But as any actor will tell you, only *no* publicity is *bad* publicity and the cartoon shown in figure 7.8 which appeared in the technical newspaper *Engineering News* in 1962 was a real boost for anyone like myself who did appear, at that time, to be alone and hanging on to a project by the fingernails. In order to tie in the subject of this chapter with those of the foregoing, however, it should be noted that 1961 was the year in which I first talked with Gilbert Sadler about crane motors and the possibility of forming the company 'Linear Motors Limited'.

Over the years that followed the formation of Tracked Hovercraft, I grew to respect very highly the brain power of the men directing the research, especially in the case of Mr Dennis Hennessey (Chairman) and Mr Tom Fellows (Managing Director). But some of the *Board's* policies (NRDC that is), the thinking behind which was never made known to me, were to give me some little insight into the BISON story, in the years to come.

Following a Press Release on 16 May 1967, describing NRDC's proposals to form the company Tracked Hovercraft Limited, the first of many surprises came with a formal invitation to attend a joint Press Conference to launch the new

Figure 7.8 Cartoon (after Mortimer) appearing in *Engineering News* (National Trade Press) 1962

company which was to take place at the Headquarters of British Railways in London on the 17 July 1967. A few days before that event, however, there were frantic phone messages to all those invited to say that the Treasury would not release the £2 million promised and that the Press Conference was cancelled. The money was in fact made available some six weeks later and the real cause of the delay was never made clear. It is possible that some Treasury official may have read one of the several critical articles on linear motors which were appearing here and there about that time, articles coming to the conclusion that linear motors were hardly suitable for *any* application, least of all transport. For example, D. S. Armstrong, writing in the *Railway Gazette* (Armstrong, 1967) came to the conclusion that: "It is therefore unlikely that linear motor propulsion can be used with existing railway vehicles, and it can only be considered for existing routes reserved for special vehicles or for completely new routes" and he showed "test results obtained with a small two-stator linear motor" having a maximum efficiency less than 40 per cent and a power factor of 0.3 at running speed (0.45 on starting).

To be fair to Donald Armstrong I always found him ready to listen, which is more than I can say for most of the industrialists whom I met in the early 70s (when it really mattered), and Armstrong himself conducted extensive experiments at Derby, on behalf of British Rail, between 1973 and 1976 using, oddly enough, three 'existing railway vehicles'! But of course we had come through a topological explosion between 1967 and 1976 and single-sided motor techniques, the use of wide slots and narrow teeth for primary windings (Laithwaite and Barwell, 1969), the concept of transverse flux (Laithwaite *et al.*, 1971) and the invention of electromagnetic joints put right all the defects that were listed in the *Railway Gazette* of 1967. I should also like to record here that Donald Armstrong was the first researcher to test electromagnetic joints at speed on real vehicles, and was the first man outside Imperial College and THL to appreciate their worth. What is more he lost no time in telephoning me to congratulate us on the concept, once he had made the initial tests.

7.5 The pace makers

News of further overseas activity had meanwhile been received. Having tested the half-scale 01 Aérotrain in 1965, Monsieur Bertin added thrust boosters and attained a speed of 345 k.p.h. (215 m.p.h.) and went on to design the 02 Aérotrain (figure 7.9), also half-scale. In Japan, the Railway Technical Research Institute reported

Figure 7.9 Monsieur Bertin's second half-scale hovering vehicle '02 Aérotrain', built 1968, ran at 263 m.p.h. (422 k.p.h.), 1969 (*courtesy of Société de l'Aérotrain*)

a linear motor experiment (with photograph) on a small, indoor, wheeled vehicle (1964). Monsieur Bertin formed his company, the 'Société de l'Aérotrain' in 1965, the first such company since that of the other French pioneer Monsieur Bachelet, in 1914, but, as yet, there were no plans for linear motor propulsion of the Aérotrain.

Other Britons had also been 'putting it together', at least mentally and, on 9 January 1964 the *Engineering News* reported on its front page: "MONORAILS TO SPEED UP UK TRAVEL" with a sub-title "Consortium formed to pioneer new transport systems." The company was to be called 'Speedover Transport Limited' and was said to be a combination of British industrial and banking concerns with former Minister of Power, Lord Mills, as President and to consist of the Hawker Siddeley Group, Sir Robert McAlpine and Sons, Philip Hill, Higginson, Erlangers (Bankers), Rapid Transport Development Company, Edger Investments and Stockland Property. Sir Miles Thomas was to be executive chairman.

I never knew what became of this collection. One name in particular was to reappear some eight years later in a different role, but one thing is certain—no-one from the consortium ever came to see *me* during the early 1960s. Nor, according to *Engineering News*, had they 'so far' been to see Christopher Cockerell, but the article pointed out that: "Two years ago Hawker Siddeley entered into a three-year agreement with the French Safège consortium of engineering and financial interests, to explore monorail possibilities." But it passed away, no doubt, in the light of what was to follow, to the relief and delight of the Civil Servants in Whitehall.

After the initial scale-model demonstration track of a Tracked Hovercraft had been set up in the garden of what had been a private house in Hythe, Hampshire, it was exhibited at 'Hovershow 66' in Browndown, Portsmouth (figure 7.10). The success of the model gave approval to move away from the Hovercraft Development Ltd's site and to set up a full-sized track 'somewhere very straight'. The site chosen was between two drainage rivers running from Earith in the South to Denver Sluice in the North, a distance of almost 20 miles. The land was loaned from the Great Ouse River Authority and although marshy, huge concrete piles were sunk down to 70 feet to meet the bedrock. The hangar to house the vehicles was to be at the Earith end and building proceeded northwards. At about the 3-mile point there was a road crossing and it was proposed to use this to demonstrate high-level track (figure 7.11) as well as low-level (figure 7.12).

One of my first tasks for THL, and one of the hardest, was to give professional opinion on two study contracts (each worth £5000) which were awarded to Associated Electrical Industries, for whom I had acted as consultant for many years, and to the English Electric Company, with whom I had had little contact. On first submission I felt bound to say that the English Electric proposal was just what I myself would have suggested at that stage of the game. It was to supply power at 200 Hz from a trackside station and to feed it 'raw' to the linear motor. The hoverpad fans needed high-speed motors for which the 200 Hz supply was ideal. Trackside generators would not be subject to the restrictions on power factor imposed by the use of the National Grid, or to the weight penalty of inverters, and on a test track of only a few miles, transmission line reactance would not rear the same size of ugly

Figure 7.10 The model hover vehicle exhibited at Hovershow 66 (*courtesy of Hovercraft Development Ltd*)

Figure 7.11 High-level track of the full-scale tracked hovercraft, Earith, 1970 (*courtesy of Shephard Hill and Co. Ltd*)

Figure 7.12 Low-level track at the southern end of the Earith track (*courtesy of Tracked Hovercraft Ltd*)

head as it would on an inter-city link which was to be the ultimate target. The English Electric scheme was a 'belt and braces' solution, sure to be overweight, sure to be of lower power factor—sure to WORK! In my view, the only mistake one could make at this stage was for the beast not to work at all, for some unforeseen reason, or to have such 'teething troubles' as would lead to a long development time and give our overseas competitors a chance to see the mistakes we had made and to profit from them. I hoped that we would have found a solution to the power/weight and power factor problems by the time the prototype was running, so that we could demonstrate a 250 m.p.h. vehicle to the World in a few months with the powerful additive: 'Of course, we now know how to do *much* better than that.'

The AEI submission could not have been more different. D.c. supply was to be fed to the vehicle by the slip tracks and fed 'raw' to the fan motors (for safety—a strong argument indeed). On board the vehicle the d.c. was to be converted to variable-frequency a.c. by solid-state inverter for the linear motor. This would allow efficient speed control, increased power/weight ratio in the motor itself and regenerative braking—a whole bagful of powerful selling points, but the system was complex, to say the least. I might have found myself in a dilemma in that I favoured the English Electric scheme but was consultant for AEI, but something happened which made my opinions of no consequence, for English Electric followed their original submission with a revised scheme (within a few weeks) and the latter was almost entirely in accord with the AEI proposals. Whether or not this decision was in any way connected with the AEI (by now 'GEC') merger with English Electric which was shortly to follow, I doubt whether I shall ever know, but THL accepted the AEI scheme and a meeting was arranged between the design engineers of that company, THL representatives and the consultants, to thrash out the more detailed aspects of the design.

The motor to be built was for THL's first full-scale vehicle (named 'research test vehicle' (RTV) 31) to have a full speed of up to 300 m.p.h. and to weigh approximately 23 tonnes. When we looked at the inverter design, the consultants and THL men looked across the table at each other in horror, for this item alone could weigh as much as 13 tonnes! It was then explained to us that the blame lay with the linear motor whose power factor was only of the order of 0.3. "Inverters don't like reactive kVA," we were told, "the motor needs 40 per cent more of this than the original estimate." "Then why," I asked, "is the inverter not 40 per cent heavier instead of 100 per cent as shown?" It was explained to me that inverters came 'in lumps', like loaf sugar. You could have one lump or two, but not 1.4, and we were just lucky with the first linear motor estimate in that it consumed marginally less than one lump could supply.

The meeting broke up in some disorder with no firm conclusions as to the future —except that we were in a Grade I mess. We were, as we now know, tasting the poisonous fruits of long magnetic circuits which could not be magnetically 'insulated'.

1966 had seen the entry of some new and formidable 'gladiators' in the shape of the Department of Transportation in Washington D.C., who gave a study contract to the Garrett Corporation (AiResearch Manufacturing Division), and of the West German firm of Messerschmitt-Bölkow-Blohm, who also began papers studies, closely followed by Krauss-Maffei. The Americans spotted the difficulties of long pole pitches and had coined a phrase that defined the 'best' frequency for a given running speed. It was to be 'one cycle per second per mile per hour', which can be translated as: when the pole pitch of a linear motor reaches 10 inches, hold it at that and increase beyond standard frequency for higher speeds.

7.6 The long pole pitch problem

The synchronous speed, v_s, of a linear motor is given by

$$v_s = 2pf$$

and if one substituted 50 Hz for f and 400 ft per second for v_s ($\approx$ 250 m.p.h. and by the gift of Dame Fortune also $\approx$ 400 k.p.h.), the disturbing value of pole pitch, p, emerges as 4 feet (with allowance for slip, something over 4 feet 6 inches ($\approx$ 1.5 metres)). Imagine now a linear motor as shown in figure 7.13, perhaps 12 inches wide and 18 ft long (4 poles) and think of the weight of copper wasted in end windings and the leakage it produces (a). Of course one could short-chord the winding viciously (say to 1/3), doubling the primary ohmic loss due to the copper in the slots but dividing it by between 2 and 3 in the end windings. One could, indeed *must*, adopt aircraft generator standards where current densities of 13 000

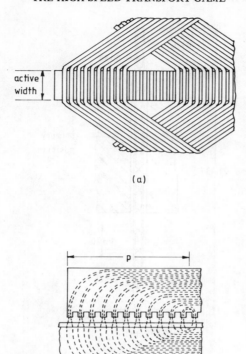

Figure 7.13 (a) The end winding problem of a mains-frequency, high-speed motor; (b) the core flux problem

amps/sq. inch are commonplace. As suggested earlier, this is not the only place where the constraints placed on designers of high-speed ground transport motors are similar to those on aircraft designers.

The real 'end of the road' for conventional, mains-fed, high-speed linear motors in 1968 was not, therefore, the primary end windings. Rather it was the *magnetic* circuit that called the tune (b). Tracked Hovercraft and Imperial College had already agreed in 1967 that double-sided motors were OUT on grounds of safety alone. There was a fantastic hollow track section of aluminium subsequently produced by one American team (shown in figure 7.14) which looked as if it would wave about in a gentle breeze, *before* the motor ever got near it. The Americans also tried to weld aluminium track in the manner in which British Rail had welded steel rails to such good effect. Aluminium does not have the strength of steel. It was worth a try, but on a cold night the track aluminium split in several places.

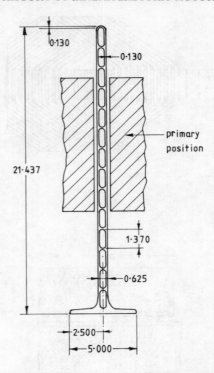

Figure 7.14 Aluminium plate section actually used in a test track for double-sided sandwich motors

To go 'single-sided', however, looked as foolish to the Americans as their welded plate looked to us, for at first sight it looks as if the disadvantages of single-sided working are monstrous—a *range* of mountains one might never cross. Listed, they are

(1) Magnetic pull between primary and track, by all the rules of rotary machine design, must increase the weight of the vehicle by several times the force of gravity.
(2) The depth of steel needed for big pole pitches must be of the order of a foot or more and therefore
(3) The track iron must be laminated. The cost of such a track is prohibitive.

What no-one in the world seemed to know was contained in a paper that Fred Barwell and I had published in 1969 (Laithwaite and Barwell, 1969) in the *IEE Proceedings*. But people do not read other people's papers—there just isn't time. But this one expanded on a concept that a colleague, (then Dr) E. M. Freeman, had first put to me some years earlier that $B^2/2\mu_0$ newtons per square metre was not the exact evaluation for the normal force across the airgap of an induction machine and that under certain circumstances, what commercial designers had long called 'unbalanced magnetic pull' (UMP) could even become 'unbalanced magnetic push' (Freeman and Laithwaite, 1968).

The full expression for normal force should read $B^2/2\mu_0 - \mu_0 J^2/2$. If typical values for B and current loading J (amps/metre) are inserted into this formula the first term dominates the whole to the tune of perhaps 97 per cent. The industrial designer is right to ignore $(-\mu_0 J^2/2)$. But the paper with Fred Barwell proved the relationship for all induction machines that

$$\frac{B}{J} \propto \frac{p}{g}$$

where p is the pole pitch and g the airgap. In small rotary motors, where p is small, or large linear motors, where g is big, the value of B is likely to be some five times smaller than normal, so that the teeth for a given tooth pitch can be reduced by that factor. This virtually doubles the available slot width and, for the same slot leakage, doubles the possible slot depth, multiplying J by a factor of 4.

The effects of this radical difference between linear and rotary motors of comparable power-handling capacity are twofold and were still not fully appreciated by those who directed top projects in the technology in 1976. The first effect is the more obvious. Teeth of 1/5 the width of those of a comparable rotary machine need only 1/5 of the core depth and since this applies to both vehicle-carried primary and to track, in a single-sided system, it reduces the track steel depth to a few inches, instead of more than a foot.

The second effect is on the normal force. Even if $B^2/2\mu_0$ represents 97 per cent of the total normal force in a rotary design and *attracts* the primary to the secondary, $\mu_0 J^2/2$ is a *repulsive* force. When B is divided by 5 and J multiplied by 4, the ratio 97:3 becomes $(97/25) \div (3 \times 16)$—that is, less than 1:12—and the large *lifting* force occurs when the values of slip are high. These facts were known and demonstrated first at Imperial College in September 1964, but were still not being believed by the majority of workers outside the College as late as 1976. Even so, a few inches of laminated steel in the track was excessive, and it was thought that flux doubling, a phenomenon peculiar to linear configurations, virtually put mains-fed linear induction motors 'out of court'.

By 1968 a world situation in 'the game' was developing. Monsieur Bertin had been converted to the usefulness of linear motor propulsion by (I think) Monsieur M. Barthalon, who was developing low-speed motors for urban mass transit systems (the 'URBA' system was the name for M. Barthalon's outfit), but had yet to incorporate it in a test vehicle. The West German firm of Krauss–Maffei had started paper studies, while in East Germany Peter Budig had published his first linear motor paper. Washington DOT had economised on track length by accelerating their test vehicle by twin jet engines up to high test speeds in a relatively short distance. On-board high-frequency turbines gave the USA an apparent lead over the UK and France.

When faced with the long pole pitch problem and the possibility of a 13-tonne inverter for a 23-tonne vehicle, emissaries from THL went to France and to the

USA to see what 'the other guys' were doing about it. In France it appeared that the old Gramme-ring winding had, like Earnshaw's Theorem, been 'taken out and dusted' and designs were proceeding using large pole pitches and 50 Hz supply. They too, had spotted *one* of the constraints—that is, long end windings—but not the other (thick, leaky magnetic circuits). The leading contender in the USA had a proposal almost identical to the AEI scheme, except that the inverter was to weigh not 13 tonnes, but 16.

Within a fortnight of the news from France, we had put together a laminated iron core 10 ft long and 5 inches × 5 inches cross-section. A few turns of thick, insulated wire were wrapped around the bar, Gramme-ring fashion, to correspond to the winding in each slot (had there *been* slots). We were going to give this winding every chance to be good by eliminating primary slot leakage, zig-zag leakage, and the like. The pole pitch was 5 ft. Opposite one of the two appropriate primary faces—that is, the one displaying the edges of the laminations—a second laminated block of the same dimensions was mounted, carrying a 1/4 inch sheet of aluminium on the face opposite the primary, allowing a 6-inch overhang on each side. There were three immediate observations:

(a) With a mere 10 amp supply the leakage field was so great that feed wires writhed and twisted in it at distances of several feet from the motor, as the current was switched on. Measuring instruments had to be at least 10 ft from the motor.
(b) A search coil around the bar indicated that the core would display marked saturation effects at only 10 amps/phase.
(c) The power factor measured was 0.05.

7.7 The 'darkest hour before the dawn'

There was deep gloom for a period of about six weeks and for myself, who had been described more accurately as its 'prophet' rather than its inventor, I felt as if I had 'sold my friends up the river'—or rather more literally down the Earith 'drain' in this case. Work had proceeded to the point where there was an empty slot up the centre of the top face of the track (figure 7.15) waiting to receive—who knew what kind of secondary metal? I must have repeated to myself a hundred times "The French made Gramme-ring windings and made it worse." Then late one night I changed the wording to "The French turned their electric circuit through a right angle and made it worse." In a flash came an obvious retort "Why not do the *opposite* and make it better?" The 'opposite' of a Gramme-ring winding was *not* obvious. But the opposite of an electric circuit *was*—the magnetic circuit.

My first effort is shown in figure 7.16 and it lingered about two minutes in my mind before I dismissed it as bad since it contained more steel than would a Gramme-ring arrangement, but the second concept was already supplanting it (all this I should add, was taking place on the top deck of a No. 72 London 'bus!). The second layout is shown in figure 7.17. This half-a-slot/pole and phase winding could of course be extended to a properly distributed winding as shown in figure 7.18.

Figure 7.15 The Earith track top awaiting its metal 'filling' (*courtesy of Tracked Hovercraft Ltd*)

But this also would leak badly, I argued, so I put the windings right on to the level of the pole faces as in figure 7.19. By this time I had arrived at my office. I rang THL at once and told the Chief Engineer, Denys Bliss, of my findings, dictating the shape over the telephone. During the following week my secretary received a letter (on 4 March 1969) that Tom Fellows had sent from San Francisco to London for typing on 24 February. It began as follows:

"Were it not for the fact that I am visiting the Garrett Corporation the day after tomorrow and do not wish to appear to have been influenced by them I should have waited to ask the question contained in this letter until my return. If there is some fundamental error in the underlying thinking perhaps you would be forgiving enough to put it in the category of the nine false trails which we have to follow for each one that succeeds.

146   A HISTORY OF LINEAR ELECTRIC MOTORS

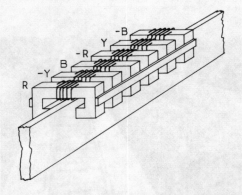

Figure 7.16 First thoughts on a transverse magnetic flux machine

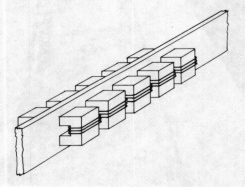

Figure 7.17 The 'Mark II' TFM

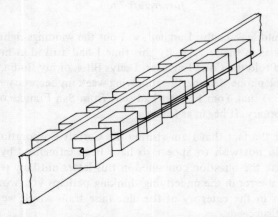

Figure 7.18 Distributed winding arrangement for TFM

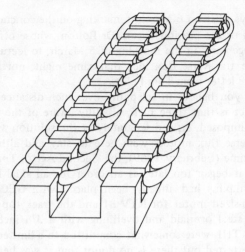

Figure 7.19 Reduction of leakage is achieved by placing windings near pole faces, as in this single-sided 'C-core' primary

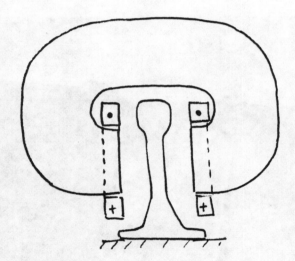

Figure 7.20 Sketch by Mr T. G. Fellows of his idea for a transverse flux configuration

"The question relates to long pole pitch motors with narrow pole faces and it is to ask if there is any point in rotating the plane of the magnetic circuit through 90° so that it lies transverse to the direction of motion? (see simplified sketch)." (figure 7.20). I was in Southampton on 4 March and read the letter on my return on 5 March.

I wrote to Denys Bliss on 6 March, remarking on the concurrence of thought in relation to my call of the previous week. Dr Bolton, whose office is next to mine, left Imperial College early in the morning of 5 March, to lecture at Runcorn, and came up with the transverse flux idea that same night, not knowing about my ideas, nor about the letter from Tom Fellows.

Whether or not you believe in telepathy over such distances is of little consequence. What is fact is that three men, all fully aware of the problem, all under the stress which it imposed, arrived separately at the solution within a few days of each other. Transverse flux motors were here to stay and although we were not aware of it at the time (February 1969), the 'Second Age of Topology' had begun. For THL however it began too late, as an order for an axial flux machine for a lower speed (150 m.p.h.) had already been placed with GEC, who went ahead and designed a full-sized motor for RTV 31 and the track top was filled with an inch of laminated steel overlaid and overhung with a 3/8 inch thick aluminium sheet (figure 7.21). THL were somewhat committed to a limited number of motor types that could be tested but there is no doubt that it was better at that time to do something than to sit thinking about it.

Figure 7.21 The 'filling' is decided, 1 inch of laminated steel is shown being surfaced with a $\frac{3}{8}$-inch aluminium extrusion (*courtesy of Tracked Hovercraft Ltd*)

7.8 The three-part problem concept of the mid '60s

About the same time, a Japanese leaflet printed in English had declared that the problem of high-speed transport could be divided into three distinct facets

(i) supporting the weight
(ii) guiding the vehicle laterally
(iii) propelling the vehicle.

It was claimed that cryogenic coils (superconducting niobium-tin in a liquid helium bath), mounted on the vehicle and carrying monstrous direct currents, would produce (i) and (ii) when the vehicle was at speed, if the coils faced an aluminium-covered track (both horizontally and vertically) while the linear motor was the obvious choice for (iii).

Other workers, of course, had their own ideas. We had put THL's money on air cushions in the first instance. British Rail were backing the Advanced Passenger Train, APT (steel wheels on steel rails) for the very best of reasons. It was not until Dr Sydney Jones (Board Member for Research of the British Railways Board) gave a stimulating colloquium at Imperial College in 1973 that I fully appreciated just how right was the concept behind APT. I cannot guarantee to remember his exact words and if I misquote him I trust he will forgive me, but they went something like this: "We have, from our ancestors, a great legacy in the form of thousands of miles of track. They made it *flat* enough but they didn't make it *straight* enough (for modern needs). So we are developing APT to take the corners faster as this is the only way open to us at this time to reduce journey times. We are making the best of what we have got, with the money available." I reminded him that the Lancashire cotton industry did just that between 1930 and 1955 and lost a national industry in a quarter of a century. (See also 'The Future of Steel on Steel', *Illustrated London News*, 1966.)

Monsieur Bertin was totally committed to air suspension and to linear motor drive for low speeds. In July 1969 his first full-scale Inter-urban Aérotrain I-80 was presented before a press conference at Le Bourget Airport and in September it achieved 250 k.p.h. (156 m.p.h.) on the 9 km track (figure 7.22). It should be noted that the propulsion system of a shrouded propeller with boosters was still retained but a second string project was now being undertaken for suburban transport and the Suburban Aérotrain 44 (figure 7.23) was linear motor propelled by a double-sided sandwich motor (figure 7.24) built by Merlin Gerin. The full-scale track, 3 km long at Gometz, with current collection system, is shown in figure 7.25. The S-44 carried 50 seats and weighed 10.5 tonnes. The 350 kW supply was 3 phase at 50 Hz and the motor developed a maximum thrust of 1.6 tonnes.

West Germany was now in the hunt too, and the firm of Krauss-Maffei was putting its money into a magnetic suspension now referred to as 'Maglev', but much confused with the cryogenic lift and guidance system which many American workers also referred to as 'Maglev'. The magnetically attractive type was first demonstrated to me at Manchester in 1946 by John West, now Vice-Chancellor of the University of Bradford. In its simplest form it consists of a sensing device (figure

Figure 7.22 Bertin's inter-urban vehicle I80 on its 18 km track. By 1971 it had run for 700 hours at 160 m.p.h. (260 k.p.h.) (*courtesy of Société de l'Aérotrain*)

Figure 7.23 The Suburban Aérotrain S44, propelled by double-sided LIM (*courtesy of Merlin Gerin*)

Figure 7.24 The double-sided LIM for S44 in its test chassis, delightfully called 'Chariot d'essais' (*courtesy of Merlin Gerin*)

Figure 7.25 The French track at Gometz. A 3-phase line can be seen alongside the inverted-T aluminium alloy secondary (*courtesy of Société de l'Aérotrain*)

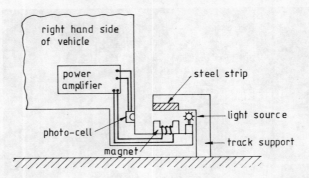

Figure 7.26 Basic system of 'attractive Maglev'

7.26), such as a light source and photocell, detecting the position of a set of electromagnets that are attached to the vehicle and feeding error signals to an amplifier that in turn energises the lifting magnets. Such a system can be stabilised and made to correct disturbances rapidly and with damping. Professor West moved to the University of Sussex and set up a research team to develop the idea. This can now be seen to have been going on in parallel with the work of the German firm.

In the USA they were doing 'some of each'. The Garrett Corporation were developing thrust units on contract from the DOT, Washington D.C., who also awarded study contracts for cryogenic levitation studies both to the Stanford Research Institute and to the Ford Motor Company. The Rohr Corporation decided to pursue Maglev ('Romag', the feedback type) and in parallel developed a prototype (150 m.p.h.) tracked air cushion vehicle (PTACV) with Bertin-type air cushions and a double-sided, Gramme-ring, fixed-frequency linear motor built by Le Moteur Linéaire with a GEC voltage control, illustrating that international exchanges of technological know-how *are* possible (despite assertions to the contrary later to be made before the Select Committee in 1973). At Pueblo there was an air cushion vehicle TACRV (Tracked Air Cushion Research Vehicle) designed for 300 m.p.h. with peripheral jet air cushions similar to those developed in the UK, and a double-sided linear motor, inverter-controlled and fed from the trackside, together with a Linear Induction Motor Research Vehicle (LIMRV), running on steel wheels, the counterpart of THL's plans for a similar 160 m.p.h. vehicle, RTV 22, which was never completed. In France the 'Tridim' vehicle was driven by mechanical rack and pinion. I reckon that if radiation pressure (by means of which stable levitation of microscopic objects can be achieved) had been only vaguely 'on' at that time, someone would have taken it up during this period of gestation when the engineers of the world saw future transport as a three-part problem and were sharpening their weapons for the real in-fighting to begin.

7.9 Global competition 1966-1972

Interest in high-speed linear motors was expanding rapidly. To recapitulate, the line-up in 1966 was as follows:

(1) The Garrett Corporation had their first contract from the DOT, Washington D.C.
(2) Messerschmitt-Bölkow-Blohm were doing studies. The Technical University at Braunschweig were also doing paper studies.
(3) Hovercraft Development Ltd had a 6 ft working model shown at Browndown in 'Hovershow 66'.
(4) Imperial College had a 7-ft diameter disc rig.

Monsieur Barthalon had joined the contenders with the URBA system in France, initially for comparatively low-speed urban transport, and had made his first tests of Gramme-ring linear motors on an air cushion vehicle, and attained 40 m.p.h. Garrett had started their full-scale linear motor programme. They had produced a computer program for the design of linear motors to include end effects, and this was being used in British firms as well as in those of other countries, in preference to what became known as 'the Manchester method' (Laithwaite, 1966). The trouble with the latter was that, as published, it contained no terms to account for either primary or secondary leakage, so got the starting conditions very wrong. Fred Eastham and I knew how to put leakage in and considered our system of calculating effects near synchronous speed as superior, for the Motor Industry Research Association's accelerator at Nuneaton, officially opened by the Minister of Technology (The Rt Hon. Anthony Wedgwood Benn) in March 1968, had a terminal speed correct to better than 1 per cent. The motor and track (figures 7.27 and 7.28) were built by AEI and by 1976 had completed eight years of continuous service.

If there was one facet of the game that was still in its infancy in 1968, it was communication between workers with common interests. At Imperial College we were 'going it alone' in the belief that we were winning. So, I suspect, were several other competitors. When overseas engineers visited Imperial College they were shown as much as we thought was good for them. They were shown the problems on one rig while the demonstrator was literally leaning on the 'answer' in the form of another, more advanced rig, to which their attention was not to be drawn.

In Germany, Messerschmitt-Bölkow-Blohm landed a systems analysis contract with the Federal Ministry of Transport in 1969 (the year we invented the TFM). Another German firm, Krauss-Maffei, began a 'study of the development of future long-haul, high-speed transportation' in 1968 and had made a static rig to prove the feasibility of attractive magnetic suspension with feedback amplifier and followed this with a 'worldwide study on the development status of high-speed transportation systems'. (In 1974 these two German firms were to combine their programmes.) MBB completed a full-scale vehicle in 1970 (figure 7.29).

In 1971 Krauss-Maffei produced a vehicle labelled 'Transrapid 02'. This machine weighed 11 tonnes, was 11.7 metres long, and with it they claimed a 'world record'

154   A HISTORY OF LINEAR ELECTRIC MOTORS

Figure 7.27 The car crash test facility at the Motor Industry Research Association's laboratories at Nuneaton (*courtesy of Motor Industry Research Association*)

Figure 7.28 The moment of impact must occur at constant speed, in this case 30.4 m.p.h. (*courtesy of Motor Industry Research Association*)

Figure 7.29 Messerschmitt–Bölkow–Blohm were first to test a full-scale feedback amplifier-type magnetic suspension (*courtesy of Transrapid International*)

Figure 7.30 'Transrapid 02' achieved a 'world record' of 164 k.p.h. in 1971 using a double-sided linear motor (*courtesy of Transrapid International*)

of 164 k.p.h. (102 m.p.h.) on a 930-metre track on a curve of 800 metres radius, with linear motor propulsion (figure 7.30). (Garrett at that time had done 94 m.p.h.)

German academic interest in linear motors began as an offshoot of magnetohydrodynamics, but not until 1971 did they do more than paper studies. In that year the West German Ministry of Research and Technology established a programme to "introduce novel technologies into the field of high-speed transportation." They concluded that "Gramme-ring windings were preferable" for large pole pitches, but tried to improve the power factor by "an additional ladder-like damper winding that reflects the magnetic flux back into the secondary." The results for a motor 1 metre long were said to be "highly satisfactory." Be that as it may, I had tried copper sheet shielding of Gramme-ring motors (see figure 7.31) in the late 1950s in Manchester and found that all you did was to convert a primary series impedance of $0.1 + j5.0$ into one of $2 + j2$, and I would hardly have called this 'satisfactory'. But the Germans and I were unaware of each other, even in 1971.

Meanwhile, in France, Monsieur Barthalon was claiming the "very first tests on a linear motor associated with an air cushion vehicle of the type URBA" in 1968. This claim was justified only to the extent that his air cushion was a vacuum type, whereas THL's was a pressure cushion. Nevertheless, in 1969 his project 'Test Wagon' using a double-sided, Gramme-ring motor, reached 180 k.p.h. on Monsieur Bertin's 3-km track at Gometz. When large organisations begin claiming world records, I reckon it can be assumed that 'the game' has become serious. I mention this here for there were many suggestions in the later 1970s, in evidence before

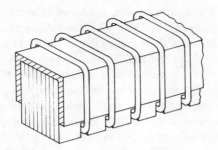

Figure 7.31 Copper sheet shielding in an attempt to reduce primary leakage flux

the Select Committee and elsewhere in Britain, that no-one overseas was really taking linear motor propulsion for inter-city transport seriously.

At this time (1971) the French firm Le Moteur Linéaire had begun collaborating with the British firm Linear Motors Ltd (note that both companies had the initials 'LML'—which was most confusing for managers and politicians alike!), to the extent that there was a joint company named 'Lintrol', the Loughborough firm's marketing trade name being 'Lintrol Systems, UK' and the French counterpart 'Lintrol Systems, France'. There were rumours of a merger of Le Moteur Linéaire with the very large industrial firm of Merlin Gerin of Grenoble, at which time the Loughborough subsidiary of Herbert Morris cut the French firm adrift. Le Moteur Linéaire made the Gramme-ring motor for Monsieur Barthalon's Gometz vehicle and went on in 1972 to make a motor for the Rohr PTACV in the USA, despite 'home competition' from the Garrett Corporation who had had the large study contract from the Department of Transportation in Washington D.C. Garrett could justifiably have felt affronted, for they probably had the best theoretical design equipment in the world at that time. Even the under-modest publicity for the French machine had a hint of some disappointment. Designed to give a starting thrust of 2100 daN, and with a synchronous speed of 77 m/s (277 k.p.h.), fed from 4160 volts at 60 Hz, the motor was only run at 3100 volts and did 163 k.p.h. when consuming 13.2 MVA (a power factor less than 0.1, but then, they still believed in Gramme-ring windings and had not heard of TFMs).

Whether this motor was the beginning of the end of 'LML' (France) is not clear but the company was closed in July 1974 and news of this did no good at all for the British 'LML' who were *not* making mistakes. From now on, whoever failed anywhere in the world, whether it was seen as an air cushion failure, a magnetic suspension failure or a bad Gramme-ring design, seemed to be receiving translation in British 'political' (by which I mean perhaps 'Scientific Civil Service') circles as 'proof of the failure of linear motors'. Whatever it was, overseas 'competition' was making itself felt not as *superior* technology but as *inferior* with the implication that if the might of the USA, German and French industries could not produce a winner, what chance had the British?

In the USA, Garrett Corporation apart, academic interest began at the Massachusetts Institute of Technology in 1971 with the 'Magneplane' project—a cryogenically suspended model on a short track (figure 7.32). At the same time 'MinTech' support (effectively from the DOT) awarded two study contracts for cryogenic suspension jointly worth a quarter of a million dollars. One grant was made to the Ford Motor Company, the other to the Stanford Research Institute. Their reports showed artist's impressions of good 'old-fashioned' double-sided linear induction motors. We knew at this time that the Americans in particular could not contemplate the possibility of single-sided motors for fear of a magnetic pull that might increase the effective weight of the vehicle tenfold. They were not alone in this, indeed in 1972 only THL seemed to know anything at all about vertical forces from single-sided LIMs, having, as pointed out earlier, taken the decision to go single-sided in 1967. This was the 'apparent' size of the British lead (in time) in 1972. The word 'apparent' is used because the Americans did know how to design motors for fixed low-frequency supply and high speed. The reluctance to manufacture this kind of machine could possibly be put down to 'national pride', without intending to cause offence, for there is a quite justifiable preference in most countries to develop their own ideas first.

Figure 7.32 The MIT model 'Magneplane' using cryogenic suspension and guidance in its V-shaped track (*courtesy of Professor Henry H. Kolm*)

American linear motor interests had progressed in General Motors and the Rohr Corporation (1970). Japanese interest had been directed to cryogenic suspension and by 1972 an article had appeared in *Japanese Railway Engineering* (Kyotani, 1972). Japanese National Railways had begun building the two-man cryogenically levitated vehicle shown in figure 7.33. THL's big vehicle RTV 31 (figure 7.34) did its first test run on 4 December 1971 and a modest 30 m.p.h. a few days later for the press; no-one could fault it as a vehicle, or as a concept, at this stage.

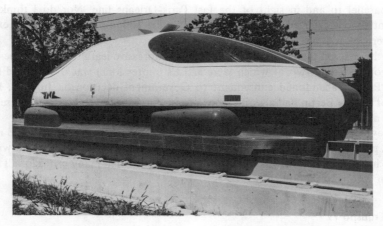

Figure 7.33 Japanese National Railways' two-man, cryogenically levitated vehicle on its test track (*courtesy of Japanese National Railways*)

Figure 7.34 RTV 31 being installed on its concrete beam (*courtesy of Ron Bailey Studios*)

In terms of speed alone, Monsieur Bertin was definitely winning in 1972. He had begun air cushion development as far back as 1957, the year of my first linear motor paper (Laithwaite, 1957). The Aérotrain concept was dated 1961 and the Gometz track was begun in 1965. By April 1966, speeds of 112 m.p.h. (180 k.p.h.) were attained, 190 m.p.h. (303 k.p.h.) by December of that year and 215 m.p.h.. (345 k.p.h.) by 1967, all without linear motors. The track was now 6.7 km long. Monsieur Bertin built a second (elevated) track 18 km long and gave rides to 'VIPs' and, as one of them told me, "I don't think it was an accident that this track was built parallel to and in sight of the fast Paris-Orleans line. Nor do I think it an accident that we started a few moments after the 100 m.p.h. crack French express and overtook it at a relative speed of 80 m.p.h."

But it suffered the fate of THL and a most impressive leaflet showing nine stages of Aérotrain (only two of which were artist's impressions) was overprinted rather sadly 'Project abandoned consequent to Government decision on July 17, 1974', but not before Bertin had a world record for ground speed vehicles with passengers, at 270 m.p.h. (430 k.p.h.).

Likewise, at the time of closure of THL, RTV 31 had achieved 108 m.p.h. (173 k.p.h.) against a 20 m.p.h. headwind, stopping and starting under its own LIM power in less than a mile. Like the Aérotrain, RTV 31 had many critics of all kinds, none of whom could do better.

7.10 Transpo 72

What happened just prior to both of these project closures reminded me forcibly of the ancient jousting tournaments which took place in England between the Normans and Saxons whenever a test of strength was called for. The supposition was that the side that won the tournament would win the battle (if there was to be one). A similar phenomenon perhaps persists, even in modern times, except that the words 'Olympic Games' replace 'Jousting Tournament', even though the results are far more disconnected from a nuclear war than were the knights and wars of old. However it was, the Americans decided to hold a 'Tournament' in high-speed transport in May 1972 in Dulles, near Washington D. C. Named 'Transpo 72', each potential 'winner of the game' was allocated a space 30 ft × 20 ft in which to demonstrate a working system. Although the outcome was totally indecisive, it was important for THL and its consultants in providing a target date for 'something'. The exercise illustrated quite clearly for us the need to *have* a terminal date from time to time, even if the goal sometimes seemed impossible. Later historians may possibly regard 1972 as the year when panic about *pollution* was at its height. In such a climate Tom Fellows suggested that we should try for a THL exhibit which was entirely electromagnetic and pointed out quickly that the elongated rectangular-plate levitator was already the basis for such a system, merely requiring a single-sided linear motor to be fitted along the centre, between the two, long C-cores.

The scale of the operation, of course, was against us. On a 30-ft run, a vehicle would look ridiculous if it exceeded 3 ft in length. To scale, this meant 4 inches wide, at most, and our levitation track was 10 inches wide. Since the levitator was only 2-minute rated as it stood, a reduction to 4 inches wide would have such a devastating effect on its performance as to make it 2-second rated! In December 1971 we made a start by the assumption (and this was my contribution, I believe) that only 'edges' mattered (see figure 7.35(a)). The edges A, B of the floating plate, I argued, could not 'see' the edges C and D of the primary, nor even E and F, so we could cut off the areas P, P' and therefore probably Q, Q' also, to leave a narrower system, as in figure 7.35(b). Then, since only *circuits* matter in electromagnetism, the system should not care where the m.m.f. was applied, so the outer sides of the exciting coils could be re-positioned *below* the C-cores as in figure 7.35(b). This reduced the width to 7 inches without loss of goodness, so we made a short length, just to be sure that it still worked—and it didn't! The wretched plate tried to tilt sideways into one or other of the positions shown dotted in figure 7.35(c). That was to be the only contribution I was to make to the proposal. I was entertaining an important visitor on the afternoon of the day we tested the first developments of the long levitator but half way through that afternoon my colleague Fred Eastham burst into my office, recognised the visitor and retired in confusion with an exclamation which began almost as a whisper and rose to almost a shout at the end of the very short sentence: "Oh, I'm sorry—but it's STABLE!" I ended my discussions with the VIP as soon as I could, without making it obvious that I wanted him to go, and rejoined Fred. "What did you do?" I said, for the system looked identical to that in figure 7.35(c). "You may remember" he said, "that in our Manchester days

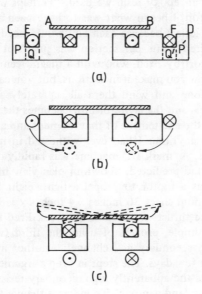

Figure 7.35 (a) Cross-section through a rectangular plate levitator. (b) Reduction of cross-section by purely 'edge arguments'. (c) Instability of a simple system

someone once connected up the long levitator with one of the coils reversed and it seemed to make little difference, so I tried it with this, just for the sake of it, and it works." The system was now that shown in figure 7.36 and, since we had used smaller C-cores than before, the levitation part of the system was down to the 4 inch width required, but only about 15-seconds rated. We could blow it hard, we argued, by putting it over a duct.

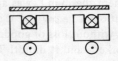

Figure 7.36 The first self-supporting, self-guiding linear motor at Imperial College, December 1971

At this point we called a meeting of both user and manufacturer and John Bell and Tony Davey of Linear Motors Ltd, Geoff Easton and Geoff Brown of THL and Fred and I sat around a table to discuss a final design which had to be built and shipped by March 1972. John Bell was first to comment, and to the effect that we would never build a 30-ft length of C-core stampings and ship it to the States in one piece, however strong an epoxy resin we used. "Perhaps we could build it in sections," said Geoff Easton. I began what was to have been a lengthy explanation as to why that was impractical because each coil would need hundreds of turns and the making of a hundred joints per section was just 'not on'—or something like that—but I was cut short by Fred, who with a magnificent authority in his voice, declared: "Not only can you make it in sections, but you can make each section no more than 4 inches long and wind them all separately and connect them—red, minus yellow, blue, etc.—and the levitator now becomes the motor."

I shall never forget the excitement of that moment and as a piece of sheer inventive genius it rivalled anything I had ever witnessed in my F. C. Williams/Tom Kilburn computer days. A 'mark III' machine was rapidly constructed and worked just as Fred Eastham had predicted. Shown in plan view in figure 7.37 it had just about enough thrust as a motor to propel a lightweight bodywork of fibreglass mounted on an aluminium sheet 36 inches × 4 inches × 1/8 inch. We would really have liked a little more thrust, and within the week Fred Eastham provided it by the, now apparently simple, process of using the third (vertical) dimension and making the aluminium secondary a U-channel. Another mountain had crumbled into a heap of dust in a few days. The steps in Fred's argument were as follows. The reason for low thrust is the apparently high secondary resistivity—an effect created by the lack of overhang 'end-winding'. To increase the width of the plate would be to destroy the stability, for which edge positions are vital. The channel idea allows both edge and extended plate width to co-exist.

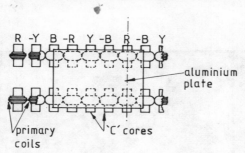

Figure 7.37 Plan view of magnetic suspension/propulsion system that became known as 'the Washington Prototype'

Still we were not satisfied. An air duct was still required for a reasonable time-rating and the air made so much noise that we might as well have had an air-bed, defeating the original objective. I argued that if vertical edges were a good idea at the extremities of the plate, perhaps they would be equally good at the centre, as shown in figure 7.38. We bolted two channels together and found it to be unstable. The pressure was really on now. The manufacturers said it was touch and go as to completion by March. Tom Fellows said we could *fly* it out, if we really believed we could still do better. One night Fred said to me just before leaving for home (we were both practically exhausted at the time), "I suppose it would be silly to try just *one half* of that double channel over *one set* of C-cores." We tried, and it took up an unstable position as shown in figure 7.38(b), but by now we were both so 'saturated' in the subject that we both shouted in unison, "That's only the effect of the other half!" We went home both happy and confident and in the morning we took a woodsaw (all experiments were mounted on wooden bases not unlike the 'breadboards' of early electronic days), and sawed it up the middle, placing the two pieces end-to-end, and we floated a half-width channel (2 inches overall) successfully. Now we were 'home'. With an increase in width to 4 inches the goodness was multiplied by 4 and a successful track was sent out to Dulles and operated throughout the fortnight of the exhibition (figure 7.39). The 'magnetic river' was born.

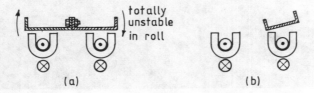

Figure 7.38 (a) Upturned centre plate is unstable in roll. (b) A single U-plate appears unstable only on a double track

164   A HISTORY OF LINEAR ELECTRIC MOTORS

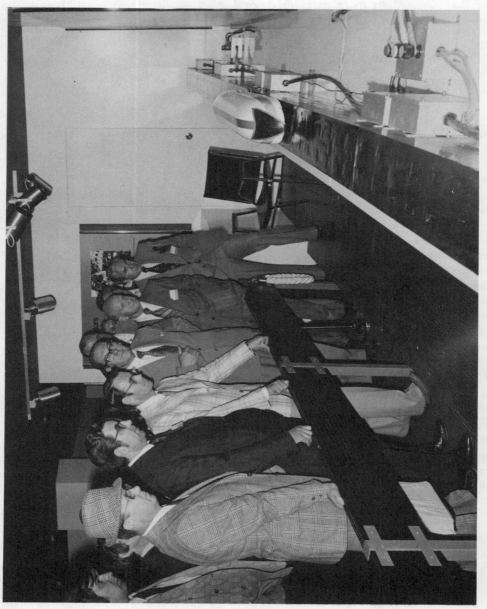

Figure 7.39 The first demonstration of a Magnetic River at Transpo 72, Dulles, Washington D.C. The spectator second from left is Mr Clinton Davis, Parliamentary Under Secretary of State at the Department of Trade, and

There was just one more bit of know-how which emerged during the manufacture of this equipment which I found particularly pleasing. For some 12 years I had believed that there was probably an absolute maximum height to which a piece of conductor could be levitated at 50 Hz for complex reasons similar to those which dictate the maximum size of bird that can fly. This belief was based largely on the work of the American W. V. Lovell who showed that over a levitation coil system there existed a region (conical if the coil system was circular, wedge-shaped if linear) within which small conducting objects would be attracted. Outside of this region they would be repelled (Lovell, 1955). It appeared that a necessary condition for stability was that the supported object must lie partly within the 'cone of attraction' (see figure 5.5).

The first test of the Dulles model showed the vehicle to be underdamped in roll, and Tony Davey had discovered that a slightly wider plate was more damped. I was shown this on a visit to Loughborough. "Yes," I said, "but you will have lost some lateral 'stiffness' will you not?" This proved to be so and I thought I had rounded off the matter by saying: "If you were to raise the supply voltage so as to lift the plate higher it will become totally unstable laterally." A higher voltage *was* available so it was used, and with increasing height the plate achieved *increased* lateral stiffness! At once it was clear that the cone of attraction had been eliminated. The single C-core operated on an expanding geometry, as shown in figure 7.40, whereby increased power supply, resulting in increased supported height, could be maintained with the same stability, so long as the plate width was increased to fit between the dotted lines, as shown—but the input power needed tends to increase at something like the fourth power of the levitated height. "Give me a place to stand and I will move the earth," said Archimedes, having appreciated the principle of the lever. "Give me Battersea Power Station for only two minutes and I will float the Taj Mahal 30 ft in the air!" I could now add, noting that both Archimedes and I were incapable of actually performing the respective feats, but both of us at least knew *how*. 'Knowing how' is what engineering is all about.

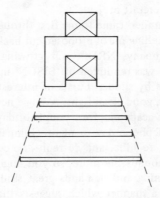

Figure 7.40 The levitation system with an expanding geometry

The real importance of this discovery was to come later when workers on cryogenic suspension were to declare that 'only cryogenic levitation' enabled 6-inch clearances to be possible between track and vehicle. Transpo 72 really 'paid off', so far as *our* efforts had been concerned.

We had no illusions about the 'magnetic river'. It was amazing that it was even possible to float, guide and propel a sheet of conductor from one and the same set of coils, but we had measured the propulsive force and the power input and knew that 95 per cent of the latter was consumed as non-mechanical-power-producing ohmic loss in the plate, so the efficiency of the machine *as a motor* was less than 5 per cent. It was a curiosity and a device to impress visitors. It was also 'upside down' in the sense that the coils were in the track, but so were most of the Washington models, for neither miniature fans nor miniature linear motors are good at producing adequate lift to support their own weight and I know of no electromagnetic system which will not invert—that is, it is either stable both ways up, or unstable both ways.

7.11 The stop–go phase

What was to be the next development? In late 1972 we saw it as more exploitation of the TFM idea. With all three dimensions now available, design parameter variations were semi-infinite in number. A full description of the thinking at this time is very complex, for although TFMs were invented specifically to solve the long pole pitch problem they could have application for lower speeds also. What happened in the high-speed transport game in 1972 and subsequently can now be seen in several countries more as political manoeuvring than as technological development, and the storm that broke in the UK in February 1973 was only one of a number of similar political/technological tangles that were to play havoc with development of the subject of linear motors as a whole.

Although a historian must constantly be reminded not to over-emphasise the immediate past of which he was a part at the expense of older events that concerned other people, there was nothing to compare with the sequence that led to the closure of the Tracked Hovercraft project.

The noise of 'distant thunder' came to me first throughout 1972 by visits from a very worried Tom Fellows telling me of proposed cut-backs in Government spending on THL. The parent company, NRDC, was growing unenthusiastic about the project, it appeared, and I was reminded of 'BISON' in 1951. We were somewhat harassed from time to time by sudden bursts of interest from the daily press who are always ready to report a good clean fight. A single newspaper headline is enough to 'stop production', for an academic, for a couple of days or more during which he is scarcely away from a telephone day or night. Going into hiding implies that you personally have something to hide, and to reply 'no comment' to every question only stimulates the news hounds the more, so you might just as well tell them all, which is both time-consuming and demands great skill, if you are not to be misquoted out of context in a manner which suggests that you have called Mr X a fool.

But there was no doubt that 'stop-go' policy existed in high places, inaccessible to an academic like myself, and I was fully aware of the size of the impending storm and the amount of cleaning up which would be necessary afterwards, quite some weeks before it broke. What I could not possibly foresee was the manner of it.

7.12 The Select Committee

The future of THL had, on the 8 February 1973, been placed before the Select Committee for Science and Technology of the House of Commons for examination. The Select Committee consisted of Members from both sides of the House and was chaired at that time by that most able and famous man, the late Mr Airey Neave. Members of the Committee were expert in various fields of science, and were all back-benchers. Thus Mr Arthur Palmer, Labour Member for Bristol Central was a fellow electrical engineer with whom I could, and in fact did, converse at the highest technical level and be assured that I was being understood. It was, for me, a most commendable organisation. It took evidence in public, which of course included the gentlemen of the Press. At the second meeting on 14 February 1973 the Committee had before them five Memoranda submitted by the following organisations

(1) The Department of Trade and Industry
(2) Tracked Hovercraft Ltd
(3) National Research Development Corporation
(4) British Railways Board
(5) Tracked Hovercraft Ltd Staff Association.

The tone of the first of these was not so much a suggestion of closure as "an indication of the type of questions being studied by the Department" (the last sentence of the first Memorandum). On that same day the Select Committee took evidence from and examined two witnesses. Mr Michael Heseltine, Minister for Aerospace and Shipping, and Mr S. N. Burbridge, an Assistant Secretary of the Department of Trade and Industry.

The first question from the Chairman was "Has a decision been reached on this project?" Mr Heseltine replied simply: "Yes." "When was it reached?"—"On the 29 January." "What was the decision?"—"That we should in fact close down the Company Tracked Hovercraft Ltd as it now operates and that we should come to agreements with two particular contractors to deal with those two aspects of the technology at Tracked Hovercraft which are of immediate interest." This shocked and angered the Committee whose members saw it as a snub, insofar as they were effectively being asked to advise on a project which was already a dead duck and there was natural resentment, particularly in view of the tone of the Department's Memorandum.

The Select Committee's immediate retort was to announce their intention to make a complete examination of the THL project, closed or not. During the subsequent weeks they took other evidence from small groups of representatives from the various interested parties, including NRDC, the Department of the Environment, British Rail and the Universities. The detailed minutes of evidence given in public form remarkably good reading, for anyone interested in seeing Government working, and perhaps even follow a pattern of confusion and error, mostly because of lack of proper communication, such as was soon to follow in other countries, notably West Germany and Canada.

7.13 References

Armstrong, D. S. (1967), 'Application of the linear motor to transport', *Railway Gazette*, Vol. 123, pp. 145-150

Bedford, B. D., Peer, L. H. B. and Tonks, L. (1939), 'The electromagnetic levitator', *General Electric Review*, Vol. 42, No. 6, pp. 246-247

Freeman, E. M. and Laithwaite, E. R. (1968), 'Unbalanced magnetic push', *Proc. IEE*, Vol. 115, No. 4, p. 538

Kemper, H. (1934), German Patent No. 643 316

Kyotani, Y. (1972), 'Magnetic levitation research vehicle', *Japanese Railway Engineering*, Vol. 13, No. 4, pp. 6-9

Laithwaite, E. R. (1957), 'Linear induction motors', *Proc. IEE*, Vol. 104A, No. 18, pp. 461-470

Laithwaite, E. R. (1966), *Induction Machines for Special Purposes* (Newnes)

Laithwaite, E. R. and Barwell, F. T. (1969), 'Applications of linear induction motors to high-speed transport systems', *Proc. IEE*, Vol. 116, No. 5, pp. 713-724

Laithwaite, E. R., Eastham, J. F., Bolton, H. R. and Fellows, T. G. (1971), 'Linear motors with transverse flux', *Proc. IEE*, Vol. 118, No. 12, pp. 1761-1767

Lovell, W. V. (1955), 'Electromagnet removes nonferrous metals', *Electronics*, Vol. 28, No. 9, pp. 164-166

Zehden, A. (1902), 'New improvements in electric traction apparatus', U.S. Patent No. 88 145

8 The world-wide game

8.1 New readers begin here!

I built my first linear motor in 1948 and wrote my first paper on the subject in 1954. The Gorton experiment took place in 1962. The first model of a tracked hovercraft was publicly demonstrated at Browndown in the summer of 1966. We abandoned double-sided sandwich motors in 1967 on the grounds of safety and with the realisation that a track of aluminium sheet backed by steel need not necessarily put large downward attractive forces on the vehicle. The 'we' referred to here consists of Fred Eastham, Hugh Bolton, other members of my staff, and of Tom Fellows and his THL senior staff at Cambridge. We met and conquered the long pole pitch problem in 1969. We were on the track of very far-reaching experiments with the emergence of a 'magnetic river' following Transpo 72 in May of that year. We were aware of the feedback amplifier type of magnetic suspension and of the cryogenic method (superconductor). We were aware of a potential customer for a linear motor transport scheme in the form of the Canadian Government who had invited 11 firms to submit schemes. Two of the 11 (Tracked Hovercraft Ltd and Hawker Siddeley (Canada)) combined effort to put in a single bid. Their solution was to use rubber-tyred wheels with no treads running in concrete guides with linear motor propulsion. The absence of treads would reduce noise levels to well below those of the tyres of ordinary cars.

By the time the Select Committee began their job on THL, the Canadian Government had shortened their list to two contestants:

(1) Krauss–Maffei of Germany who proposed a feedback amplifier type of 'Maglev' plus linear motor propulsion.
(2) The THL/Hawker Siddeley (Canada) proposal.

Just how all this occurred was explained to me by a man who stood as close to the Toronto project as I did to THL. I will tell his story here, for much of it can be grouped under the heading 'The perils of having a *System*'! But first I will digress merely to illustrate just what I mean by that phrase.

Many academics believe that 'Benevolent Dictatorship' is the ideal form of Government, for it is only when a 'SYSTEM' is set up that anomalies begin to occur.

In the summer of 1975 the French Government had promised the apple farmers a minimum price for their fruit—a very reasonable thing to do. But that year there was a glut. The only sensible thing to do would have been to reap what was needed, pay the farmer for the estimated total of *all* apples, let the remainder drop from the trees and rot and roll them in for fertiliser. But there would have been an outcry about the waste of public money.

They were paid for, so they had to be seen to be reaped and carted away—at considerable expense. Then came the problem of what to do with them. You could not dig a hole and bury them because someone would have to do the tipping and the digging and the filling in. And he would tell the Press and the cat would be out of the bag again.

So they found an elegant solution. A fruit canning factory had a pulping machine and was located near a river. So the lorries delivered the surplus apples to the factory (to be canned, everyone thought). But only two people knew that the output pipe of the pulping machine went straight into the river. That was nice and tidy.

But alas, they made a river of pure cider that poisoned all the fish. The dead fish floated to the surface and the conservationists wanted to know who killed them, and not one but *two* cats flew out of the bag at once.

Now for the punch line: while all this was going on, France was importing apples!

To the layman this looks like stupidity in any language, but try putting yourself in the place of any *one* of the people involved and you will find yourself doing your best and doing exactly as they did. You *must* import apples or you won't export your wine, and so on. The fault belonged to no one person, but to the social system as a whole.

If the actions in the UK over the Tracked Hovercraft project seemed illogical, it was comforting to see the same kind of 'circus' taking place in Toronto about the same time—in fact the two systems had considerable effect on each other. The following account is one man's view of what went on in Toronto. Only the portions in brackets are my 'asides'.

8.2 The Toronto Urban Transit Scheme

In the 1960s there was public concern over pollution, soon to be replaced by the 'Energy Crisis'. It was proposed to build a North to South Transit System through Toronto and many 'action groups' were formed to see that something was done. The most important ingredient was that the Provincial elections were near (1972). A political issue was made of the idea of stopping the building of 'expressways', and building an Urban Transport System instead, even though no-one had any firm ideas as to what form it would take.

In February 1971, a group was formed to look into what kind of system would be best, not only for Toronto, but elsewhere in Ontario also. Outside authorities were to be encouraged to bring in their own expertise. Ten firms were invited to submit their own versions and anyone else was invited to volunteer proposals. Of the 200 applications in this second group, all were rejected and the 10 invited submissions became the 'finalists', only one of which (the Toronto Transit Commission) was a 'local'.

The objective was to transport 20 000 passengers an hour at, perhaps, no more than 20 m.p.h. Basically three ways of doing it were possible:

(1) Do it yourself
(2) Go out to experts—as was in fact done
(3) Do it conventionally.

Of the nine 'outsiders', five were from the US, two from France, one was British and one German. The rules were not specified at the time of the invitations but, a year later, a short list of three was funded for more detailed study with a view to demonstrations in 1973. During this period the number of staff employed by the Canadian Transportation Department to assess the projects exploded from 5 to 200; some were 'rooting for air cushions', some for cryogenic suspension, and so on, and as one worker put it: "An awful lot of guts got spilt on the floor—after all, how do you compare where the water closet is situated in rival systems?"

Toronto became 'a central meeting place for junior engineers from the Colonies'. The short list of three and their systems were as follows:

(1) Krauss–Maffei (Germany) with three linear motor bidders including SPAR Aerospace Products Ltd in Canada itself, proposing feedback amplifier magnetic suspension, and linear induction motor propulsion.
(2) A combined Hawker Siddeley (Canada)/Tracked Hovercraft Ltd (UK) project using rubber-tyred wheels with no treads (and therefore very quiet) with single-sided linear induction motor propulsion.
(3) Ford Motor Company proposing cryogenic magnetic suspension and linear motor propulsion.

Then Ford pulled out, presumably because they had done less experimentation than either of the other two.

A few weeks later the Toronto contract was awarded to Krauss–Maffei. In Toronto there was a great clamour for an answer to the question 'Why not a Canadian system?', to which the ready answer was the urgency and shortage of time. Krauss–Maffei had available studies for Taiwan, Fiji and other metropolitan areas.

The first test was to be carried out in a place to which nobody went and then moved to a public area when seen to be good. Trees were uprooted, conservationists got cross, but a system with points and four stations was built, the stations alone costing 18 million dollars. Things began to go seriously wrong at the start of 1974. Parts were not delivered on schedule, schematics were not finalised, the requirement for 16 passenger seats was increased to 22 seats. This added to the on-board power supply and, by the summer of 1974, the original 7-tonne proposal had increased to a 13-tonne vehicle (almost 1 tonne/passenger).

In the end there was disaster—in simple terms it just did not work. All faces had to be saved, of course, and in order that *some* public confidence should remain it was claimed that the exercise had given 'a feel' for the problem. [Rather like putting one's hand into liquid helium and discovering just what it can do.] They tried to make the best of what there was and no official details of why or where it failed were given. The Canadian Government would carry on their studies, but with less money spread over a longer time. Putting wheels on was thought to be rather a good idea—especially when they could be treadless to reduce noise, since the linear motor could be used for propulsion. (The end of the personal view of a Canadian academic.)

What a tragedy for THL—Toronto was proposing to use their system after all. The 'one man's view' in the foregoing nine paragraphs did not reveal the nature of the failure but talks were arranged between myself and an engineer from Krauss-Maffei when cards were laid on the table and, since there was no confidential correspondence and no money changed hands, there is no reason why I should not disclose the nature of the conversation as part of the history of the subject.

It appeared that at some stage it had been decided to 'go single-sided' on the LIM, and that no proper investigation had been made of the vertical forces likely to be produced by a single-sided motor. In the event no tests on full-scale devices were made until Maglev and LIM were 'married' on a disastrous day when the LIM was found to deliver an enormous upthrust on starting that banged the suspension magnets against the track. In this condition the carriage skidded down the track until the cross-over point of the lift/attract characteristic was reached when it peeled off the rail entirely under the LIM's magnetic attraction. Even this might be thought to have been retrievable, but the *fundamentals* were wrong. The engineers had scaled up a sparrow to the size of an ostrich and found that it could not fly (they were unaware of the 'great Divide' (Laithwaite, 1973))—and for the same reason. After magnetic saturation sets in, *magnetic* levitation forces (as opposed to *electromagnetic* forces involving induced *current*) can increase only in proportion to pole-face *area* (L^2). Weight increases as volume (L^3). The same is true of the lift/weight considerations in birds. To continue to increase the size of such creatures is to reach a point of no return. It seemed incredible that in the face of such fundamental restriction, grants from public funds in the UK were still being awarded in 1976, and Japan Air Lines were interested in pursuing the same form of Maglev for high-speed transport.

8.3 Earith and after

Eight of the redundant engineers from THL went to Canada and joined the SPAR company. This was immediately followed by an announcement that SPAR had "bought the entire linear motor design team from Cambridge (THL)." This, however, did not imply that Britain had lost its expertise, for a full-scale longitudinal flux motor driving RTV 31 had been designed and built by the AEI/GEC/English Electric group, and the 1/5 scale transverse flux motor tested at Cambridge on the vehicle 'RTV 41' had been built by Linear Motors Ltd, at Loughborough. But then, the most amazing things were being said in various places and I had a telephone call from the technical director of a manufacturing company which began thus: "Good morning, - - - here, we're taking over your research on linear motors." I'm afraid he got the very rude instant reply: "The hell you are!" Mistakes through lack of communication were being made in several places, and not only in Britain. SPAR, however, were not slow to exploit their gains in personnel and soon had an experimental track (figure 8.1). One felt that a little more of this in the UK would not have gone amiss. The most common mistake was to assume that hardly any work on linear motors had been done at Imperial College, or for that matter anywhere else, but the Select Committee did a good job on unravelling all the loose ends and getting at the facts. Their approach was really tough. A minister who said he did not intend to answer a particular question was severely rapped by one of his own back benchers who replied, "This happens to be a Select Committee and I am asking the questions. I do not want reproof from you. I will ask my question, if the Chairman finds it out of order he will speak to me, not you."

The Committee told a Board Member of NRDC that he should resign. The leaders of THL were almost made to appear the villains of the piece. A member of the Select Committee suggested that I myself was being 'bloody-minded', and I replied that in the circumstances I regarded that as a compliment! The Minister of

Figure 8.1 The SPAR track in Canada, testing single-sided motors mounted on a wheeled carriage (*courtesy of SPAR Aerospace Ltd*)

Transport (Mr Peyton) made his almost famous remark about the APT having seats in it which, for many, indicated his lack of knowledge of the facts, but for me was one of the wisest statements made before the Committee. I cast my mind back to the Gorton days and to Fred Barwell's equally wise remark that we were doing the tests (which mostly consisted of giving rides to VIPs) to get money to build a bigger one.

The Select Committee announced their intention to visit Imperial College to see the linear motor work. This gave me another 'target' date and once again under pressure Fred Eastham produced another 'world beater'. He had a magnetic river primary constructed within three weeks, but instead of single coils on single C-cores, he wound each of the two rows of projecting iron core faces with a properly distributed winding, as shown in figure 8.2. This was a new concept. It was exciting, for it lifted the magnetic river at least halfway out of the Ivory Tower, possibly towards the workshops of the world, for it was now clear that it could be regarded as purely a pair of parallel linear induction motors and as such it was possible to do a *power balance* horizontally, in the direction of motion, for if F was the force in this direction and v_s the synchronous speed

input power = Fv_s + (primary ohmic loss) + (primary iron loss)

The vertical lift and lateral guidance forces were obtainable for *zero input power*! There was, ideally, no movement vertically or laterally so no power *should* be called for. We were in a new world in which you allow the forces of induction to act in quite separate planes and if you 'pay' for your thrust in one plane (payment in the

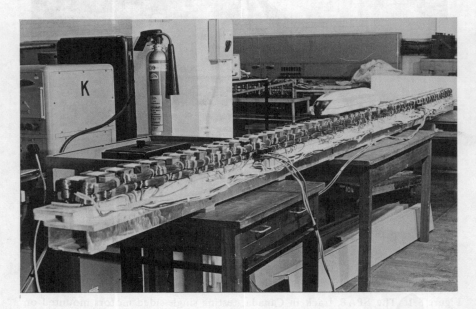

Figure 8.2 The first distributed-winding Magnetic River at Imperial College, 1973

form of losses) you do not have to pay for it in the others. Once again in the 'bright light of hindsight' it was obvious, for unbalanced magnetic pull in a conventional rotary motor can be enormous in relation to the useful tangential thrust, but there are no watts attached to the radial forces, otherwise no rotary motor would have an efficiency greater than 10 per cent, except by good fortune.

If the work of the Select Committee had done no more than stimulate this single concept, it was a job well done. The odd thing is that in 1976 there were few engineers, either in the UK or anywhere else, prepared to believe these relatively simple facts, the others arguing that, at standstill, all the input producing the lift goes as loss. You point out that this is also true of a purely propulsive, conventional, rotary induction motor and they fail to see the relevance. The fact is that you can either elect to say that your levitation costs you 1 kW per lb lifted, and you get your propulsion force for zero power, or that you need 1 kW per lb thrust at 250 m.p.h. and you get all the lift for nothing.

But you don't lose twice!

8.4 The Consortium of British Universities

The letters of sympathy flowed in at a 'personal' level following the THL closure. It was quite an interesting situation for I was able to discover my enemies in other spheres of activity very quickly. On the closure day itself, for example, a colleague whom I was accustomed to passing frequently without exchanging words, crossed the main road especially to say with a great beaming smile "Sad day for you, eh?" Of course it wasn't sad at all. But it *was* a call for action or our accumulated stack of patents (all the property of THL) might well be of no use to British Industry. What was important and yet not realised by the 'man in the street' was that the Government had closed the project but not the *Company*. So the patents were safe and the track was still standing and, what is more, was used for some months as a noise-testing track for experiments carried out on behalf of the Department of Trade and Industry. There was a natural desire on the part of the senior staff of THL, who knew enough about the state of the project and about the state of the 'game' in the world generally, to want to salvage the track and to limp along on a skeleton staff. On the other hand, there were clearly people who wanted to see that mile-and-a-bit of track put under the bulldozer at the earliest possible date. Who these people were I never knew; perhaps they did not really care all that much, but the Press whipped up what they could of a fight and the actions which followed were quickly described by journalists as 'the Universities' fight to save Earith'.

What was fact was that Professor Sir Hugh Ford, the Head of the Mechanical Engineering Department of Imperial College, took time off from his many other duties to interest himself in the work that we were doing in the Electrical Engineering Department and to do what he could to ensure a continuation of our researches and of those in other universities which were directed towards the coveted 250 m.p.h. 'goal'. It was clear that British Industry was not over-enthusiastic about high-speed linear motors. That being so, the Department of Trade and Industry could not be blamed for taking little more than a passing interest and nobody was willing to put up money for development—not in the six months or so after the closure of THL. Sir Hugh Ford pulled together the interested parties among the academics and formed a policy Steering Committee.

An application was made to the Science Research Council for financial support to maintain the high-speed track and to turn the site and buildings at Earith and Cambridge into a Centre for High Speed Transport Research. The Science Research Council responded by appointing an *Ad Hoc* Committee to look into our claim and a formal meeting was arranged in May 1974.

And it was not entirely wasted. Of course our application was declined. So was a last-ditch personal application that I made to SRC for £52 000 to re-surface the track top with a 3/8-inch thick aluminium sheet, the full track width, backed by a $\frac{1}{2}$-inch sheet of boiler plate steel, somewhat narrower, and to build a full-sized 'Magnetic River' motor (with no vehicle) that would be able to accelerate at over $2g$ and attain 250 m.p.h., self-guiding and levitating within a mile. The objective was to have been a demonstration to the delegates of an international conference planned by the Institution of Electrical Engineers for October 1974 (I was the organising chairman). The reason for declining was that SRC funds were not available for public demonstrations.

But by this time I would have needed a very strong belief in the inherent benevolence of Mankind not to suspect that from where I stood, very powerful decisions had been made (at very high level) that could not be reversed. Whether we ever discover who made them and why they were made is doubtful and probably not important.

A possible result of the formation of the SRC *Ad Hoc* Committee, although I cannot be certain for I was not a member of it myself, involved the Company of GEC in a counter-proposal. A letter from the then Chairman of the Select Committee on Science and Technology, Mr Airey Neave, to Professor Arthur Ellison of the City University made the first reference to these counter-proposals:

"After our meeting with Lord Beswick, Arthur Palmer and I saw Professor Edwards, Chairman of the Science Research Council, at the House of Commons last week.

"Professor Edwards explained that the view of the Council was that they should support the principle of the University scheme for research into electric surface transport. They did not, however, think the Earith Track was suitable, being only a mile in length. They suggested that this should be moth-balled.

"They have in mind to build a circular track on land adjoining the Rutherford Laboratory in my Constituency, but have not, I think, worked out details of the finance."

Only when we learned later that GEC were involved did we realise that the 'circular track' was a moving track! GEC were in fact proposing to build, not so much a track, as a legendary Great Wheel, 15 metres in diameter and 3 metres axial length, to test full-scale linear motors, nearly all of which was to be paid for out of public funds. By the time a meeting had been arranged at GEC's Hirst Research

Centre, to which 'outsiders' such as myself, Professor B. V. Jayawant of Sussex University and Dr R. G. Rhodes of Warwick University had been invited, SRC had apparently completed a 'first stage assessment' and appeared to be proceeding to a second stage. To my surprise, members of the SRC *Ad Hoc* Committee turned up in force to the Hirst Centre meeting and what must surely qualify for a famous Record Book was the fact that the roles of the industrialist/civil servant and the academic were reversed in respect of linear motors. The former were making the proposals and the latter were throwing the bricks. It was a refreshing experience!

The biggest 'brick' was the fact that members of the Department of Industry's Linear Motor Programme Committee (LMPC)—one of the better outcomes of the THL closure—including academics, industrialists and civil servants, had been to Grenoble to inspect a wheel test rig of not dissimilar proportions, built by the French company of Merlin Gerin. It was suggested prior to the trip that British interests might 'buy time' on the rig or, since French enthusiasm for the project had waned, buy the thing outright and rebuild it in the UK. When the facts were known, the peripheral speed attained had never exceeded 100 m.p.h. by very much on account of failure to balance the wheel. Yet the motors that propelled it were of the double-sided sandwich type, giving every chance to attain balance. The GEC/SRC wheel was for *single-sided* motors known to be capable of exhibiting enormous radial forces that changed with a change in speed.

One of the reasons voiced in those days why a drum rig of this type was clearly superior to a test track like Earith, was that linear motors could not be given a heat run on only a few miles of track. Yet on a drum rig of 15 metres diameter doing 400 k.p.h. at the periphery each part of the drum re-enters the field zone for a another heat injection every 0.424 second which is not at all the situation on a real track. Nor could track aerodynamics be simulated on a drum rig.

Consultancy contracts, meanwhile, were changing hands in a most curious and interesting way. For example, at the time the staff of Earith were given just two days' notice to quit, the consultants to THL were given the necessary three months' notice of termination of contract—all, that was, except me. My services were retained. I was not particularly surprised to receive notice of termination of my contract in February 1976 because the world as a whole was in a state of financial recession and many 'luxury' activities were under the axe, and not only in Britain. But life is full of surprises and an article in the *New Scientist* (New Scientist, 1976), based on an interview with Dr Kenneth Spring of British Rail, began with the amazing revelation that "British Rail now believes that magnetic levitation may be better than traditional railways for the underground and suburban lines of the future." I would be less than fair if I failed to include the fact that there were two schools of thought on means of support for transport of the future in the UK which had remained quite independent of THL, British industry and the Universities' Consortium, even though they themselves were based at universities.

One of these, at the University of Sussex, had made a concentrated study of the feedback amplifier type of Maglev, under the direction of Professor B. V. Jayawant. The other, at the University of Warwick, led by Dr (later Professor) R. G. Rhodes, was likewise engaged on a study of the other kind of Maglev—the cryogenic magnet suspension. Funding for these projects had come from the Nuffield and Wolfson Foundations.

One wondered how many people who gave evidence before the Select Committee and spoke of 'saving the technology' in respect of Maglev and linear motors were aware

(a) that there were two forms of Maglev
(b) that both were already researched in the UK
(c) that a great deal more money had already been spent abroad on the same subject
(d) that one 'attractive Maglev' system, namely that of Krauss-Maffei on the Toronto job, had been a demonstrable failure
(e) that the magnetic river was quite different from either of the two recognised forms.

Now we were told in the *New Scientist* that "This dramatic reversal in BR thinking comes after three years of research which BR believes has cracked the problems in magnetic suspension. BR now claims a world lead in mag-lev (magnetic levitation). The big remaining problem, according to Dr Kenneth Spring, BR's head of research, is the relatively poor efficiency of the linear induction motors required by mag-lev systems. Previously this was thought to be the easier half of the problem."

An interesting point to note is that by 1973, the feedback-amplifier, magnetic form of Maglev was being commonly known as 'attractive Maglev' in the USA while the cryogenic type was, by contrast 'repulsive'. When one considers the average technological make-up of the 'decision makers' whether in Government, the Civil Service or big industrial concerns, and one thinks of the everyday meaning of the words 'attractive' and 'repulsive', one is perhaps entitled to ask how much weight was given simply to the subtle use of words. After all, who wants anything that is 'repulsive'! Another thought that occurred to me about the time of the Toronto project was how interesting and useful it would have been to have had an open debate between the extremists of the day—the engineers of Krauss-Maffei, committed to building a system 'for real' and the engineers at Stanford Research Institute and at the Ford Motor Company, both in the USA, whose $\$\frac{1}{4}$ million studies of cryogenic suspension on behalf of the Department of Transportation in Washington D.C. came out so overwhelmingly in favour of the 'repulsive' system (Coffey *et al.*, 1973; Reitz *et al.*, 1973).

Again for the record, a magnetic river is, within the above terminology, a 'repulsive' system.

8.5 'Landspeed'

As it emerged later, the Universities' Steering Committee by no means wasted its time in meeting the SRC *Ad Hoc* Committee, for its voice was at least heard in a sector of British Industry prepared to give it a try. A lot of the credit for this must go to Mr Geoffrey Hart who had acted for Tracked Hovercraft in matters of public relations. The firm of Brian Colquhoun and Partners, civil engineering consultants and holders of the Queen's Award to Industry, were prepared to finance a separate company whose directors included members of their own Board and members of the Universities' Steering Committee. Thanks to foresight by Geoffrey Hart, 'Landspeed' was available as a registered name and fitted well into the gap left by Lockheed's 'Airspeed' and British Rail's 'Seaspeed'. Lord Kings Norton, who had long been anxious to help in linear motor development (he had been for some years President of the Royal Institution, during which time I was an external Professor of that famous establishment), agreed to be Chairman of Landspeed and the announcement of the formation of the Company appeared in the Press on 5 June 1975. The Universities' Consortium then formed a subsidiary company under the name 'Landspeed University Consultants Ltd'.

Now perhaps is the time to examine what was going on elsewhere.

8.6 The world scene up to 1976, country-by-country

Rather than attempt to write a chronologically correct account of developments in places thousands of miles apart, by teams of engineers largely unaware of each other's existence, this section will be divided into a few introductory pages, showing the interleaving of various projects that *did* have some small effects on each other (most of which have already been discussed in the context of the political/industrial scene), and a main section of the developments, country-by-country. The large number of photographs showing advanced vehicles will give an immediate indication of the amount of investment in the high-speed transport game, a fact to be borne in mind when considering the amount of fuss made in the UK about the relatively small sums of public money spent there (£$5\frac{1}{4}$ million by NRDC on THL, £0.3 million by British Rail and small amounts by Hawker Siddeley, GEC and Linear Motors Limited).

In the early days of a new technology the failure of any one project is inclined to put the fear of God in those holding the purse strings for all similar projects throughout the world. Only success succeeds in such a climate and to this extent those of us involved in the high-speed game were all dependent on each other, and some made mistakes and did the others no good at all. The historian cannot say, in such cases, how far the ripples from any one event spread, or he will end up in court. Even to state the facts that followed in the same paragraph may be construed as linking the two events, and therefore as a direct assault on individuals or on companies.

Let a new paragraph therefore be started and let me begin by saying that there were many actions in the mid 1970s for which I could find no logical reason. Sometimes there was support, sometimes the lack of it, in the most unusual places.

When the Garrett Corporation had done so much work on linear motors with DOT support, including a linear motor design programme that was accepted worldwide, why was the Rohr vehicle's motor entrusted to an overseas company? The much publicised 'not manufactured here' doctrine did not seem to apply in this case. Had British Rail gone a stage further than Gorton in 1963, the motors supplied to the Rohr Corporation would have been *British* motors. Why did an organisation as effective as the Merlin Gerin Company, with some of the best brains in the business working on linear motors projects (Monsieur Victorri and Dr Sabonnadière, for example), suddenly terminate all interest in linear motors? There were conflicting reports of this company taking Le Moteur Linéaire on board, but having seen the intricacies of industrial politics in action in the UK I cannot hope to untangle that particular overseas manoeuvre.

It is easy to excuse mistakes on the grounds that there is insufficient time to keep pace with events elsewhere and the bibliography in this book will more than amply support such a statement. Of *course*, engineers do not have time to talk to each other, but failure to do so can result in enormous waste of money and effort and ultimately can be made the scapegoat for failures. With some aspects of technology there is usually a 'no-man's land' of knowledge where all failure can be hidden, but with Maglev the situation is more akin to a ball game where one side *must* lose if the other wins. If your test vehicle fails in urban transport, you can always put wheels on it and pretend that such an act represented 'phase 2' of the development, but in the high-speed game you put all your money on one square.

A similar dichotomy exists for the induction/synchronous motor choice that was offered on the assumption that cryogenic Maglev is the one that succeeds. It will be pointed out—has *been* pointed out—that induction machines are lossy, have low power factor, are unproven as guiders and levitators, and it has been stated in several situations that transverse flux machines offer no solution to these problems. So the protagonists of synchronous linear motors, who were possibly at the same stage of development in 1976 as were those of LIMs in the mid 1960s, might reasonably have been asked to prove their claims to full-scale test vehicles. The sad state in 1976 was that there were the engineers with the courage to do this, but their accountants did not allow it, and from the accountant's viewpoint, quite rightly so. The spirit of adventure was fading, alas, in the high-speed game, in the face of inflation and other world problems, such as energy resources, hunger, pollution—survival! But I am an optimist and I have no doubt that someone must ultimately win the High-Speed Game. I cannot see 'the beloved motor car' dominating transport for ever, nor yet the omnibus.

Examination of the distance run by the several countries as witnessed in the photographs in this chapter alone should surely be enough to convince the fiercest critic that we weren't ALL wrong. Could *all* the engineers who designed the vehicles shown in figures 8.4 to 8.15, as well as that in figure 7.34 have been misguided, over-enthusiastic cranks?

8.6.1 USA

The start of the High Speed Ground Transportation (HSGT) studies can be seen to have been initiated by President Kennedy at the start of the 1960s. By 1964 there was a Department of Transportation in Washington D.C. A 1965 report from the Massachusetts Institute of Technology (MIT) was directed towards what became known as the 'North East Corridor Project', an area of highly urbanised land along the Atlantic coast that included Washington D.C., New York and Boston. In the same year the HSGT Act was passed. Among its requirements was that the Secretary of Transportation was to report at least once annually to the President and to Congress. Another direct outcome was the funding of a high-speed test centre at Pueblo. The MIT report recommended that a development programme be set up to examine the use of Linear Induction Motors (LIMs) for HSGT. Two of the three major vehicles that were subsequently built and tested were to be seen as a part of the implementation of the report. They were as follows.

(1) A linear motor research vehicle (LIMRV) using wheel-on-rail support and guidance, and planned to test linear motors up to 400 k.p.h. (250 m.p.h.). The track length was to be 10 km and the principal contractor was the Garrett Corporation (AiResearch Manufacturing Division). The LIMRV programme began in 1966 as a paper study at Garrett. In 1967 motor building began, to be followed by building the vehicle to house and test it in 1968. The official 'roll-out' was made at Los Angeles in 1969. After a year of standstill and low-speed testing on 400 metres of track in California, the vehicle was transferred to the now-established Transportation Test Center (TTC) at Pueblo in California. Figure 8.3 shows the state of the layout in 1974. By May 1971, LIMRV had given a ride at 150 k.p.h. to the Secretary of Transportation. Figure 8.4 shows the original vehicle. A lot of money was spent on the TTC. What is also of interest is that in the 8th Report on the Railroad Technology Program (1974) there is a paragraph specifically devoted to single-sided linear motors:

"A research phase on single-sided linear electric motors (SLEM) has been successfully completed. Because of the positive results of analytical and laboratory work at Polytechnic Institute of New York (PINY) and TTC, as well as in Europe and Japan, the FRA (Federal Railroad Administration) is now planning a development program for SLEMs... The most important potential advantage offered by SLEMs over the present double-sided linear motors is that they eliminate the protruding vertical reaction rail (about two feet, i.e. 610 mm, tall) and use instead a horizontal reaction rail structure that is flush or nearly flush with the track. Elimination of the vertical rail represents a major simplification of construction and maintenance, as well as vehicle switching from one track to another."

THL came to the same conclusion in 1967. I wondered how long all these excellent organisations would take to 'discover' the excellence of TFMs and of Magnetic Rivers in particular? I guessed quite a long time, for in the 8th Report, one of the advantages listed for the single-sided motor is that it inhibits "track hunting because of attraction forces between the vehicle and track."

(2) An air cushion vehicle with peripheral jets running in a trough section track initially using aero propulsion. Called the Tracked Levitated Research Vehicle (TLRV) the contract was given to Grumman in 1969. A linear motor was also designed and a 'second generation' linear motor in 1970 was commissioned under contract to the Garrett Corporation. This incorporated an on-board frequency changer (usually referred to in the US as a 'power conditioning unit'), with current collection from trackside. In 1971 the Garrett Corporation obtained a further contract for current pick-up experiments.

A HISTORY OF LINEAR ELECTRIC MOTORS

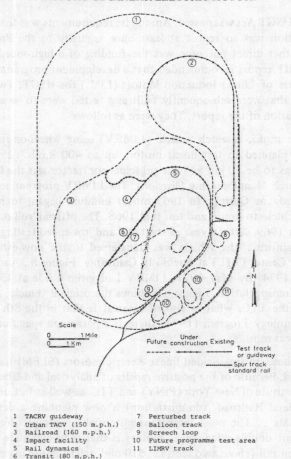

1 TACRV guideway
2 Urban TACV (150 m.p.h.)
3 Railroad (160 m.p.h.)
4 Impact facility
5 Rail dynamics
6 Transit (80 m.p.h.)
7 Perturbed track
8 Balloon track
9 Screech loop
10 Future programme test area
11 LIMRV track

Figure 8.3 Plan of the DOT Test Center at Pueblo, California

This second generation LIM developed 22 200 N (over 2 metric tonnes) of thrust continuously rated, and this could be augmented 50 per cent on overload. The motor under normal operation consumed nearly 3 megawatts from 4120-volt, 3-phase supply. The variable-frequency supply was designed to reach a maximum of 165 Hz, corresponding to 480 k.p.h. With an airgap (entrefer) of 1.5 inches (38 mm) in a double-sided system, the efficiency was 80.5 per cent and the power factor 0.575. Water-cooled tubes were used as primary conductors. The thrust/weight ratio was 1.13 and the power consumption some three times that of the first generation LIM. Both the LIM and its power conditioning equipment were designed in two halves so that, among other things, the motor and its supply unit could be tested at full power, at standstill, as a back-to-back test in the best tradi-

Figure 8.4 The original Linear Induction Motor Research Vehicle (LIMRV) before the boosters were fitted (*courtesy of US Federal Railroad Administration*)

tions of the Victorian 'greats', Hopkinson (for rotary d.c. machines), and Sumpner (for transformers).

The TLRV weighed over 12 tonnes, but carried $8\frac{1}{2}$ tonnes of equipment. It was designed to stop in 3.2 km from 480 k.p.h. (a 'fierce' deceleration of 0.28 g average). Not wishing to add the weight of a transformer to the rest of the power conditioner, the frequency changer was to operate at 8.2 kV and to use a synchronous (rotary) capacitor. This had special relevance in the UK in 1976, where low power factor was condemned nation-wide in industry and in some universities alike as the 'death knell' of LIMs. (It all seems to depend, for my money, on the willingness to win.) The size of the capacitor is clearly not intolerable in vehicles of the size of TLRV.

The initial current collection tests were carried out on a rocket sled at China Lake, California, and led to the design by Garrett of the most sophisticated current pick-up system to date. Figure 8.5 shows a view along the inside of the track. The end view of the collectors gives only some idea of the futuristic brush carrier. Figure 8.6 shows the carrier alone, described by Keith Chirgwin of Garrett (Chirgwin, 1977) as a 'double-ended, missile-like collector'. Missile-like indeed at 480 k.p.h.! Brush pairs pinched the copper rails so that reduced pressure on one side arising from misalignment and inertia implies increased pressure on the other and contact is never lost.

The vertical bounce on TLRV was over a foot in amplitude. One cannot help wondering how the 'attractive Maglev' protagonists see themselves either improving on this, or coping with it. Incidentally, in the same reference quoted above, Chirgwin came out strongly against active track systems with linear synchronous motors, a sentiment I personally endorsed, and he concluded with most ingenious suggestions

Figure 8.5 The Tracked Levitated Research Vehicle (TLRV) current collection system (from the inside) designed to pick up 17 MW at 480 k.p.h. (300 m.p.h.) (*courtesy of The Garrett Corporation*)

Figure 8.6 The spacecraft-like brush gear for the track in figure 8.5 (*courtesy of The Garrett Corporation*)

of a linear Kando phase multiplication and frequency changer such as was used in rotary form on the electrified railways of Europe (as opposed to those in the UK) at the start of this century. Chirgwin also proposed transverse flux, but unfortunately only referenced an early patent of 1904 by Andrée in this context.

The TLRV suffered a similar, though not as cruel a fate as RTV 31. Because of shortage of funds, only one-half of the LIM/inverter was built and installed in 1973. Had the second half been built it was planned to replace a hot water sink that had been used in connection with the cooling system by an air-to-water exchanger at the front of the vehicle "to make room for the second half of the power conditioner." Even this statement is not without its implications, of course. An onboard frequency converter is clearly an object of considerable size when this kind of 'room' has to be made for half of it.

The *official* line on the TLRV was that it would be diverted to a programme of lower speeds on 'more conventional' vehicles.

TLRV and LIMRV were parallel projects of course, and meanwhile the latter had been run at high speeds, only to demonstrate that the LIM could not attain the highest speeds required in the track length available. Such 'failure' would have been justification for demolition in the UK, but such was the determination of the experimenters on the other side of the Atlantic that jet engines were fitted to assist the LIM (figure 8.7) and in 1974 a world record for LIM propulsion with wheel-on-

Figure 8.7 LIMRV after the fitting of thrust boosters (*courtesy of US Federal Railroad Administration*)

rail was claimed at 411 k.p.h. (255.4 m.p.h.). Braking was accomplished with the aid of a 2000 kW resistor on board. This was assisted by two air doors, friction brakes, a set of parachutes and finally an emergency arresting cable installation. Travel at 400 k.p.h. is a different world.

A third vehicle known as the Prototype Tracked Air Cushion Vehicle (PTACV) was originally intended to serve an airport and travel at a maximum speed of 150 m.p.h., but was diverted to Pueblo when the airport plans fell through. This vehicle was notable in that it used a double-sided LIM and was guided by the reaction rail. In order to do this, the aluminium rail had to be continuous and despite earlier attempts to use the welded rail technique which had failed, the PTACV track was pre-heated and welded and survived the extremes of temperature that occur in Colorado. The 5 km track was completed in 1974 at a cost of $206 000 per mile. An extension of 4 km was proposed in the 1974 report at a cost of $352 000 per mile "due to inflation."

The contractor for PTACV was the Rohr Corporation who seemed to have ridden out the storm of earlier disappointments when the French firm Le Moteur Linéaire built them a Gramme-ring LIM whose power factor was rumoured to have been 0.03. Their latest vehicle (figure 8.8) was certainly one of the most science-fiction-like structures to date. But the PTACV project has now been stopped. Perhaps they put too *many* seats in it!

Co-operation eventually began between the US and other nations on the subject of HSGT. An agreement was signed in 1974 between the Federal Railroad Administration and the West German Ministry of Transportation and the Ministry of Research and Technology, for an exchange of R and D data. The Germans have

Figure 8.8 Full-scale fabrication of the last-proposed Rohr vehicle (*courtesy of Rohr Industries, Inc.*)

provided information on Maglev, the Americans in return provided LIMRV noise test results and a noise demonstration at Pueblo—a tentative beginning but better than nothing. Without doubt one of the most impressive figures in the 8th report of the DOT is that in the financial year "FY 1975, the budget authority in the combined HSGT and Railroad Research and Development Appropriations is $40.2 million." I would have settled for less than a quarter of 1 per cent of that amount to test a Magnetic River at Earith in 1974! Nevertheless, the pace, even in the US, began to slacken thereafter. The Pueblo test centre was being maintained, LIMRV was testing single-sided LIMs, but DOT funding for the MIT Magneplane stopped and the TLRV aims were made short-term, after having operated successfully at 100 m.p.h. (160 k.p.h.).

During a sabbatical year, 1975–1976, Professor Barwell spent considerable time in a number of countries, studying opportunities for linear motors. The countries included India, Japan, Taiwan, Canada, USA, France and Germany. In his visit to the Office of the Secretary of the Department of Transportation in Washington D.C. he learned that they were "unimpressed by the British system because Eric Laithwaite had stated that there was a poor power factor on his demonstrations"! It is a bit like the whispering game at Christmas. You say—'Here is a new invention. As a device it gets better as it gets bigger.' The Establishment says—'prove its value to us for £500.' You have to make a small model for that amount of money. It works. They say 'What is its power factor?' You say '0.2, but it gets better in bigger sizes.' The last part goes unheard. Professor Barwell found the same insistence at the DOT that lift forces from linear motors must *of necessity* involve loss of power. His failure to quote kW/tonne was taken to mean that it must be unacceptably high.

8.6.2 Canada

Following the abandonment of the original Krauss-Maffei project, a re-grouping of forces was necessary and an account of this activity is well reported in the *New Civil Engineer* dated 15 July 1976 (Hayward, 1976). Not all was lost in Toronto it appears. "Fortunately someone had the sense to include a cancellation protection clause in the £9m contract with the Germans, and when Krauss-Maffei lost German government backing and pulled out of the deal in November 1974, Ontario lost more time than money. Apparently the principle of using a linear induction motor to transport a 7.5 m long vehicle along an elevated guideway worked well. The no-moving parts motor and the frictionless propulsion were genuine assets." One feels that the last two sentences of this quotation were written with some surprise! The same author continues:

"The contract had collapsed just six months short of scheduled completion of a 2 km prototype loop in Toronto and the government had been bequeathed not only the technical information but also the use of 30 German engineers to help develop the principle further. Ontario, however, did not relish the possibility of its fingers being burnt twice, and all 'magnetic levitation' thoughts were ignominiously banished to the shelves."—And who shall blame them?

"However, the principle of 'intermediate capacity transit' was to survive and the UTDC (Ontario's Urban Transportation Development Corporation) was the obvious body to pick up the pieces of the Krauss–Maffei failure. Adopting the approach of 'find what the user wants before producing anything' the Corporation's committee included eight transit operators and, in announcing the birth of the Advanced Light Guideway Transit system, it is confident that government aid to develop the programme will follow.

"If all goes well, the 12 m-long cars incorporating the best of the proven technology from the Krauss–Maffei competition, will be running off the production line by 1980. Steerable axles should take care of the previous tight curvature problems and the linear motor, with its high acceleration, built-in braking facilities and low maintenance costs, seems once again the most suitable elevated guideway powerbox."

This description makes an amusing (even if tragic) comparison with a description of the same kind of motor quoted in the SRC report *Advanced Ground Transport*: "The linear motor with its inherent low power factor and efficiency..." It all depends which way one *wants* to look at an object. Even a rosebush, described as 'an irregular assortment of branches, covered in thorns that tear the flesh, terminating in sex organs (stamens) distorted and made gaudy by man's interference with Nature...' could be made to appear the least desirable plant in a garden.

In 1973 three Canadian universities were allocated Government money (from the Transport Department Agency) for transport projects and in particular Queen's University, Kingston, Ontario, used some of it to build a drum rig in which the 8-metre diameter, 1.5-metre long wheel carried the wave winding of a synchronous motor to co-operate with stationary cryogenic magnets around the periphery. It was also possible to replace the windings with a continuous aluminium strip. Maximum peripheral velocity was 100 k.p.h., which suggests that the apparatus was intended for Urban rather than Inter-city transport research. The DOT in Washington D.C., it appears, had arranged to spend $400 000 using the Canadian wheel rig at Queen's University and testing was due to start in the Spring of 1977.

The Canadians were aware of what failed in Toronto and had the courage to persevere with the linear motor as "the best of the technology." But whereas this latter phrase was probably the most overworked of them all in the UK in 1973, it was paid little more than lip service in Britain.

8.6.3 *Japan*

Japanese industry earned itself, in pre-war days, the reputation of being the greatest copier of other people's ideas. That this is no longer true is evidenced by the systematic way in which it has taken over successively the electronics industry, the motor cycle business and now the motor car itself (surely 'sacred' to the West?). Whose ideas did they copy in building the Tokaido line with its 150 m.p.h. trains?

Japanese commercial linear motor work can be traced back at least as far as 1964 and my only critical comment on their progress is that whether or not they were alerted to the potential of the linear motor by what they saw in Europe, they developed, along with the technology, that strange inability to communicate with each other that has crippled linear motor development in North America, wounded it in Germany and all but arrested it entirely in the UK.

I refer specifically to the 'tug-of-war' between the feedback amplifier type of 'Maglev' and the cryogenic type, for while Japanese National Railways poured a great deal of money into the latter and planned a 40 km track to accommodate cryogenic suspension and linear synchronous motor propulsion. Japan Air Lines arranged a Press demonstration of Maglev that was almost a replica of the Krauss-Maffei Canadian system and thereby inherited the same problems, at which time they appear to have sought help from the UK!

Nevertheless, the Japanese R and D outlook was very much more in line with that of Isambard Kingdom Brunel than that of present-day UK industry and they extended a railway to an island by building a tunnel as long as the Channel Tunnel might be, with virtually no publicity or cheering. Our own technological history should tell us that such an outlook has a habit of succeeding. If it begins with poor technology it will put it right on the way, and as the US effort on high-speed transport seemed to be following the same sad pattern set by France and the UK, the Japanese effort seemed to be growing. For example, Japanese National Railways had been working towards a 7 km test track in Kyushu to be operational by April 1977 and in December 1976 it was reported from American sources to be "on schedule." The experimental sections of track of Krauss-Maffei in Allach (Munich) and of MBB in Manching, near Ingolstadt were said (Tasch, 1975) to be "lined with Japanese 'train spotters', who have filmed the trials in exhaustive detail." Figure 8.9 shows Japan Air Lines' experimental vehicle and track. In January 1976 the

Figure 8.9 Japan Air Lines first high-speed test vehicle 'HSST-01' at rest on its track at Kawasaki, near Tokyo (*courtesy of Japan Air Lines*)

first vehicle HSST-01 was demonstrated to the public on a 200 m track in Yokohama. Its target speed was said to be 300 k.p.h. to be attained in only 1.3 km. The vehicle was comparatively small, being only 4.2 m long and weighing 1 tonne. The clearance between magnets and rail was allowed a tolerance between 1.0 and 1.5 cm. Propelled by LIM, the peak thrust was 750 lbs f—modest, but impressive.

It was at this point that I ceased to regard France, Germany and the UK highspeed linear motor developments as separate efforts, and for several reasons.

8.6.4 Europe

Being closer geographically and more inter-related historically, European Nations do seem to be able to communicate rather better among each other than is possible on an East-West global scale. The formation of the European Economic Community will do—has already done—much to amplify this most important aspect of technological development. Britain's finances force its R and D into very low key, but West Germany has money. In November 1976, Britain was the first country to be visited by two representatives of the EEC in Brussels. These gentlemen were making a survey of all research into high-speed transport within the Community, for it appeared that there was to have been a high-speed inter-city network of guided land transport in Europe where vehicles in one city are effectively 'aimed' at the capital city of another country and 'fired' like a bullet in a near-straight trajectory to that city at between 250 and 300 m.p.h. In THL days there was a feeling of competitiveness with regard to the type of track that would emerge as the 'standard'. It is merely an updating of the problem of railway line gauges, with greater complexity, but as far back as 1971 we believed (at THL) that there were forward-looking plans within German industry to make the European track a German track. I know that it was a great disappointment to the Management of THL that they failed to persuade Hawker Siddeley (Canada) to adopt their box-beam construction in the submission for the Toronto job. Had we won the Toronto job, the European track might have been a British track, built by Dowmac. 'If only. . .'.

What is particularly interesting to me personally about the visit of the EEC representatives is that they learned of my work through a man in the Department of Industry whom, although we have probably met, I would not recognise if we passed each other in the street. The Brussels men also had on their list of people to visit, my then 'right hand man' Dr Freeman, and they did not have time to discuss their other visits at the time they first visited me. The result was that a few weeks later, they returned to find that the said Dr Freeman had an office next to mine!

I almost pleaded with them to persuade the German workers that they must abandon the screened, Gramme-ring wound stators that only look promising in small sizes. More than that, I begged them to believe that we were no longer 'competitors' in the high-speed game, having been torpedoed by some person or persons unknown in 1973. Co-operation between engineers—not politicians—is something that could take place *now*. I told the EEC men to urge the German authorities to make that all-important decision in 1977 to *build* 56 km of track at Donnaureid rather than to withdraw.

Figure 8.10 The banked track at Erlangen built by the powerful consortium of AEG Telefunken, Brown Boveri, and Siemens (*courtesy of Projektgruppe Magnetschwebebahn*)

Figure 8.11 An experimental vehicle on test on the Erlangen track (*courtesy of Projektgruppe Magnetschwebebahn*)

The 400 million Deutschmark track at Erlangen completed in 1974 (figures 8.10 and 8.11) was possibly the leading research facility in the world in 1976. It was an example of the kind of facility that our University Steering Committee proposed to the SRC, since it was not a 'one-system' track and was equally capable of testing cryogenically-levitated or wheel-supported vehicles.

A project known as 'Cabinentaxi' had a $2\frac{1}{2}$ km track at Hagen (figure 8.12). The linear motor drive was controlled purely by voltage adjustment at constant frequency. Similar systems were in use on a 600 metre track in a hospital and on a short mountain railway. Although stepping out of geographical location for a moment, it seems appropriate to point out that there were equivalent linear-motor-driven systems in the USA, as figure 8.13 shows. In Dallas, George Scelzo put linear motors on the Astroglide carriages (figure 8.14) and lost 1 tonne of equipment per coach (Scelzo, 1976). The change was made in 1973, being fully operational by 1974. Having run successfully for over a year, it was proposed to convert all the LIMs to TFMs by 1980.

Academic work in West Germany might be said to have been 'moderately funded' by Government. For example, at Braunschweig under Professor Weh, Federal Government supported research which had advanced as far as the levitation and propulsion of a 1-tonne vehicle on a 30 metre track. The primary winding was in the track and was fed from variable frequency. The airgap was 10 mm and the research was directed towards a 'City-Bahn' (urban) transport system.

Figure 8.12 The 'Cabinentaxi'—a joint PRT project between DEMAG and Messerschmitt-Bölkow-Blohm (*courtesy of DEMAG Fördertechnik*)

THE WORLD-WIDE GAME

Figure 8.13 A fully automated PRT system by Transportation Technology Inc. (now part of the Otis Elevator Company) of Denver, Colorado, used single-sided linear induction motors (*courtesy of Otis Transportation Technology Division*)

Figure 8.14 The 'Astroglide' PRT system in Dallas uses magnetic attraction of the LIMs to relieve the wheels of weight (*courtesy of PRT Systems Corporation*)

Figure 8.15 'Transrapid 04', using double-sided LIM propulsion, which achieved a 'world record' for passenger-carrying Maglev vehicles of 204.3 k.p.h. in 1975 (*courtesy of Transrapid International*)

Following the combination of the two major German groups in 1974, the vehicle Transrapid 04 (figure 8.15) was designed for double-sided linear motor propulsion up to 400 k.p.h. Later the motor was rewound for 200 k.p.h. because of excessive side forces at very high speed. The later vehicles of this consortium carried the name 'Komet'.

At Imperial College we continued to develop the Magnetic River. We tried to make as many people as possible aware of a third magnetic suspension system which is neither tensile nor compressive 'Maglev'. We hoped that other research teams would ask our advice before they 'went big', for we knew from bitter experience that a failure of a linear motor project anywhere in the world would immediately reflect back as *our* failure.

8.6.5 USSR

It is always difficult to know exactly what is going on in Russian technology at any one time. The monorail at Kiev was reported in 1969 (*New Scientist*, 1969) and in March 1966 *The Guardian* reported the design of a 300 m.p.h., linear-motor-powered, wheel-less train. The source of this information was the official journal *Soviet Military Review* which stated that the train was to have magnetic suspension (without saying which kind) and would "be carried by the magnetic waves passing through the airgap between the cars and the rails." Synchronous linear motor and cryogenics—or Magnetic River?—I would think the latter. Russian research on linear *induction* motors has always been of interest to me personally and I think this latest development may be no exception. That it combined levitation and propulsion in one unit seems certain.

8.7 References

Chirgwin, K. M. (1977), 'Linear electric propulsion systems', Chapter 2 of *Transport Without Wheels*, edited by E. R. Laithwaite (Paul Elek)

Coffey, H. T., Colton, J. D. and Maher, K. D. (1973), 'Study of a magnetically levitated vehicle', Stanford Research Institute, *Report prepared for DOT, No. PB 221696*

Hayward, D. (1976), 'Toronto adds two new people movers', *New Civil Engineer*, 15 July 1976, pp. 21-24

Laithwaite, E. R. (1973), 'Magnetic or electromagnetic? The great divide', *Electronics and Power*, Vol. 19, No. 14, pp. 310-312

New Scientist (1969), 'Monorail for Kiev airport', 13 March 1969, p. 571

New Scientist (1976), 'BR gives big boost to mag-lev', 22 April 1976, p. 163

Reitz, J. R., Borcherts, R. H., Davis, L. C., Hunt, T. K. and Wilkie, D. F. (1973), 'Preliminary design studies of magnetic suspensions for high speed ground transportation', Ford Motor Company, *Report prepared for DOT, No. FRA-RT-73-27*

Scelzo, G. P. (1976), 'Astroglide—the advanced automatic guideway transit system', *Proc. 2nd International Conference on Hovering Craft, Hydrofoils and Advanced Transit Systems, Amsterdam*, pp. 29-35

Tasch, D. (1975), 'Hovertrain breakthrough imminent', *The German Tribune*, 5 June 1975, No. 686, p. 8

9 The second age of topology

9.1 A voice from the past

The Royal Society of Arts was simply 'The Society of Arts' in Victorian England. The word 'art' can be used to cover a much wider field than painting and sculpture and in the context of the following quotation, it may safely be taken to mean 'engineering'. The occasion is a paper by Professor Osborne Reynolds and the date is 1883. He has a message for academic and industrialist alike as we approach the 21st Century. It is a message all too often overlooked by politician, accountant and industrial manager alike.

"I have to deal with facts, and I shall try to deal with nothing but facts. Many of these facts, or the conclusions to be immediately drawn from them, may appear to bear on the possibilities—or, rather, the impossibilities—of art. But in the Society of Arts I need not point out that art knows no limit; where one way is found to be closed, it is the function of art to find another. Science teaches us the results that will follow from a known condition of things; but there is always the unknown condition, the future effect of which no science can predict. You must have heard of the statement in 1837, that a steam voyage across the Atlantic was a physical impossibility, which was said to have been made by Dr Lardner. What Dr Lardner really stated, according to his own showing, was that such a voyage exceeded the then present limits of steam power. In this he was within the mark, as anyone would be if he were to say now that conversation between England and America exceeded the limit of the power of the telephone. But to use such an argument against a proposed enterprise, is to ignore the development of art to which such an enterprise may lead."

In this remarkable paragraph he defended Dr Lardner, clearly forecast the transatlantic cable and ended with the most relevant remark in the context of the 'High-speed Transport Game'. If you fill in football pools your chances of winning may be slender, even over a lifetime, but if you *never* fill them in, it is *certain* that you will never win. The hazards of linear motor development were far less likely to prevail in the later 1960s–early 1970s, but those who refused to give them a try ignored 'the development of art' to which they may have led.

9.2 Can no-one be right?

As stated previously, a world financial recession brought governments into conflict with technological innovation in linear motors in the mid 1970s. Looking back from the next century it will seem amazing that at a time when millions of pounds-worth of commercially manufactured linear motors had been sold and had proved their worth, everyone was so slow to appreciate their value in the transport scene, knowing that bigger, faster motors would have enormously superior characteristics to those used for sliding doors, travelling cranes, conveyor belt drives and the like. Although the bibliography and detailed text of this book were effectively 'frozen' at the start of 1975 to avoid escalation of the narrative, the final pages were penned as late as 1985 and there is still no outright 'winner' in the High-speed Transport Game. Yet Japan Air Lines, Japanese National Railways, Transrapid (in West Germany) and British Rail all made advances in their own particular versions of Maglev and linear motor propulsion in the mid 1970s. Let it also be recorded that exciting activities in university departments continued into the 1980s and a great deal of this was an extension of the topological developments of the 1960s. Surely the point of no return was passed at this time? There could not have been a continuing stream of wrong answers from the research departments of the world such as was forecast by the prophets of doom of the late 1960s.

9.3 The lateral dimension

The legacy of rotary machine design can be seen, in part, as an inhibition of linear motor experimentation, even as far as the 1970s. In rotary machines, the tangential direction was the *thrust* direction and the axial direction was simply a means of increasing power output. Three-dimensional thinking was, in some ways, more advanced in the Victorian era, so far as machine innovation was concerned. At the risk of accusations of self praise, the Second Age of Topology can be seen as having had its beginnings in the demand for high-speed propulsion, the problem of the long pole pitch and the resulting development of the TFM concept.

Maglev systems went 'transverse flux' in an almost identical manner to that of the induction drive, U-cores replacing single electromagnets. The Maglev of Transrapid (MBB and Krauss–Maffei) used the system shown in cross-section in figure 9.1(a). This is a true TFM with inverted U track. A longitudinal flux system, but still using U-shaped magnets under a simple steel beam, as in the cross-section view of figure 9.1(b), was investigated by Tracked Hovercraft and subsequently developed by British Rail. Plan views of the poles of both systems (in their primitive form) are shown in figure 9.2. The German system (a) used a U-core magnet almost as long as the vehicle, along both sides. The THL/BR system (b) was a series of alternating poles. Both systems were virtually d.c. fed (apart from adjustments of the control system).

A broken pole system, as in (b), will induce considerable eddy currents into a solid steel rail so the rail was laminated in the THL/BR track. The disadvantage of the long magnet (whose eddy current effects are negligible) is that it is heavier than

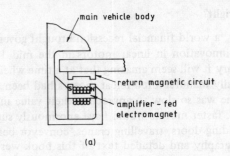

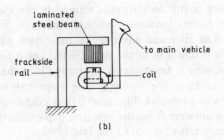

Figure 9.1 Cross-section through the lifting magnets of tensile Maglev: (a) the West German system developed by Krauss–Maffei and Messerschmitt–Bölkow–Blohm; (b) the British system developed by THL and British Rail

the longitudinal flux machine and it is not easy to bend a U-shaped rail for track curves. Bending a laminated beam is no problem.

Both systems were improved by staggering successive magnets laterally as shown in (c) and (d) of figure 9.2. In the German system (c) the entire block was divided into a number of alternately displaced sections, as shown. In the British system alternate magnets were displaced (d). If the alternate pairs are now energised selectively, the natural lateral stiffness (which was already considerable) was strengthened and was well worth while, even though some lift power had been sacrificed. Further advantages and disadvantages emerged as development proceeded. The windings on the pole faces of the British system were exposed to natural air cooling and could generate a higher m.m.f./weight, but the German system might prove better for heavier, faster vehicles because of its longitudinal extensibility which, in the case of the longitudinal flux system, might produce unacceptably high iron losses. At the time of writing, Maglev is still at the crossroads.

Back in the 1950s the Manchester team were often pleasantly surprised at the performance of their experimental motors (even at standstill). Theory had advanced to the stage where many workers claimed to be able to predict the force of a given machine on the basis of $(B \times J)$, yet repeatedly the real machines exceeded it. There was much argument about whether the Russell and Norsworthy factors for the overhang portions of a secondary sheet were sufficiently detailed. It was some years

THE SECOND AGE OF TOPOLOGY

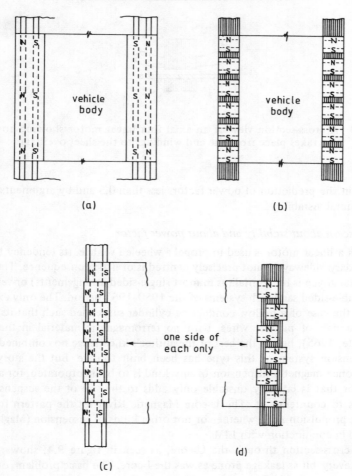

Figure 9.2 Plan view of the pole faces of tensile Maglev systems: (a) primitive transverse-flux system; (b) primitive longitudinal-flux system; (c) and (d) staggered versions of (a) and (b) for improved stability and damping

before the realisation that the end windings of primary and secondary were coupling was to be implemented in the developments of the TFM. Figure 9.3 shows the flux around the end regions as displayed in a cross-sectional view of a simple, single-sided motor (now called an I-core arrangement in the light of what follows). From this diagram can be seen not only the tendency of primary-generated, end-winding flux to couple with a wide secondary plate, but also the leakage flux from the same primary end windings which led to very high leakage reactances in machines of large pole pitch. Political decisions relating to the role of linear induction motors in the High-speed Transport Game were often backed by technical

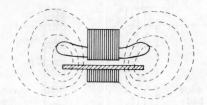

Figure 9.3 A cross-section view of an axial flux linear motor showing how induction takes place from the end windings to the sheet overhang

expertise in the prediction of power factors less than 0.3 and by arguments relating to basic lateral instability.

9.3.1 Concern about stability and about power factor

So long as a linear motor is used to propel a wheeled vehicle, its tendency to throw the secondary sideways if not precisely centred is of little consequence. This is true whether the airgap is horizontal (as in most single-sided arrangements) or vertical, as in the double-sided sandwich systems of the 1950–1965 period. The only exception occurs in the case of a hollow conducting cylinder suspended such that its axis lies between a pair of parallel wires, with no ferromagnetic material in the system (Laithwaite, 1965), but to the best of the author's knowledge no combined propulsion/suspension system of this type has been built to date. But the story is very different once magnetic suspension of any kind is to be incorporated, for a propulsion motor that is laterally unstable only adds to the size of the suspension unit which has to counteract it. The U-core Magnetic River set the pattern for a self-stabilising propulsion unit, whether or not other kinds of suspension (Maglev) were to be used in conjunction with LIMs.

But a cross-section through the U-core, as seen in figure 9.4, shows that the system is every bit as leakage prone as was the I-core. The basic problem, of course, is as old as the subject of transformer design, and from the latter context it should have been obvious that wherever there is a 'U' there can always be an 'E'! But it was not until Professor Eastham's suggestion late in 1969 that the E-core shown in figure 9.5 was built and tested. Even then, it took a week or two for the full impact of the arrangement to be realised. Not only is all the copper 'trapped' within steel

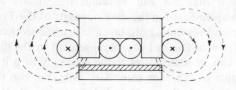

Figure 9.4 Cross-section through a U-core Magnetic River, showing that end-turn leakage is as dominant as in the corresponding I-core

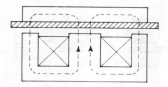

Figure 9.5 Cross-section through an E-core TFM in which almost all useful flux is confined by the iron circuit

boxes, but such an arrangement of a linear motor uses the flux twice! While it can be argued, and rightly so, that the return of half the main flux across the working airgap on either side of centre limb is done at a ratio of width/pole pitch only half that of the centre core, it is still a 'gift horse', since the original intention had been to improve power factor.

But now came another surprise. An E-core, whether used as a Magnetic River or merely as a propulsion unit, is basically unstable laterally. By 1974 it appeared that the only alternatives for a linear induction motor were a low power factor and lateral stability, or a reasonable power factor and lateral instability.

Again it took far too long to realise that the technique of coupling the end windings of the primary by embracing them with steel limbs could be applied to the U-core, just as it was to the I-core. The result was a four-limbed primary core as shown in figure 9.6. Given the name ξ-core (since Xi is the only letter that could be said to have four spurs) these machines were seen to be stable and to have low magnetic leakage. Subsequently a theory was developed showing that, to a first approximation at least, the stability criterion was extremely simple: odd numbers of polar projections, in a lateral sense, were basically unstable; even numbers were basically stable but could still be rendered unstable by poor design. The criterion was found to work, even though the polar projections were not connected magnetically, and systems such as those shown in figure 9.7 were shown by experiment to be stable even though each half itself carries an odd number. Clearly one must count *all* lateral projections whether or not magnetically connected.

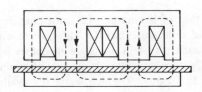

Figure 9.6 Cross-section through a four-limbed Xi-core which can be used as a Magnetic River

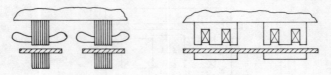

Figure 9.7 Twin E-core or even twin I-core motors can be laterally stable

It is now a good time to look back at the last sentence of the Osborne Reynolds quotation with which this chapter began. The development of the art to which enterprises led are to be seen, in the case of linear induction motors, as the Magnetic River, the Xi-core, the transverse flux tubular motor and the 'electromagnetic joint' (the last two of which are about to be described), and all of these were the result of spending £5.25 million on a Tracked Hovercraft. Who knows what other developments might have been forthcoming if the project had been continued?

9.4 Tubular motors

Tubular motors have never been popular, even though a range of short-stator versions was commercially developed. It is easily appreciated that one of the difficulties in making either a short-rotor machine or a moving-primary, short-primary version is inaccessibility to the inner member, in the case of the first, and inability to suspend a long secondary, in the latter case (see figure 9.8). But there is a second difficulty with tubular motors. All the radial (useful) flux generated by each pole must pass longitudinally along the secondary core. With large pole pitches the problem becomes acute and leads to excessively massive and expensive centre rods.

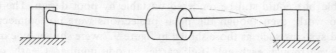

Figure 9.8 The difficulty of the bending of a long tubular motor secondary

A third difficulty is that of laminating the rotor core. For small tubular motors the core can be of solid mild steel, but it is difficult to obtain a goodness factor greater than unity for such a system. In larger machines where high G is needed, lamination becomes obligatory and there is no way of laminating except by blocks of stampings as shown in figure 9.9 which waste the available cylindrical surface because of the contracting rotor geometry. Bundles of iron wire will not do, for the flux must jump from wire-to-wire in penetrating it. It is a topological law that although a straight flux path can be contained in steel which may be laminated in any plane that contains the path, no non-planar flux can be contained in lami-

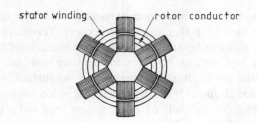

Figure 9.9 Lamination of a tubular motor using parallel-sided blocks

nations that are not divided into differently-oriented blocks, which is not practical in a tubular motor secondary.

The latter two difficulties were swept away in 1971 when the tubular equivalent of the flat TFM was invented (Eastham and Alwash, 1972). The principle is to wind the primary with helical coils as shown diagrammatically in figure 9.10. This is a primitive winding having effectively only 'half a slot per pole and phase' (had there been slots!). The ingenious part of the invention is shown in figure 9.11 where the negative phase currents are interlaced by taking each helical phase coil to the end of the motor remote from the supply, passing it diametrically across the end and returning it midway between the outward helix back to the starting end. When all three phases have been so treated, a star point is made at the same end as the terminals, as shown in figure 9.11.

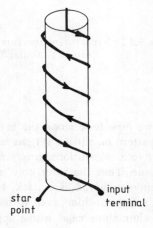

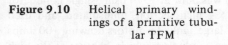

Figure 9.10 Helical primary windings of a primitive tubular TFM

Figure 9.11 Reversed phases interlaced in a tubular TFM (one phase only shown for clarity)

If we now draw a diagram of an instantaneous flux pattern, as in figure 9.12, we can see just what kind of a flux pattern to expect. Travelling along the surface parallel to the cylinder axis (as along PQ) we see a succession of N and S poles, each of pitch p. Every line such as PQ gives an identical resultant field to that of a conventional tubular motor. But to travel around the surface in a diametral plane (the path R) is to see alternate N and S poles just as in a conventional rotary motor. The helical travelling field is made up of a rotating field and a linear field, and an unrestrained rotor would receive both torque and axial thrust. The inventors pointed out at once that the rotating component could be cancelled and the linear component doubled by superimposing a second layer of helical winding of opposite handedness, as shown in figure 9.13.

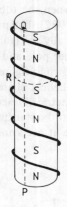

Figure 9.12 Instantaneous flux pattern in a tubular TFM

Figure 9.13 A double-layer-wound tubular TFM produces only longitudinal thrust

If we now look along the axis of such a motor and observe the instantaneous field pattern on a plane at right angles to the axis (figure 9.14) we see what is tantamount to a conventional rotary motor flux, in which no lines need enter or leave the plane. Thus the rotor core, so vital in tubular motors, can consist of circular punchings mounted as a stack. In fact, the rotor core becomes that of a conventional rotary machine, even to the extent that helical slots can be milled to contain a cast aluminium 'cage' whose appearance is similar to that of a sheet of extruded aluminium, used for radiator grilles and the like, wrapped around a cylinder.

A motor of this type (single layer only) was built in 1974 using single conductor primary windings of copper bar, 1 inch $\times \frac{1}{4}$ inch section, uninsulated, and used 'inside out' to stir molten aluminium in a large tank. In motors drawing 100 amps at 5 volts, line, there is no point in insulating conductors which, because of the high

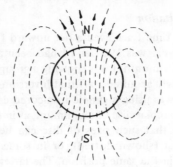

Figure 9.14 Cross-section through a tubular TFM showing instantaneous flux pattern

working temperatures then possible, are covered in insulating oxide. The turbulence caused by the rotating flux component was utilised for the mixing in of additives and new aluminium swarf, while the linear movement provided the overall stirring action.

Further research into this type of motor revealed what should have been an obvious phenomenon, but which at the time was experienced as a surprise. If a solid cylindrical secondary conductor is used in conjunction with a single-layer helical primary, and restrained from axial movement, it will spin up to near synchronism. If now the axial constraint be removed, there will be no axial force. Conversely, if a missile is allowed to accelerate axially it will only experience spin torques until it approaches the linear synchronous speed ($2pf$), by which time virtually all the torque will have disappeared. *If it had been otherwise the development of spherical motors would never have progressed as far as it did in the 1950s.* An angled-field motor with rotor constrained to move at 45° would appear to have an input current component $I_1/\sqrt{2}$ working on the rotor in the direction of motion and producing losses appropriate to a normal slip. At the same time the cross component should 'see' the rotor at standstill, producing enormous I^2R loss to no purpose. That spherical motors angled at 60° and beyond produced efficiencies over 70 per cent should have enabled us to predict the phenomena of the helical motor. More than that, it should have given us 'ammunition' to use at the time of the Tracked Hovercraft closure, when there was widespread disbelief that a combined lift and propulsion unit consumed zero power in achieving the lift (see chapter 8, section 8.3).

9.5 The vertical dimension

Fear of the effects of magnetic pull between primary and steel-backed secondary sheets inhibited the designs of several large propulsion units in the 1960s and the 'double-sided sandwich' arrangement was retained as far as full-scale experimental vehicles. The subjects of linear motors and electromagnetic levitation did not come together until the late 1960s although the latter subject had its roots much further back in time.

9.5.1 Electromagnetic levitation

I remember well the first time I achieved a net upward force from a steel-backed, single-sided system. The year was 1964 and I had been appointed to my Chair at Imperial College less than a month. I was showing a group of postgraduate students and one or two of my newly acquired staff that inversion of a levitation system did not affect its stability. I was using a simple iron-cored coil as shown in section in figure 9.15. With a small circular disc of aluminium, as in (a), the system is unstable laterally, despite the fact that such sizes of disc can be stabilised by means of a second coil inside the first (shown dotted), or in some cases, merely by a thick copper cylinder inserted in the same position. The latter acts as a shading ring to produce the inward-travelling fields necessary almost as the first ingredient for stability. One must always be on guard against over-simplification of analysis, and inward-travelling fields are not an absolute prerequisite for lateral stability. The same man who discovered the 'odds and evens' rule (page 201), Nigel Greatorex, also showed that by suitable choice of arrangement and dimensions, lateral stability could be obtained when outward-going fields, as detected by search coils, were present. With a further different arrangement he was able to demonstrate the reverse—that is, lateral *in*stability with a system of inward-travelling fields (Greatorex, 1978).

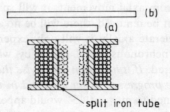

Figure 9.15 A levitated aluminium sheet is unstable over a single coil, whether it is (a) a disc of about the same size, or (b) a much larger sheet

With a large plate of any shape, as in figure 9.15(b), the single coil primary is *never* able to stabilise it in suspension, this statement being almost the only generalisation I have so far been able to make in this difficult aspect of electromagnetism.

When this system of figure 9.15(b) is inverted, as in figure 9.16, and the floating coil fed from loose coils of untwisted, stranded wire to reduce the mechanical forces inevitably applied thereby to a minimum, the coil will float in a tilted position as in figure 9.16(a), if the coil is small and/or the secondary sheet thin (an 8-inch diameter coil with an inner bore of 4 inches will tilt on a $\frac{1}{8}$-inch thick aluminium sheet) or in level suspension, as shown in figure 9.16(b), if dimensions of coil and secondary thickness are somewhat greater. (Increasing the plate thickness to $\frac{1}{4}$-inch in the example just quoted will suffice to produce level suspension.)

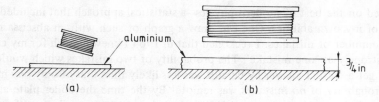

Figure 9.16 Self-levitation of an iron-cored coil over an aluminium sheet: (a) if the sheet is thin or the coil small; (b) if the sheet is thick or the coil large

Stable as such systems appear superficially, it is only the supply leads that hold them central over a very wide plate and, given complete freedom, self-levitated coils will always move towards the point on the edge of the secondary which is nearest, even though the definition of 'nearest' may include displacements less than a thousandth of an inch (other geometry, such as the coil dimensions being assumed 'perfect'). What is more, when the floating coil approaches the edge it tilts in such a manner as to suggest that it is 'falling upwards' (see figure 9.17). During this tilting, rapid acceleration off the plate can be observed.

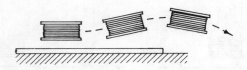

Figure 9.17 'Falling upwards' illustrated by a levitated coil near the edge of a sheet

It was such phenomena that the 'New Professor' was demonstrating to his learned colleagues in 1964, when one said, "What would happen if you put steel below the plate?" The reaction to "I don't know," was a quick comment from my Experimental Officer, Barry Owen, to the effect that we could easily try, and he was off to stores to bring a piece of boiler plate. As he was doing so I told my captive audience that we ought to be good enough to be able to *predict* the behaviour of such a simple electromagnetic arrangement. "It was, after all," I argued mentally, "basically a yes/no-type problem. It would either float higher or clamp down solidly."

Now I knew enough in 1964 to beware of any obvious answer, which in this case was that it would clamp down—was this not, after all, the basic reason (in 1964) why none of us favoured a single-sided motor for a propulsion system for traction application? I also knew that electromagnetic systems were fickle and that in a yes/no situation one was more likely to get the wrong answer by sticking to the well-trodden path. Looking back, I suppose that my apparent 'wild guess' was founded on the best of modern physics—a statistical approach that included assessment of my own abilities. One can draw a graph of such with an abscissa marked off in number of mistakes. I reckoned that in 1964 I knew enough for my curve to peak strongly at 'one mistake'. The probability of two mistakes which would allow me to get a correct answer by default was less likely than that of only *one* mistake. The probability of *no* mistakes was remote! By the time the boiler plate arrived I had decided to embark on a technique which I was to use much more frequently afterwards—work out the answer to the very best of your ability, then take the opposite! That this applies 'right across the board' in life in the 1970s and 80s is a piece of advice I would give to all those who are occupied in politics, local government, foreign affairs, bringing up children—etc. It stimulated me to write an article on 'The about-face age' (Laithwaite, 1978).

In the case of the boiler plate incident, the ferrous sheet was of smaller dimension than the aluminium sheet, as shown in figure 9.18(a), a fact which made the result still more interesting. I declared my prediction as 'higher levitation' and got it right! More than that, the coil could 'see' the edges of the steel through the aluminium, and was rapidly ejected from over the nearer edge of the steel, as shown in figure 9.18(b). The sight of an electromagnet (even if it *is* fed with a.c.) repelling a piece of ferrous material was not a common sight in 1964.

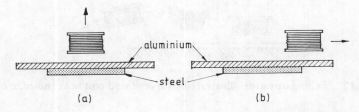

Figure 9.18 An iron back plate produces greater repulsive forces both in (a) vertical and (b) lateral directions

The thinking by which I reached the wrong answer, enabling me to triumph before my newly won staff, went something like this: if the coil is very small, the iron thick and the aluminium thin, the coil is sure to clamp down (magnetic things get better as they get smaller). But if the coil is large and the aluminium thick, then any amount of backing iron must improve the goodness factor, induce more current and result in more lift, since electromagnetic things get better the bigger they are and they will overhaul the attractive magnetic force which is not increasing even in proportion to L^3. Now the only thing to decide was whether an 8-inch diameter coil on a 0.25-inch thick aluminium plate was *big* enough to be *good* enough to provide extra lift. It seemed to me almost obvious that it was not, so I had reasoned that it would clamp down, took the opposite and got it right. The one mistake was that 8 inches *was* big enough.

9.5.2 The topology of track joints

As in many previously discussed situations, the bright light of hindsight revealed the solution to the track discontinuity problem as nothing more than a 90° plane shift. The use of aluminium channel as the secondary of a Magnetic River had long been known to enhance the propulsive force (figure 9.19). The folded-down edges were seen as a wide plate, so far as Russell and Norsworthy factors were concerned, but the folding enabled the primary to 'see' the folds as *edges* so far as lateral guidance forces were concerned.

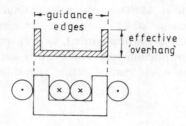

Figure 9.19 Cross-sections through U-core Magnetic Rivers showing how plate overhang can be both 'eliminated' (from the point of view of lateral stability) and retained (for purposes of propulsion)

As far back in time as the Gorton Loco Works tests, the effects of discontinuities in secondary conductors were known. An indication of the nature of the effects was given in chapter 7, section 7.2. Preoccupation with the effects of long pole pitches had led to an examination of the old Gramme-ring method of winding (notably in France) and possibly this was in the mind of Tom Fellows when he suggested to me early in 1974 that a solution to the track joint problem might lie in wrapping the track conductor right around the steel backing, as shown in figure 9.20, immediately adjacent to a joint, on either side.

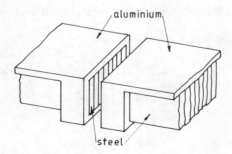

Figure 9.20 Gramme-ring technique applied to the track joint problem

In a linear motor designed only for propulsion, the track joint problem is not serious. Temporary reduction in thrust, not occurring as a step function so much as a dip with gentle slopes, would hardly be noticed, but in a system using the linear motor for lateral guidance or in a Magnetic River using all three axis forces, the result of joints might be disastrous.

We discussed the possibility and concluded that the same leakage flux that bedevilled the French motors would stifle circuital currents around the secondary steel. But on the following day I did some more topological thinking and came up with the idea shown in figure 9.21. As I contemplated the effect of a current path on the left-hand half of the joint, flowing, so far as the plan view of figure 9.22 was concerned, in paths such as ABCD, as if no discontinuity existed, there was the marked increase in adrenalin which always accompanies a new thought at the instant of conception, for the vertical current, as it disappears at D and re-appears at A from the folded-down aluminium would either *assist* similar currents in the right-hand half or the pair would mutually stifle each other: a moment's anxiety— and all was well!

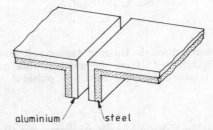

Figure 9.21 Folded-down edges in a lateral plane

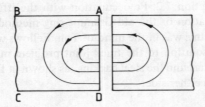

Figure 9.22 The obviously 'impossible' ideal current pattern at a joint (in plan)

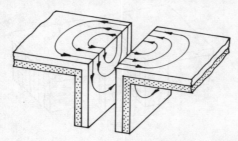

Figure 9.23 Vertical-plane currents in the folded-down joint cancel each other

Figure 9.23 shows that the two semi-circuital currents in the vertical plates will be equal and opposite, as in a transformer winding with single turn coils. What is more, the system can be made into an even better transformer by folding down the steel core also, as shown in the figure, for the effective area of the 'transformer' core is now large enough to offset the fact that it has an inherent airgap (the expansion gap, as shown).

An early test on the system by British Rail at Derby showed that a vertical dimension of only 5 inches on a track used by a primary with a 9-inch pole pitch sufficed to stifle the transients to the extent shown in figure 9.24. Another piece of pure topology took its place in the tool chest of the linear motor designer.

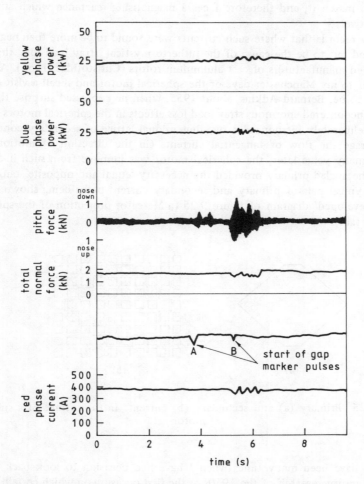

Figure 9.24 Pen-recorded transients at a track joint with and without folded-down edges. (A) corresponds to a folded-down joint, as does (B) for a simple gap between adjacent sheets

9.6 The longitudinal direction

It is common practice, in considering the action of conventional rotating machines, to regard the slot currents as the 'useful bit' and the currents in the end windings as the 'wasted bit'. Induction motor end rings, in particular, are a means of getting current from the bars under an instantaneous N-pole to the bars under the adjacent S-poles and little more, unless the designer decides to opt for a rotor resistivity higher than pure copper and puts the necessary impurity into the metal of the end rings only. The flow of current in the direction of motion which looks as if it should flow in cast rotors in which the aluminium is brought into contact with every rotor lamination, is stifled by the fact that it finds no opposite primary current to match it, and therefore faces a magnetising reactance which stifles it very effectively.

History records that where such currents were found to be more than negligible they turned out to be the cause of the hitherto mystical 'stray load loss' that had long baffled manufacturers of cast aluminium rotors (Christofides, 1965). Again I can return to my Manchester days of the spherical motor and recall a visit of my colleague to be, Bernard Adkins, about 1955, when he expressed surprise that we had not encountered enormous stray load loss effects in the spherical motors where we had deliberately introduced a two-dimensional lattice of secondary conductor to encourage the flow of tangential currents (in the direction of motion). As emerged many years later, the spherical motor was immune from such ill-effects because the angled primary provided the necessary 'equal and opposite' tangential paths, a typical pair of primary and secondary current paths being shown in the 'double-developed' diagram of figure 9.25 (a Mercator projection of the spherical surface in fact).

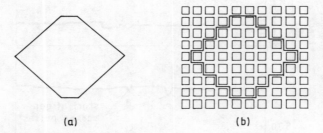

Figure 9.25 Primary (a) and secondary (b) currents in the surface of a spherical motor

There have been many times when I have had occasion to look back on the spherical motor research of the 1950s as the first occasion on which certain effects were encountered and studied, and I am forcibly reminded of the powerful words of Osborne Reynolds with which this chapter began: " . . . the development of art to which such an enterprise may lead." What began as a simple disc experiment (Williams and Laithwaite, 1955) turned out to be a mighty splash that sent ripples to all parts of the earth where machines research is pursued.

9.6.1 Surface current topology

During the development of Tracked Hovercraft and before the concept of the Magnetic River, attention had been focussed on the need for lateral stabilisation, if possible, using the linear motor whose currents were primarily designed for propulsion—side-to-side currents therefore, in relation to the direction of motion. It had long been realised, after all, that when the secondary conductor is a continuous sheet of conductor lying totally in the 'airgap' between primary and secondary steel, currents in any direction could be utilised. Once one has 'paid the price' for increased magnetising current (lower G) necessitated by airgap-located conductor, one cannot go on 'losing' *all* the time!

In a paper devoted almost entirely to what might now be called 'surface current topology' (Eastham and Laithwaite, 1973), a number of novel arrangements were suggested in which longitudinal currents played an active part. Figure 9.26, 9.27, 9.28, 9.29, 9.31 and 9.35 illustrate a number of 'preferred embodiments' (to use patent language).

The simple herring-bone pattern shown in figure 9.26(a) can be designed either as a surface winding or as a more conventional winding in slots. (The conventional technique of rotor slot skewing to prevent magnetic locking is like one half of a slotted herring-bone winding with a large obtuse angle.) This arrangement can be seen as a direct descendant of the spherical motor for the following reason. A developed diagram of one primary block of a pre-skewed spherical motor is as shown in figure 9.27, where alternate poles are marked corresponding to one particular instant of time. The rotor takes the distance y' as its effective pole pitch. In the lateral direction, however, the pole pitch is seen to be only a half of the block width, but the half pole is differently located for different positions along the block. Thus, at positions such as ab, cd, the primary behaves as a C-core TFM, while at positions such as ef it is clearly an E-core. At all intermediate positions it behaves as an asymmetric E-core whose actions have never been studied as a separate topic. But it is interesting to note that for C-core sections, for example, the slot geometry of the spherical motor was distributed, as it always is in machines with conventional, cylindrical geometry, and there is no call for a large central slot that is always associated with what might (in the 1980s) be called 'conventional C-core TFMs'.

There were two ideas incorporated in the 'herrring-bone' winding arrangement shown in figure 9.26 (Eastham and Laithwaite, 1973). The first was to exploit the distributed slot principle, the second was to eliminate the unilateral forces that would occur at all but small values of slip. The two arrangements differ in that the same pole is continuous across the centre line in (a), while the polarity changes at the centre in (b). The latter is clearly a TFM, exploiting the technique especially when the pole pitch, p, is large compared with the half width $w/2$. The arrangement in (a) is, like the spherical motor, a curious hybrid of longitudinal and trans-

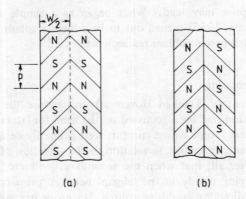

Figure 9.26 'Herring-bone' patterns of instantaneous poles on the surface of a linear motor: (a) longitudinal flux motor; (b) TFM

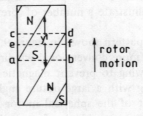

Figure 9.27 Instantaneous pole pattern on a spherical motor surface

verse flux, being a simple I-core in particular sections, an E-core proper in other particular sections and an asymmetric E-core for all other intermediate sections.

A part of the original paper on linear induction motor topology (Eastham and Laithwaite, 1973) was devoted to discussion of the ways in which magnetic flux can be 'directed' along desired paths. For example, lamination places certain constraints on flux directions because steel is a conductor of electricity as well as of magnetism and any flux crossing from lamination to lamination penetrates the sheets, inducing circulating currents that oppose the primary current components parallel to the plane of the laminations. At the same time there are magnetic gaps between adjacent laminations which are filled with paper, varnish or merely natural oxide (rust). In a stack of punchings a metre thick, the total effective 'airgap' in a direction normal to the lamination plane may amount to several centimetres. Both of these effects can be deliberately amplified by the insertion of layers of conducting foil between each pair of adjacent, insulated laminations. This technique has been used for many years in d.c. machine rotors for the discouragement of slot-leakage flux, which flux produces inferior commutation.

A less obvious, but nonetheless effective technique consists simply of differential dimensioning, using the natural reluctance of the iron circuit (which must be of the same order as that of the total airgap, however, for this method to be effective). In the example shown in figure 9.28 there will be a tendency for more flux to pass laterally from N_1 to S_1 and N_2 to S_2, despite the constraints imposed by the features described in the foregoing paragraph, than will pass from N_1 to S_2 and N_2 to S_1, but especially so if the depth of the stack in a direction normal to the plane of the diagram is of the order of the half width $w/2$ or less. In such cases, the use of high airgap flux densities would produce saturation of the core in the longitudinal or p-direction which would lower the permeability and further increase the reluctance in this direction. If the ratio of the permeability in the p-direction to the w-direction is μ_p/μ_w the ratio of longitudinal (p-direction) reluctance to transverse reluctance is $(2p^2/w^2)(\mu_w/\mu_p)$ which, for $\mu_w/\mu_p = 20$ and $p/w = 10$, makes this ratio equal to 4000. This powerful topological tool can be used in the design of herring-bone primaries.

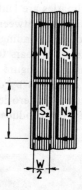

Figure 9.28 Flux-directing by suitable choice of dimensions

9.6.2 Superimposed skewed windings

What emerged as a much more profitable topological exercise was the splitting of the diagram of figure 9.26(a) along its centre line and the overlaying of one half on top of the other. This produces a diamond-shaped mesh of alternate poles, draught-board fashion, whose pattern is seen to move longitudinally and to have merged the end windings completely into the slot-contained winding. If now an actual winding diagram for such a machine is attempted, it takes the form of figure 9.29(a). This would appear to be a very difficult winding to construct until it is realised that it is identical to the arrangement of simple, diamond-shaped coils shown in figure 9.29(b).

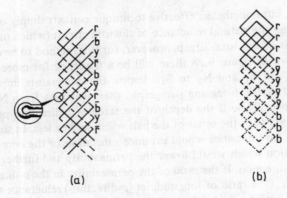

Figure 9.29 The overlaid skew-windings (a) are equivalent to simple, overlapped diamond coils (b)

There is now only one further topological stage to complete a 'circular argument'. If two conventional linear motor primaries are laid side-by-side as in figure 9.30, a suitable choice of coil pitch and spacing between the blocks allows a complete physical fit between the inner sets of end windings, as shown, so that a red-phase current carried, for example, by one set appears to continue *in the same position* —that is, in the top layer—exactly as in the figure 9.31 schematic. In other words, the latter diagram could have been arrived at by consideration of a simple U-core TFM with its 'active' conductors shrunk to zero length! If further proof is needed that the end windings of the E-core TFM produce useful thrust *pro rata*, here it is.

Figure 9.30 The end windings of two, double-layer, conventional linear motors fit each other to produce the double-skew pattern of figure 9.31

Figures 9.32 to 9.34 show a full-scale wooden model that was constructed to illustrate how the motor that propelled RTV 31 might have been replaced by a TFM that produced three times the thrust and weighed less. The fittings were made as replicas of those already existing in the vehicle, including air cushion feeds. The method of designing this machine was to begin with standard 11 kV windings and fit the steel around it, lest it be said that the whole exercise was purely academic and that windings of this shape could not be built. This defensive attitude of the academic was a direct consequence of the reaction of industry to change which had been so much in evidence during the development of the Tracked Hovercraft.

But years after the large wooden model had been built it was realised that a surface-wound arrangement of the overlapped diamond coils shown in figure 9.29(b) would have been the simplest machine to build, and a 3 ft × 1 ft primary was built and tested at Imperial College. There can be little doubt that if the Tracked Hovercraft project had been continued, this form of winding would have emerged as the most advantageous.

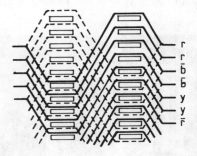

Figure 9.31 Schematic winding diagram of a U-core TFM with cross-over windings

Figure 9.32 View of pole surface of a full-scale wooden model of an E-core TFM, designed to deliver 4 MVA

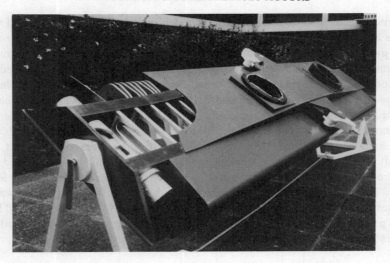

Figure 9.33 Back view of the TFM in figure 9.32

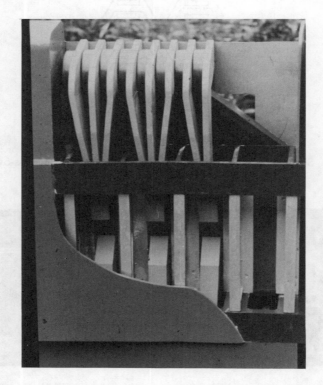

Figure 9.34 Cut-away view of the TFM to show the small quantity of backing steel needed

Other topologies might have followed. If we return to figure 9.26(a), for example, total overlap of the two halves could be replaced by partial overlap, as shown in figure 9.35, in which the centre, diamond-form system is flanked by herring-bone lateral stabilisers. But until there is a real need to make a large machine of this type it will be almost impossible to say whether or not such an attempt at lateral stabilisation would be successful.

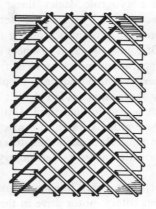

Figure 9.35 A central, diamond pattern flanked by partial herring bones

Once again, in the 'bright light of hindsight' it is obvious which arrangement held the most promise. If high-speed motors of large pole pitch are the target, then the significance of large airgaps shrinks and the induction motor designer has a newly found freedom, especially if he is given the 3 or 4 inches that separated the motor of RTV 31 from its track, or the comparable gaps that are contemplated by the designers of superconducting Maglev, possibly combined with linear synchronous motor propulsion. The designers of synchronous and d.c. machines long enjoyed this freedom and never used it until Professor E. J. Davies produced his paper on airgap windings for alternators (for which I believe I was an enthusiastic referee) (Davies, 1971). The basic idea behind large airgaps in these machines was to stifle 'armature reaction', so that, for example, when used as generators of a.c. and d.c. respectively, the terminal voltage would not fall appreciably on load. The phrase 'making a stiff machine' was often used in this context. In the case of alternators the philosophy that 'd.c. amp-turns are cheap' was pursued with vigour in 20th Century designs and airgaps of up to 7 inches were used (a full-scale design for a multi-megawatt alternator with a 14-inch gap was once completed but the machine was never built).

The message of 1970 is now clear. Of *course* we should investigate machines with both primary and secondary conductor in the 'entrefer'. Of *course* the arrangement shown in figure 9.29(b) is the most elegant. Of *course*, no-one has had the courage to try one yet. No-one has the money!

9.7 Linear synchronous motors

Whether it was the result of encouragement provided by the at least 'partially successful' progress of linear motors in the 1960s, the desire to *compete* with induction machines for the sake of competition or the search for a solution to the problem of 'wasted flux' in a cryogenic Maglev, linear induction motor-propelled system could form the subject of a hot debate—the fact is that linear synchronous motors came more and more into prominence during the 1970s as an alternative drive for levitated vehicles. Of course the rapid advances in solid state inverters were crucial to this subject. A synchronous motor on fixed frequency will not self-start. Variable-speed running *demands* variable-frequency supply, for the motor is locked-on to the supply as faithfully as are all mains-fed electric clocks.

Any type of rotary machine can be linearised. Until the 1950s all adjustable-speed motors of large size were either commutator machines or slip-ring induction motors wasting losses by the tankful (in liquid resistors). Linear versions of either of these would be fairly horrific pieces of technology! But linear synchronous motors could be different.

Before embarking on specific proposals it is worth asking the old question: why, for every h.p. of synchronous rotary motor drive in the world, are there 100 h.p. of rotary induction motor? The answer is not quite as straightforward as it might at first appear for

(1) solid state inverters ought to solve the problem of self-starting synchronous machines
(2) unity power factor (simplifying solid state supply design enormously), can be assured in synchronous motors.

When it comes to *linear* motors, however, there are new considerations which militate further against the use of synchronous, as opposed to asynchronous drives. The most fundamental of these are

(a) There are two types of synchronous machine: the d.c.-excited type and the reluctance motor. One could even add the hysteresis motor as a third type, which will run-up asynchronously in a most desirable manner.
(b) Of these three, only the d.c.-excited machine is 'electromagnetic'. The others are ill-suited to very large power levels.
(c) By definition, the electromagnetic machine has m.m.f. on *both* sides of the airgap and therefore demands a 'live' track, even if the m.m.f. is provided by a line of permanent magnets.

9.7.1 Claw-pole motors

There have been two attempts to develop linear synchronous machines as drives for high-speed transport. In the first, the opinion expressed in (b) above (and at other places in these pages) has been effectively refuted and a linear form of a 'claw-pole' motor has been proposed (Balchin and Eastham, 1979). It is interesting to compare the rotary and linear versions of this configuration, for the two are not identical. In the first instance the rotary claw-pole motor was designed to exploit powerful permanent magnets such as were developed between the wars and during World War II. Permanent magnet materials of high BH_{max} value are inevitably brittle and virtually unmachinable. The intricacies of multi-polar magnetic circuits must therefore involve ingenious arrangements of soft iron pole pieces. Figure 9.36 illustrates the topological steps in assembling a rotor. Mild steel sheet is punched out to form 'teeth', as shown in (a), with a central hole for bolting right through the permanent magnet. The teeth are then bent over, as shown in (b), to produce interlaced 'fingers' of alternate polarity. When spun inside a stator consisting of individual coils wrapped around teeth such that there are as many teeth as there are fingers on the rotor, the machine is extremely useful at quite low speeds and is widely used as a bicycle dynamo. The requirements are right for this permanent magnet

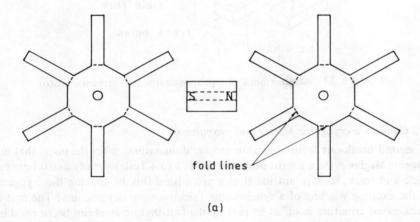

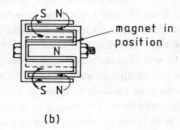

Figure 9.36 Construction of a 'claw-pole' (bicycle) dynamo: (a) original steel punching; (b) result of folding two punchings

machine—small, low-speed and therefore multi-polar and robust—and yet the machine is still electromagnetic.

The track of the linear reluctance motor is as shown in figure 9.37 in which the E-core primary (on the vehicle) completes its magnetic circuit first to the right of centre, then to the left, and so on, as the fingers of track steel alternate, side to side.

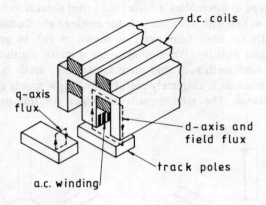

Figure 9.37 Arrangement of a proposed linear reluctance motor

9.7.2 Combined cryogenic Maglev and propulsion

The second break-out from induction motor 'domination' takes the point that in a cryogenic Maglev system a most powerful field (3 to 4 Tesla) already exists between vehicle and track, so why not use it as a propulsion flux by making the cryogenic coil the exciting winding of a synchronous (electromagnetic) machine? The cost of a track-based armature is offset in part by the fact that no steel can be, or need be, used in such a high flux density and the armature winding in its simplest form can consist of nothing more than a single conductor bent into the shape shown in figure 9.38, in which the size of a cryogenic 'pole' is shown.

The inversion of the field/armature roles in this arrangement has one inherent drawback however. Even though track cost may be met, no company could afford to energise miles of track all the time. Nor would it be desirable, anyway. If more than one vehicle or train of vehicles is to occupy the same track at any one time, it is unlikely that both will require precisely the same frequency of supply and therefore the track must essentially be subdivided into relatively short sections, each of which is fed from its own variable-frequency, switchable inverter, and this makes the installation very expensive. To return to a moving armature, stationary field system is unthinkable in terms of superconducting track members.

But it must be said at the end of a 'Second-Age-of-Topology' chapter that ingenuity has not yet run out, and development of these and many other forms of drive is likely to proceed well into the next decade.

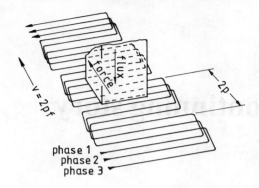

Figure 9.38 Zig-zag track conductor for a linear synchronous motor

9.8 References

Balchin, M. J. and Eastham, J. F. (1979), 'Characteristics of a heteropolar linear synchronous machine with passive secondary', *IEE Electric Power Applications Journal*, Vol. 2, No. 6, pp. 213-218

Christofides, N. (1965), 'Origins of load losses in induction motors with cast aluminium rotors', *Proc. IEE*, Vol. 112, No. 12, pp. 2317-2332

Davies, E. J. (1971), 'Airgap windings for large turbogenerators', *Proc. IEE*, Vol. 118, Nos 3/4, pp. 529-535

Eastham, J. F. and Alwash, J. H. (1972), 'Transverse-flux tubular motors', *Proc. IEE*, Vol. 119, No. 12, pp. 1709-1718

Eastham, J. F. and Laithwaite, E. R. (1973), 'Linear-motor topology', *Proc. IEE*, Vol. 120, No. 3, pp. 337-343

Greatorex, N. (1978), 'A study of transverse flux linear induction motor performance', Ph.D. Thesis, Imperial College (University of London)

Laithwaite, E. R. (1965), 'Electromagnetic levitation', *Proc. IEE*, Vol. 112, No. 12, pp. 2361-2375

Laithwaite, E. R. (1978), 'The about-face age', *Electrical Review*, Vol. 203, No. 18, pp. 33-34

Williams, F. C. and Laithwaite, E. R. (1955), 'A brushless variable-speed induction motor', *Proc. IEE*, Vol. 102A, No. 2, pp. 203-210

10 A continuing story

10.1 General

It was the original intention of the author to end the narrative, the reporting and the references at the end of 1975. But the manuscript went through a traumatic experience between 1978 and 1984 (in fact, several!) and was later revised after adoption by a new publisher.

It was suggested that the last chapter might contain an update, at least of the High-Speed Transport Game, if only in summary form, and since the whole subject was expanding more rapidly than it had ever done in 1976 it did seem that the reader deserved some indication as to where the subject was going—history book though it is supposed to be!

First then, a few facts and figures about the development of non-transport-oriented developments might be of interest, especially if condensed into a mini *Guinness Book of Records* form.

The research director of Linear Motors Ltd told me in the late 1970s that he had then listed over one thousand *different* applications for linear motors. By this he meant that motors had been manufactured and sold for that number of different jobs. The most common applications included sliding doors, travelling cranes and conveyors.

The items that were moved varied from 0.1 mg in weight to over 5 tonnes. In the case of the lower end of this range the particles moved were themselves the whole secondary members of the motors.

Within such a wide variety of types and sizes there was bound to be overlap with other topics, one of the most notable being that of liquid metal stirrers. 'Face plate' motors, in which the conductors were laid radially on a steel backing plate and rotated a copper disc secondary, are clearly not strictly 'linear' but they were 'sheet-rotor motors' and they did have airgaps much longer than those of conventional machines. So they were a natural spin-off or secondary product for a firm building strictly linear machines because the design techniques were extremely similar. Electromagnetic levitation of both solid and liquid secondaries is obviously a closely allied subject and at least one professional conjuror is presenting as 'magic' something which we know is as ordinary as the mechanism of a hair-drier—and yet I shall always believe the force of induction to be sheer magic in its own right! From the levitation of liquid metals and the pumping of these came applications in powder metallurgy, in the scrap metal industry (for sorting non-ferrous metal from non-metal trash) and in the glass industry.

Linear motors for industrial plant applications 'came of age' between 1976 and 1986, even though their general acceptance was still slow.

10.2 'Stop-press' on HSGT

The question 'What happened next in high-speed transport?' must of course be answered, as far as possible on a world-wide basis.

Air cushion vehicles in the USA, France, West Germany and the UK were mostly abandoned in the early 1970s, including the American TACV (Tracked Air Cushion Vehicle).

West Germany branched out into yet another new topology with a large-scale demonstration of the 'M-Bahn' system for urban transport. Permanent rare-earth magnets were used to provide the lift from the underside of a ground rail. Guide wheels were used to control the gap. The philosophy here was simple—better to provide a lifting force of 120 per cent of the vehicle weight and run the wheels on a 'ceiling', thus reducing wheel-rail pressure from 100 per cent weight to 20 per cent weight, the latter being necessary to avoid fluctuations in track etc. from 'unhooking' the vehicle from its magnetic suspension.

The field of the permanent magnets was then used as excitation for what was effectively a very long stator synchronous motor, with active track, in which it was proposed to switch on one kilometre at a time.

West Germany was also actively engaged in a quite different system—along with Canada—in which the lift was provided by cryogenic magnets, and lateral guidance was effected by the use of closed, figure-of-eight loops of conductor laid along the central section of the track between the twin, parallel levitation strips.

The Japanese cryogenic system that came into favour used a powered winding along each side of a U-shaped guideway, cross-connected in such a way that any lateral displacement produced circulating currents that interacted with windings on the vehicle to produce restoring forces. A recent Canadian design also makes use of this topology.

As if to leave no stone unturned or perhaps to emphasise that *any* rotary machine can be 'linearised', one Japanese venture involved the use of a linear commutator motor.

The above examples are chosen to illustrate the intensity of the thinking, determination and ingenuity that continued into the 1980s and will doubtless still continue. A summary of events, country-by-country, now follows. This material is taken from an international conference held at Solihull, England, in October 1984 (Institution of Mechanical Engineers, 1984).

10.2.1 USA

All Government support for research into basic concepts was withdrawn in 1975. A study of a possible high-speed link between Las Vegas and Los Angeles, carried out by the Budd Company, recommended an attractive Maglev system as the most viable.

10.2.2 Canada

A study was made for the Toronto–Ottawa–Montreal 'corridor' and it was concluded that Maglev was better than wheels, but so far it remains a paper study only.

A paper on economic proposals for the corridor project contained only five references—all to Canadian papers.

10.2.3 Japan

Japanese National Railways claims record speeds of 517 k.p.h. for a single vehicle on a T-shaped track at Miyazaki in 1979, 305 k.p.h. for two coupled vehicles and 221 k.p.h. for three coupled vehicles. It is said to be aiming at a 500 k.p.h., Tokyo–Osaka link. A paper on this topic lists 13 references—all to Japanese publications.

10.2.4 West Germany

Work at Erlangen was suspended in favour of attractive Maglev, which has been developed over a period of ten years jointly by the firms Krauss-Maffei and Messerschmitt-Bölkow-Blohm. A full-scale demonstration of the attractive Maglev system was given at an International Transport Fair in Hamburg in 1979, using a linear synchronous motor as propulsion unit.

A new test centre is being started in Emsland with a proposed track 31.5 kilometres in length for a 200 passenger, twin-bodied, Maglev vehicle for speeds up to 400 k.p.h., but this project is still only at the planning stage. Work to put the site at Emsland into service, however, began in 1983 and the vehicle known as Transrapid 06 has run there. The work is described in a paper containing 16 references—all to German papers.

One thing emerges crystal clear at this point. The main obstacle to progress is a lack of communication between what are obviously very deeply committed research teams. They pursue their own researches without looking left or right and what has come to be known as the 'Not Invented Here Syndrome' operates at all levels. Belief in what one is doing is, of course, vital to any project. But there must surely be room for a wider look at a whole relatively new technology. The extent of the literature on it is evidenced by the bibliography in this book.

10.2.5 UK

British Rail's Birmingham Airport–Birmingham International Rail Link has been completed and carries fare-paying passengers. There are now many papers on this project, each containing references mainly to papers from the UK, although at the Solihull conference on 9/10 October 1984 Donald Armstrong of British Rail wrote a fascinating world review paper which included the topology figure reproduced here as figure 10.1. It is interesting to compare this with the topology classification with which this history began (chapter 1, figure 1.6 to 1.8). The various layouts depicted in figure 10.1 refer to the following systems.

A — Permanent magnets on the vehicle mounted underneath a steel reaction rail on the track to give more lift than the weight of the vehicle so that guide wheels run on a ceiling. This is the 'M-Bahn' system described in the third paragraph of section 10.2—said to be 'suitable for low-speed, light-weight vehicles'.

B — A combination of permanent magnet and electromagnet attractive Maglev. The permanent magnets are represented by the vertical lines between adjacent poles.

C — Straight cryogenic lift system needing wheels at low speed (credited originally to Powell and Danby).

D — A split-track cryogenic lift system attempting lateral guidance.

E — Split, inclined track plates attempting lateral stability for a cryogenic system.

F — Inverted-T system for cryogenic lift.

G — Straight cryogenic system with wound coils in the track replacing the conducting sheet.

H — Multiple coil/coil system for combined lift, guidance and propulsion (the proposed Japanese National Railways system).

I — Ferro/null-flux system due to Danby, Jackson and Powell in which levitation is achieved by cryogenic coils being attracted to a steel plate. Adhesion is prevented by active repulsion loops fixed to the track.

J — Horizontal flux from coils on the vehicle reacts with horizontal conductors on the track. Said to achieve combined lift and propulsion with either iron-cored or air-cored cryogenic coils on the vehicle. The saddle coil is intended to produce lateral stability.

K — The so-called 'mixed μ' system comprising a combination of a superconducting energised coil, a superconducting sheet screen and a ferro-magnetic rail (research at Bangor University).

L — Attractive Maglev with longitudinal flux.

M — Attractive Maglev with transverse flux.

N — The Hamburg IVA demonstration system, 1979. Attractive Maglev with secondary air suspension and linear synchronous motor propulsion.

P — 'Magnetic River'.

Q — Homopolar synchronous motor. The track has staggered pole pieces.

R — The Zig-Zag linear synchronous motor due to McLean, Williams and West (1980) of Manchester.

S — A heteropolar synchronous machine.

This is clearly a classification based more on systems and their functions than on the straight topology of magnetic and electric circuits of figures 1.6 to 1.8. But it illustrates the facts that world opinion on the subject is by no means crystallised in the 1980s, that ingenuity continues unabated and that the engineers thus engaged all have enough faith in their own systems to spend considerable research money on prototypes and demonstration vehicles for the world market place—hopefully!

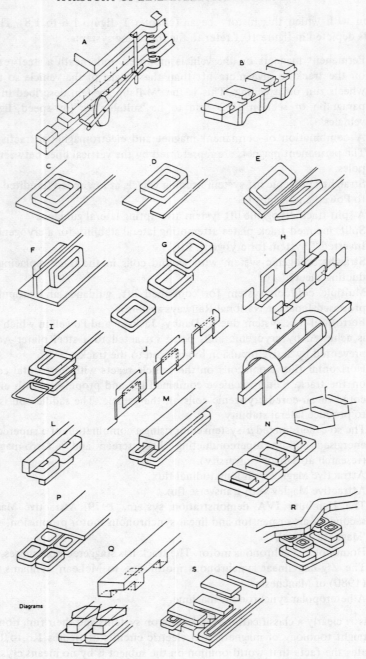

Figure 10.1 Donald Armstrong's topological classification of linear motors (*Reprinted by permission of the Council of the Institution of Mechanical Engineers from 'Maglev Transport—Now and for the Future'*)

10.3 What have we learned?

10.3.1 Wisdom

It is possible that in the whole of science there is all Knowledge—if we could ever have access to it. But that does not imply that there should be at least a grain of Wisdom. Wisdom comes of experience, the experience of living with both our fellow men and with machines. Wisdom can be found by reading history and the great literature of the world. The Bible tells us: "Wisdom is the principal thing; therefore get wisdom: And with all thy getting get understanding" (Proverbs 4, vii). Understanding?—now *there's* a paradox. We do not understand electromagnetism—our basic building block in electrical machine design. We put iron and copper together in shapes of remarkable ingenuity and intricacy, on the understanding that whenever we have done it before, the resulting behaviour produces forces in the directions and of the sizes we would both expect and desire. There is no further Truth, no Understanding. But Wisdom remains, even in an 'earthy' subject such as electric motors; after all, is this not a *history* book?

One can trace the history of a technology such as that of digital computers and arrive at the conclusion that Jacquard and Babbage were a century or more ahead of their time. They seemed to be waiting for advances in basic technology to give them the tools and the materials to open the truly 'Golden Door'. They needed the diode, the triode, the pentode, the transistor and possibly the thyristor to make their inventions blossom. But it is not the same with linear motors.

10.3.2 Courage

It could be argued that curiosity has always been the spur to advances in technology, but I am sure that whether you are a primitive man making an arrowhead out of flint or a director of a company making wheelchairs, personal greed has been at least an ingredient in your success, if successful you are seen to be. This is not a bitter remark, merely a factual one. I am sure that the increase in the rate of rise of technological know-how has had a large slice of its origin in the acceleration of greed with which the 20th Century is cursed. There are many, at first sight, anomalies that under close inspection are not anomalous at all, but merely cold, hard fact, however sad. Greed advances technology faster than hunger. War advances surgery more than does compassion alone. In the famous film, *The Third Man*, Orson Welles takes Joseph Cotton to a high place from where they can see people as tiny moving dots and asks him if he really cares whether or not one of the dots were suddenly to stop moving. He goes on to point an almost accusing finger at the Swiss as a nation for having enjoyed centuries of peace during which they invented—"the cuckoo clock"! Seen as a shocking [sic] philosophy with Orson Welles as the villain (at the time the film was made) there is an Orson Welles living in almost every expensive house today, and again I insist, this is not emotion, neither is it political and certainly not 'left wing'.

The one thing that neither success nor wealth bring is a feeling of security. Richard Lovelace's "Stone walls do not a prison make, Nor iron bars a cage" applies every bit as much to keeping others out as to being oneself confined. Yet this is almost without exception what every successful technologist effectively does—he builds a wall of lesser men around him to fend off anyone who looks as if he might pose the slightest threat to his 'empire'. It is an animal instinct after all: the territorial instinct, even though the 'territory' may be no more tangible than a theory of the Universe.

What the aspiring inventor (and to an even greater degree, the developer) needs most in such a world is courage—by the bucketful! Yet the history of technology over the last two centuries surely shows an increasing number of missed opportunities? Whether this be true or not the factors militating against the individual or even the small company to 'go it alone' with a new idea have certainly increased in the present century. The rise in accountancy as a controlling influence, the growth of very large 'empires', which inevitably favour mediocrity and the wealth decline in the 1960s on an international scale have all contributed to make it harder for a modern Brunel or Telford or Edison to emerge triumphant, no matter *how* much courage they have. The domination of the computer and lately the silicon chip have even begun to work against a latter day Clerk Maxwell, for they imply that if a computer cannot solve a problem, it is insoluble—and worse, if a computer is *not* involved in a project, that project must be of low key. One has only to watch any popular science programme on television to hear the emphasis and the reverence with which the commentator will announce: 'All this data is collected and processed by a COMPUTER,' or 'This bulldozer is completely controlled by a tiny SILICON CHIP.' As one who was at least an 'assistant midwife' at the birth of the computer* I can perhaps be allowed to say that the youth of today are being brainwashed by the media so that when they become young development engineers they may well tend to use a silicon chip instead of their common sense and to use a computer without any assurance that it has been fed the right equations that are relevant to the problem.

Courage alone will not be enough to withstand the onslaught.

So far as linear motors are concerned, however, the largest manufacturer of commercial machines was founded in the 1960s on the basis that the control unit was more important than the motor, and how right that turned out to be.

10.3.3 Theory, design and evolution

From a good look at the history of technology we ought to be able to accelerate the rate of development. In part, and particularly on the grand scale, I suppose we do. But in narrower fields I suspect that we do not. Specialisation brings its own form of 'negative feedback'. A person who teaches well, does it with authority and this can easily be translated by the student into 'This is the only way of doing it.' (Particularly is this true of theoretical analyses.) It is a form of 'brainwashing'. But there is no brainwashing half so deadly as self-brainwashing. The engineer who dis-

---

*Secretary of the Inaugural Conference of the Ferranti Mk I, the world's first commercially built computer in 1951.

covers a new technique in manipulating the quantities of his art will therefore use it on everything, often without any kind of justification.

A recent book of physics, beautifully written (Lewis, 1972) contains a chapter on the teaching of the subject and begins with an example on rate of cooling. An East African schoolboy, making ice creams in the school kitchen fridge, discovered that if you put two beakers of milk (or water) into a fridge together, the one at 100°C and the other at 20°C, the hot liquid turns into ice first! No-one believed him, of course. He was ridiculed, told to 'learn his Newton's Law of Cooling' and so on. But he impressed a visiting professor who promised to try it himself when he returned to Dar-es-Salaam. He there asked his technician to repeat the experiment and was told later in the day that the hot liquid froze first. But the technician added, 'But we'll keep on trying until we get the right result.'

'The Theory' is the most sacred of cows, perhaps the greatest ever brake on the advance of technology, rapid though the latter now is. In discussion on a recorded television show I once retorted "Well man created God in his own image" and it finished up on the cutting room floor, presumably either because: (a) they thought I had got my tongue in a twist, or (b) it would upset the Bishops (it *was* about 1964), or both. The 'Laws of Physics', the 'Fundamental Particles', the 'Striving after the Truth' are all expressions of the attempts to unravel the secrets of the Almighty—as seen through the eyes of Men, of course! But I digress, as often.

Theory is closely allied to design. When you know Ohm's Law and the theory of temperature coefficient of resistance you can improve on the original concept of a Wheatstone bridge. University undergraduate courses in electromagnetism are heavily weighted in theory—Maxwell's equations for breakfast, dinner and tea— and happily it comes as a rude shock to many to find that industrial firms do not approach machine design in *quite* this way!

The first thing that a machine designer does when confronted with a customer's requirements for a 'new' motor is to search through his 'humanoid data banks' (he sits and thinks). After time t_1 he gets up and physically searches through his filing cabinets for time t_2, pulls out a file in triumph and returns to his seat to study it. Next he writes down the stator slot dimensions. The new graduate, observing all this is disturbed. How can he *know* the slot dimensions?—He doesn't. Why did he begin there? 'It is easier,' he is told. If the newly fledged is quick of thought, he may deduce, after following through the whole process of this one design, that he has been witnessing a process of Evolution, no more and no less. The designer had first 'ransacked his sprawling inventory' (as an overseas correspondent once wrote to me) for an earlier customer who had wanted something very like it, and the new design was the same as the old one with the odd 'mutation' here and there. What is more, it works well as a technique. But it is a far cry from James C. M. or even from Oliver Heaviside.

10.3.4 Invention

It is only given to a few to make a real *leap* forward and how often one finds that it is an intuitive leap, rather than a cold hard deduction born of pure learning. I can think of no better example than the invention of a new generation of reluctance motors by my ex-research student and long-time friend Peter Lawrenson. 'Traditional' (a word so often used as an excuse for doing nothing) reluctance motors had been designed on the most obvious of grounds. The theory said you must make the shape of the rotor steel such as to give the greatest difference in reluctance between in-line and quadrature positions in relation to the axis of the stator m.m.f. *Of course* they should be of the shape shown in figure 10.2(a). Professor Lawrenson's shape is shown in figure 10.2(b). At first sight it seems fundamentally wrong to put the airgap opposite the primary axis, as wrong as hot liquid freezing faster than cold. But look at the two corresponding quadrature positions shown in figure 10.3. The conventional design in (a) has enormous gap lengths. But look at the *area* available for the flux. Reluctance is not just gap length alone. The gaps in (b) may be

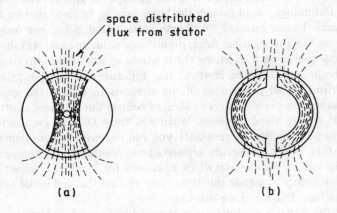

Figure 10.2 Alternative geometries for a reluctance motor rotor

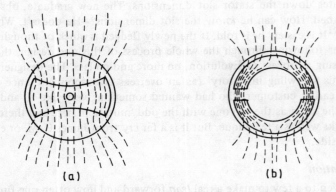

Figure 10.3 Quadrature fluxes are suppressed in the new form of rotor shown in (b)

only $\frac{1}{5}$ of those of (a), but their effective area is only $\frac{1}{20}$.

We return almost to where we began, or more strictly to where *I* began. *"Nobody* makes *linear* induction motors. They're so terribly inefficient, awful power factor, far too expensive—they've got very large airgaps, you see."

10.4 References

Institution of Mechanical Engineers (1984), *International Conference on 'Maglev transport—now and for the future', Solihull, UK* (IMechE Conference Publication 1984-12)

Lewis, J. L. (1972), 'Teaching school physics' (Penguin Books—Unesco)

Bibliography

There is no claim that any section of this bibliography is complete. All that can be said is that an honest attempt has been made to list most of the major papers in the various topics up to the end of 1975.

There are problems whenever one attempts to cover such a vast field as this. Would the reader prefer an alphabetical list of authors?—or an alphabetical list of subjects? How should the topic be subdivided to make reference easier? As soon as subdivision is made there are sure to be papers that overlap more than one section. There will also be papers that do not really 'fit' any one subdivision, so there is almost bound to be a 'General' section, which many classifiers condemn as a mark of failure. An attempt has been made *not* to duplicate a paper in two or more sections but to decide, on balance, which is the more relevant section. The main overlap occurs, of course, between 'Transport' and 'Levitation', and readers are warned to check carefully that an apparently 'missing' paper is not in the other section.

The order of presentation is fairly arbitrary. 'General' could equally well have come last. The other problems as to the order of listing were finally solved by declaring: 'This is a history book, the papers shall be listed in order of year of publication'. In that way, the order at least makes a vague attempt to follow the text.

The only thing that is claimed is that this is probably the most comprehensive list of publications to date on the subject of linear motors and allied topics.

Order

General
Industrial Applications
Levitation
Transport
Theory

GENERAL

1957   LAITHWAITE, E.R.: 'Linear induction motors', Proc. IEE, 1957, 104A, No.18, 461-70

WILLIAMS, F.C., LAITHWAITE, E.R. and PIGGOTT, L.S.: 'Brushless variable-speed induction motors', ibid., No.14, 102-18

1958   LAITHWAITE, E.R.: 'Induction machines - some new ideas', Technology (publ. by The Times), June 1958, 2, No.4, 110

VOL'DEK, A.I., VIL'YAMYAE, G. Kh., SILLAMAA, Kh. V. and TIISMUS, Kh. A.: 'Proposed compensation windings in a linear induction device', Trudy Tallinskogo Politekhnicheskogo Instituta (USSR), 1958, Ser. A, 131. In Russian

WILLIAMS, F.C. and LAITHWAITE, E.R.: 'Unusual induction machinery', Proc. Manchester Assoc. of Engineers, 28 Feb. 1958, 83-104

1959   NORTH, G.G.: 'Linear induction motors', M.S. Thesis, University of California, Berkeley, Dec. 1959

1961   MARINESCU, M.: 'On a new electric motor with oscillatory motion', Rev. Gen. Elec., 1961, 70, No.9, 453-9. In French

THORN, K. and NORWOOD, J. Jr.: 'Theory of an electromagnetic mass accelerator for achieving hypervelocities', NASA Technical Note D-886, 1961

1962   HAUS, H.: 'Alternating current generation with moving conducting fluids', J. Appl. Phys.(USA), 1962, 33, No.7, 2162-72

IDELBERGER, K.: 'Geradeaus-Beschleuniger arbeitet nach dem Prinzip des Induktions-Motors', VDI-Nachrichten, 1962, 16, No.6, 2

JAYAWANT, B.V.: 'A new linear oscillating motor', Elec. Rev., 1962, 171, No.20, 767-72

LAITHWAITE, E.R.: 'Oscillating machines, synchronous and asynchronous', Proc. IEE, 1962, 109A, No.47, 411-4

LAITHWAITE, E.R.: 'The linear motor', Discovery, 1962, 23, No.11, 17-22

LAITHWAITE, E.R. and MAMAK, R.S.: 'An oscillating synchronous linear machine', Proc. IEE, 1962, 109A, No.47, 415-26

MARINESCU, M., GAVRILIU, N. and GRÎNARU, I.: 'Asupra efectului de "histereza electromecanica". Un nou proces de transformare a energiei electrice in lucru mecanie şi aplicarea la realizarea de noi tipuri de maşini electrice', Stud. Cercet. Energ. Electrotehn. (Rumania), 1962, 12, No.3, 337-61

WEST, J.C. and JAYAWANT, B.V.: 'A new linear oscillating motor', Proc. IEE, 1962, 109A, No.46, 292-300

YANES, Kh.I., TIISMUS, Kh.A., VESKE, T.A., LIIN, Kh.A. and TAMMEMYAGI, Kh.A.: 'Proposed compensation windings in linear induction device', Trudy Tallinskogo Politekhnicheskogo Instituta (USSR), 1962, Ser. A, 215, 197. In Russian

1963
JAYAWANT, B.V. and WILLIAMS, G.: 'Analogue of a new linear oscillating motor', Control (GB), 1963, 6, No.60, 97-9

LAITHWAITE, E.R.: 'Recent developments in linear induction motors', 18th Electronics Convention, Manchester College of Science and Technology, 13 July 1963 (Institution of Electronics)

LAITHWAITE, E.R.: 'The linear induction motor', Machine Design Engineering, 1963, 1, No.5, 35-8 ('The principle of the linear machine'), ibid., No.6, 39-43 ('Applications')

LAITHWAITE, E.R. and TUSTIN, A.: 'An oscillating synchronous linear machine', Proc. IEE, 1963, 110, No.8, 1494-5 (Correspondence)

'Linear motors. Review of main features', Elect. Engr. (Australia), 1963, 40, No.4, 47-9

STUHLINGER, E.: 'Electric propulsion', Electrical Engineering (USA), 1963, 82, No.7, 459-65

1964 IDELBERGER, K.: 'The linear induction motor - a new form of the induction motor', Elektrotech. Z. (ETZ) B, 1964, 16, No.5, 105-8. In German

SEQUENZ, V.H.: 'Electrical machines with linear motion', Elektrotechnik und Maschinenbau (Austria), 1964, 81, No.17, 421-31. In German

USAMI, Y.: 'Linear motor', Japanese Railway Engineering (Japan Railway Engineers' Assoc.), 1964, 7, No.1, 3198-200

1965 BARWELL, F.T. and LAITHWAITE, E.R.: 'Linear induction motors', Mining Electrical and Mechanical Engineer, 1965, 45, No.532, 195-9

IVANOV-SMOLENSKII, A.V. and TAMOYAN, G.S.: 'An experimental investigation of the processes occurring in a layer of conducting liquid in the "air" gap of an induction motor', Elektrotekhnika (USSR), 1965, 36, No.2, 5-9. English trans. Soviet Elect. Engng., 1965, 36, No.2, 8-12

MACCHIAROLI, B.: 'Linear induction motors', L'Energia Elettrica, 1965, 42, No.1, 21-7. In Italian

MARINESCU, M.: 'Despre transformarea directă a fortelor electromagnetice pulsatorii in forte alternative. Perspectivele dezvoltării motoarelor liniare de tip electromagnetic', Stud. Cercet. Energ. Electrotehn. (Rumania) 1965, 15, No.1, 63-72

MARINESCU, M.: 'On the direct transformation of pulsatory electromagnetic forces into alternating forces. Development prospects for electromagnetic type linear motors', C.R. Acad. Sci. (Paris), 1965, 260, No.11, 3021-4. In French

MARINESCU, M.: 'Unmittelbare Umwandlung pulsierender elektromotorischer Kräfte in alternierende Kräfte. Entwicklungsaussichten linearer Motoren elektromotorischen Typs', Rev. Roum. Sci. Tech. Electrotech. Energet., 1965, 10, No.1, 99

TAKAHASHI, M.: 'An experimental study on the two-phase linear induction motor', Reports of the Faculty of Engineering, Yamanashi University, Takeda, Japan, Dec. 1965, No.16, 104-8

USAMI, Y.: 'Linear motor, the second report', Japanese Railway Engineering (Japan Railway Engineers' Assoc.), 1965, 8, No.1, 3821-3

1966   LAITHWAITE, E.R.: Induction machines for special purposes (London: Newnes) 1966, (New York: Chemical Publishing Co.) 1966

LAITHWAITE, E.R.: 'Linear induction motors', Engineering, 29 April 1966, 201, 835-40

LAITHWAITE, E.R.: 'New forms of electric motor', Science Journal, 1966, 2, No.2, 38-43

LAITHWAITE, E.R.: Propulsion without wheels (London: English Universities Press) 1966, (New York: Hart Publishing Co.) 1968

LAITHWAITE, E.R.: 'Self-oscillating induction motors', Elec. Rev., 1966, 179, No.10, 346-8

1967

BĂLĂ, C.V.: 'Predetermination of the load characteristics of the electromagnetic linear oscillating motors', Rev. Roum. Sci. Tech. Electrotech. Energet., 1967, 12, No.1, 55-73 In German

BERTINOV, A.I. et al: 'The movement of a conducting piston in a magnetic field', Magn. Gidrodin. (USSR), 1967, No.14, 149-52. In Russian

CAMPANARI, E.: 'Linear asynchronous motors', Elettrotecnica, 1967, 54, No.9, 716-27. In Italian

CERINI, D.J. and ELLIOTT, D.G.: 'Performance characteristics of a single wave length liquid metal MHD induction generator with end loss compensation', 8th Symposium on Engineering Aspects of Magnetohydrodynamics, Stanford, 1967

IDELBERGER, K.: 'Der Linear-Motor läuft geradeaus statt zu rotieren', Elektro-Jahrbuch (Zurich), Dec. 1967, 48-56

KUNTE, J.: 'Der Linear-Induktionsmotor', Industrie-Elektrik und Elektronik, 1967, 12, No. 15/16, 321-2

LAITHWAITE, E.R.: 'A tubular reluctance motor "shotgun"', Elec. Rev., 1967, 180, No.22, 836-7

LAITHWAITE, E.R.: 'Linear electrical motors', Engineering Materials and Design, 1967, 10, No.1, 83

LAITHWAITE, E.R.: 'Linear induction motors', Project, Autumn 1967, No.5, 8 & 23

LAITHWAITE, E.R.: 'Magnetic power in new forms', The Times, 28 April 1967, V.

LAITHWAITE, E.R.: The engineer in wonderland (London: English Universities Press) 1967

1968 BRANOVER, G.G.: 'Experimental investigation of the velocity distribution in the flow of conducting liquid in rectangular ducts situated in a transverse magnetic field', Izv. Akad. Nauk SSSR, Mekh. Zhidkosti Gaza, 1968, No.1, 79-83. In Russian

KALE, K.S. et al: 'Magnetic track circuit - a hypothesis', J. Instn. Engrs. (India), 1968, Pt. ET, 48, 713-9

LAITHWAITE, E.R.: 'A self-propelled magnetic circuit', Elec. Rev., 1968, 182, No.17, 622-3

LAITHWAITE, E.R.: 'Linear induction motors', Electrical India, 1968, VIII, No.4, 27-30

LAITHWAITE, E.R.: 'Linear line-up in force', The Times, 19 April 1968, 23

POLOUJADOFF, M. and PELENC, Y.: 'Recent developments of the linear induction motor', Ingénieur (Canada), 1968, 54, No.237, 14-9. In French

RÉMY, E.: 'Why the linear motor?', Flux (Revue des Anciens Élèves de l'École Supérieure d'Électricité), 1968, No.51, 12-5. In French

TIMMEL, H.: 'The travelling-field linear motor - a remarkable special form of the induction motor', <u>Elektrie</u>, 1968, 22, No.10, 398-402. In German

1969 ASTROP, A.W.: 'A linear micro-stepping motor', <u>Mach. & Prod. Eng.</u> (GB), 1969, 115, No.2966, 465-7

BARTHALON, M.: 'Reciprocating electric motor', Patent UK 1152209, 8 Sept. 1966; publ. 14 May 1969; prior. 24 Sept. 1965, France 32529

DAS, J.K.: 'Use of single-phase linear induction motor as a synchro substitute', <u>J. Instn. Engrs. (India) Elect. Engng. Div.</u>, 1969, Pt. EL2, 50, No.4, 39-42

DE COSTER, F.A.: 'Linear motor control system', Patent US 3435312, 14 Feb. 1966; publ. 25 March 1969, USA 527257

FUJIE, C., TAKETOMI, A., NAKAJIMA, H., KOJIMA, T. and MITOMI, T.: 'Feasibility study of linear motorized car retarder accelerator (Report 2)', <u>Quart. Rept. Railway Technical Research Institute, JNR</u>, 1969, 10, No.4, 221-3

GRIFFITH, J.T. and JONES, T.L.: 'Developing the d.c. linear motor', <u>Engineer</u>, 7 Feb. 1969, 227, 197

HOLM, H.: 'Linear induction motor', Patent UK 1142108, 4 Aug. 1967; publ. 5 Feb. 1969, UK 35862/67

JOHNSON, J.L. and STEPHENS, H.C.: 'Linear motor velocity detection apparatus', IBM Corp., Patent USA 3470399, 17 June 1968; publ. 30 Sept. 1969, USA 737438

KEPPERT, S.: 'Das magnetische Feld eines Planstators', <u>Elektrie</u>, 1969, 23, No. 1/2, 116-8.

LAGIER, J.C.: 'Oscillating linear motor', E.R. Squibb & Sons Inc., Patent USA 3475629, 30 March 1967; publ. 28 Oct. 1969; prior. 30 March 1966, Switzerland 4602/66

LAITHWAITE, E.R.: 'Linear induction motors', Physics Education, 1969, 4, No.5, 309-11

LAITHWAITE, E.R.: 'Linear motors', Encyclopaedic Dictionary of Physics, Supplementary Vol. 3 (Pergamon Press) 1969, 191-4

LAITHWAITE, E.R.: 'The linear motor moves on', Spectrum (British Science News), 1969, No.67, 9-11

PALMERO, A.: 'Reciprocating linear motor', Superior Electric Co., Patent USA 3441819, 18 April 1966; publ. 29 April 1969, USA 543219

PELENC, Y.: 'Linear motor with stationary field structure', Merlin Gerin SA, Patent USA 3482124, 13 Nov. 1967; publ. 2 Dec. 1969; prior. 24 Nov. 1966, France 84919

PIERRO, J.J.: 'Linear electric motor', North American Rockwell Corp., Patent USA 3456136, 26 Sept. 1966; publ. 15 July 1969, USA 581946

RÉMY, E.: 'The linear motor: I Prospect of the linear motor', Rev. Gen. Elec., 1969, 78, No.4, 357-61. In French

RICHARDS, T.L.: 'The linear motor', Copper (GB), 1969, 3, No.3, 2-4

ROUVEROL, W.S.: 'Shiftable rotor variable speed induction motor', Patent USA 3460016, 30 Aug. 1967; publ. 5 Aug. 1969, USA 664371

SAAL, C.: 'Study of an oscillating biphase synchronous linear motor: II', Bul. Inst. Politeh. Iasi (Rumania), 1969, 15, No.1-2, 77-84. In English

TIMMEL, H.: 'The linear induction motor', Maschinenbau, 1969, 18, No.4, 156-61. In German

VICTORRI, M.: 'Linear induction motors', Elektrotech. Z. (ETZ) B, 1969, 21, No.23, 535-9. In German

WALTKE, G.: 'On the performance of the MHD converter and the linear induction motor', Ph.D. Dissertation, T.U. Braunschweig, 1969. In German

WEH, H., WALTKE, G. and APPUN, P.: 'Induction phenomena in MHD converters with constant and travelling magnetic field', Energy Conversion, 1969, 9, 31-8

1970 CHIRGWIN, K.M.: 'Linear induction motor research programs' Energy 70 Intersociety Energy Conversion Engineering Conference, Las Vegas, Nevada, 21 Sept. 1970

'Elongated electric motor', Borg Warner Corp., Patent UK 1188147, 1 Aug. 1967; publ. April 1970, UK 35353/67

HARZ, H.: 'Linearmaschine mit Repulsionswirkung für einphasigen Wechselstrom', German Patent 2029462, 1970

LAITHWAITE, E.R.: 'New wave in electric machinery', New Scientist, 1970, 46, No.706, 570-3

LAITHWAITE, E.R.: 'Rack-and-pinion motors', Electronics & Power, July 1970, 16, 251-2

LAITHWAITE, E.R. and HARDY, M.T.: 'Rack-and-pinion motors: hybrid of linear and rotary machines', Proc. IEE, 1970, 117, No.6, 1105-12

'Novel reciprocating electrical drive', Elect. Engr. (Australia), 1970, 47, No.3, 45-6

RASHCHENKIN, A.P.: 'The excitation in the gap of a linear induction machine with variable travelling magnetic field velocity', Magn. Gidrodin.(USSR), 1970, No.1, 116-20. In Russian (English trans. in Magnetohydrodynamics (USA))

REITZ, J.R.: 'Forces on moving magnets due to eddy currents', J. Appl. Phys. (USA), 1970, 41, No.5, 2067-71

SAAL, C.: 'Theory of the reactive biphase synchronous oscillating motor', Electrotehnica (Rumania), 1970, 18, No.1, 1-7. In Rumanian

WEH, H.: 'Linear motors', VOE Fachber (Germany), 1970, 26, 37-43. In German

YUN JONG LEE and DAL HO IM: 'The travelling field of two phase linear induction motor', J. Korean Inst. Elect. Engrs., 1970, 19, 1-10. In Korean

1971   BOLTON, H.: 'Progress in linear motor techniques in Great Britain', Elektrie, 1971, 26, No.6, 213-4. In German

DAVIS, M.W.: 'Concentric linear induction motor', Patent USA 3602745, 27 March 1970; publ. 31 Aug. 1971, USA 23317

DAVIS, M.W.: 'Developments of concentric linear induction motor', IEEE 6th Annual Meeting of Industry and General Applications Group, Cleveland, Ohio, USA, 18-21 Oct. 1971, 85-93

BIBLIOGRAPHY

DENIS, M. and MAY, P.: 'A linear stepping motor for numerical control', Ing. & Tech. (France), 1971, No.255, 23. In French

GERRARD, J. and PAUL, R.J.A.: 'Rectilinear screw-thread reluctance motor', Proc. IEE, 1971, 118, No.11, 1575-84

GREEN, C.W. and PAUL, R.J.A.: 'Performance of d.c. linear machines based on an assessment of flux distributions', ibid., No.10, 1413-20

HÜHNS, T. and KRATZ, G.: 'The asynchronous linear motor as driving element and its peculiarities', Elek. Bahnen (Germany), 1971, 42, No.7, 146-51. In German

HÜHNS, T. and KRATZ, G.: 'The linear motor as a drive component', Elektrotech. Z. (ETZ) B, 1971, 23, No.19, 449-50. In German

'Induction motor', NRDC, Patent UK 1256091, 14 May 1969; publ. 8 Dec. 1971; prior. 14 May 1968, UK 22799/69

ISHII, T.K.: 'Impulse motor for electric propulsion', IEEE 22nd Annual Vehicular Technology Conference, Detroit, Mich., USA, 7-8 Dec. 1971

LAITHWAITE, E.R.: Linear electric motors (London: Mills and Boon) 1971

LAITHWAITE, E.R.: 'Linear induction motors', Canada - UK Trade News (Journal of the Canada - UK Chamber of Commerce), May 1971, 11,13,15,24

LARONZE, J. and FAURE, A.: 'Linear motors', Tech. CEM (France), 1971, No.81, 3-11. In French

LINDSLEY, J.C.: 'Multipole closed end linear motor', IBM Tech. Disclosure Bull. (USA), 1971, 13, No.12, 3682-3

'Linear induction motor', VEB Kombinat Elektromaschinenbau, Patent UK 1233208, 29 Aug. 1968; publ. 26 May 1971, UK 412477/68

'Linear motor', NRDC, Patent UK 1224470, 4 July 1968; publ. 10 March 1971; prior. 10 July 1967, UK 31694/67

'Linear motor system', AEG-Elotherm GmbH, Patent UK 1236119, 24 Sept. 1969; publ. 23 June 1971; prior. 20 Feb. 1969, Germany P1908457-2

LINKE, H.: 'Operation and control of a special-purpose linear motor', Maschinenwelt Elektrotech. (Austria), 1971 26, No.8, 202-4. In German

PATEL, T.R.: 'Flat coil linear motor', IBM Tech. Disclosure Bull. (USA), 1971, 13, No.11, 3385

PELENC, Y. and RÉMY, E.: 'The future of the linear induction motor', Rev. Gen. Elec., 1971, 80, No.2, 138-42. In French

POLOUJADOFF, M.: 'Linear induction machines, I. History and theory of operation', IEEE Spectrum, 1971, 8, No.2, 72-80

'Polyphase linear induction motor', Merlin Gerin SA, Patent UK 1249800, 30 Aug. 1968; publ. 13 Oct. 1971; prior. 1 Sept. 1967, France 119804

'Stretching the linear motor pole pitch with transverse flux', Elec. Rev., 1971, 189, No.27, 936-7

TANAKA, H.: 'The linear motor', JEE (Japan), 1971, No.60, 34-40

WATTS, J.L.: 'Linear motors', Mach. & Prod. Eng. (GB), 1971, 118, No.3057, 909-12

1972

ALGER, P.A. and WILSON, C.: 'Linear reciprocating electric motors', Mech. Technol. Inc., Patent USA 3643117, 3 Sept. 1970; publ. 15 Feb. 1972, USA 69267

BAUSCH, H. and NOWAK, S.: 'Performance of synchronous linear motors', Arch. Elektrotech. (Germany), 1972, 55, No.1, 13-20. In German

BEATSON, C.: 'Inducing more pull into the linear motor', Engineer, 1972, 234, No.6044, 36-8

BOLL, K.F. and STEPHENS, T.R.: 'Linear bidirectional servomechanism', IBM Tech. Disclosure Bull. (USA), 1972 14, No.8, 2333-4

BOYD, M. and HU, P.Y. 'Linear electric motor', ibid., 15, No.3, 1023-4

CANTEMIR, L.: 'Some problems concerning the linear induction motor whose armature has a circular motion', Bull. Sci. Assoc. Ing. Electr. Inst. Electrotech. Montefiore (Belgium), 1972, 85, No.3, 133-6. In French

CHAI, H.D. and PAWLETKO, J.P.: 'N-phase linear stepper motor concept', IBM Tech. Disclosure Bull. (USA), 1972, 15, No.2, 489-90

DAVIS, M.W.: 'Development of concentric linear induction motor', IEEE Trans., 1972, PAS-91, No.4, 1506-12

EASTHAM, J.F.: 'Linear induction motor stator', Tracked Hovercraft Ltd, Patent USA 3644762, 22 Feb. 1971; publ. 22 Feb. 1972, USA 117461

EASTHAM, J.F. and ALWASH, J.H.: 'Transverse-flux tubular motors', Proc. IEE, 1972, 119, No.12, 1709-18

FIREFEANU, V. and STANCIU, D.: 'Experimental study of the magnetic field and power of a linear induction model', Electrotehnica (Rumania), 1972, 20, No.2, 53-7. In Rumanian

GIRARD, A. Ch.: 'The linear motor runs from success to success', Inter Electron. (France), 1972, 27, No.47, 46-8. In French

JOSSE, B.: 'Linear induction motor', Patent UK 1279670, 21 Nov. 1969; publ. 28 June 1972; prior. 21 Nov. 1968; 27 Oct. 1969, France 174788; 6936802

LAITHWAITE, E.R.: 'How a low-speed brushless a.c. motor was invented', Elec. Rev., 1972, 191, No.7, 221-2

LAITHWAITE, E.R.: '"Rack and pinion" motors', ibid., No.22, 749-50

LAITHWAITE, E.R.: 'The development of linear electric motors, I and II', Elect. Engr. (Australia), 1972, 49, No.4, 21-2, 24-6 and No.5, 14-8

LAITHWAITE, E.R.: 'The development of the linear motor', Rev. Polytech. (Switzerland), 1972, No.1306, 987-99. In French

LAITHWAITE, E.R.: 'The shape of things to come', Proc. IEE, 1972, 119, No.1, 61-8

LAITHWAITE, E.R. and BOLTON, H.R.: 'Linear induction motors', NRDC, Patent USA 3648084, 16 June 1970; publ. 7 March 1972; prior. 31 Aug. 1967, UK 39933/67 (1226966)

NONAKA, S., HAYASHI, K. and YOSHIDA, K.: 'Operating characteristics of double sided linear motors driven by the three phase PWM invertor', Technol. Rep. Kyushu Univ. (Japan), 1972, 45, No.6, 834-40. In Japanese

NONAKA, S., YOSHIDA, K. and YAMADA, M.: 'Characteristics of double sided linear motors with Fe-Cu alloy secondary conductors' ibid., 841-8. In Japanese

PADMANABHAN, S.: 'Linear DC motor', IBM Tech. Disclosure Bull. (USA), 1972, 14, No.11, 3300-1

PAUL, R.J.A.: 'Flat single-sided linear helical reluctance motor', Proc. IEE, 1972, 119, No.12, 1693-7

PETRECCA, G. and VISTOLI, I.: 'Tubular motor design and performances', Elettrotecnica, 1972, 59, No.11, 1186-94, In Italian

ROUBICEK, O.: 'Controlled low-frequency linear oscillatory drives', Elec. Rev., 1972, 190, No.21, 727-9

RUMMICH, E.: 'Methods of braking linear induction machines', Elek. Bahnen (Germany), 1972, 43, No.12, 273-7. In German

RUMMICH, E.: 'Synchronous linear machines', Bull. Assoc. Suisse Electr. (Switzerland), 1972, 63, No.23, 1338-44. In French

SUCHODOMSKI, M.: 'Electric crank and servo-motors, types ELS and EWS', Pomiary Autom. Kontrola (Poland), 1972, 18, No.2, 83-5. In Polish

'The straight line to linear motion is a linear motor', Engineer, 1972, 235, No.6091, 49

ZARIPOV, M.F. and KHAKIMOV, Kh. Kh.: 'A noncontacting linear selsyn', Izv. Akad. Nauk USSR Ser. Tekh. Nauk, 1972, No.1, 18-20. In Russian

1973
BUDIG, P.K., MAGERL, R. and RIEDEL, C.: 'Piezoelectric linear stepping motor', Elektrie, 1973, 27, No.8, 423-7. In German

CRAIG, J.P., SELIN, K.I., KAUSS, W.: 'Operating characteristics of a two-stator induction motor', 1973 SWIEEECO Record of Technical Papers, Houston, Texas, USA, 4-6 April 1973 (IEEE), 166-71

de ZEEUW, A.: 'Linear short gap motors, I and II', Polytech. Tijdschr. Elektrotech. Elektron. (Netherlands), 1973, 28, No.25, 820-8 and No.26, 872-9. In Dutch

DELEROI, W.: 'The starting process for the asynchronous linear motor', Elektrotech. Z. (ETZ) A, 1973, 94, No.8, 488-93. In German

EASTHAM, J.F. and LAITHWAITE, E.R.: 'Linear motor topology', Proc. IEE, 1973, 120, No.3, 337-43

GERRARD, J. and PAUL, R.J.A.: 'Dynamic performance of rectilinear screw-thread motor', ibid., No.1, 73-8

GUTT, H.-J.: 'Viewpoints pertaining to the application of modern linear and moving field motors', VDI Z. (Germany), 1973, 115, No.16, 1259-65. In German

LAMB, C. St.J.: 'Parallel connected linear motor', Elect. Engr. (Australia), 1973, 50, No.7, 16-8

MATSUMIYA, T.: 'On the stationary characteristics of two-phase linear induction motors', Trans. Soc. Instrum. & Control Eng. (Japan), 1973, 9, No.1, 29-36. In Japanese

ONISHI, K., TAKAHASHI, T. and ISHIZAKI, K.: 'Characteristics of new type transistor-controlled dc micromotor', Hitachi Hyoron (Japan), 1973, 55, No.7, 23-8

SCHIEBER, D.: 'Principles of operation of linear induction devices', Proc. IEEE, 1973, 61, No.5, 647-56

'Tubular motor could combine linear and rotary motion', Elec. Rev., 1973, 192, No.1, 25

WEH, H. and LANG, A.: 'The linear motor with a ring winding and leakage-field damping', Elektrotech. Z. (ETZ) A, 1973, 94, No.2, 97-102. In German

1974 ANCEL, J.: 'A test stand for experiments on linear motors', Tech. CEM (France), 1974, No.91, 26-9. In French

ANDRESEN, E.: 'Linear induction motor with adjustable secondary and discontinuously arranged stators', Elektrotech. Z. (ETZ) A , 1974, 95, No.2, 69-74. In German

BASAK, A. and OVERSHOTT, K.J.: 'The ferrite field d.c. linear motor', IEE Conference on Linear Electric Machines, London, 21-23 Oct. 1974, 39-44

BROUGH, J.J.: 'An introduction to the linear d.c. motor', Electronics & Power, 1974, 20, No.15, 628-32

BROUGH, J.J.: 'The linear d.c. motor improves control flexibility', Elec. Rev., 1974, 195, No.17, 593-4

BUDIG, P.K.: 'Some remarks on the device of three-phase linear motors for low synchronous speed', IEE Conference on Linear Electric Machines, London, 21-23 Oct. 1974, 25-30

CHEPELE, Yu. M.: 'A study of cylindrical linear induction motor', Elektrotekhnika (USSR), 1974, No.6, 56-8. In Russian

CHURA, V. and DVORAK, F.: 'A square linear motor', Elektrotech. Obzor (Czechoslovakia), 1974, 63, No.4, 211-3. In Czech

DELEROI, W. and HUBNER, K.D.: 'Effect of the armature winding arrangement on the operation of asynchronous linear motors', Elektrotech. Z. (ETZ) A, 1974, 95, No.11, 601-6. In German

EASTHAM, J.F. and BALCHIN, M.J.: 'Pole-change linear induction motors', IEE Conference on Linear Electric Machines, London, 21-23 Oct. 1974, 9-14

GUZMAN, A.M. and LAWES, H.D.: 'Rotary/linear motor', IBM Tech. Disclosure Bull. (USA), 1974, 27, No.5, 1306-7

JUFER, M. and CASSAT, A.: 'Static and dynamic characteristics of electromagnetic transductors. Applications to a linear step motor', Feinwerktech. & Micronic (Germany), 1974, 78, No.4, 151-60. In German

LUDA, G. and POHLSEN, K.H.: 'Control and regulation of linear motors', Elektro-Anz. (Germany), 1974, 27, No.24, 515-8. In German

MEYER, H.: 'Linear motor', Elektrotech. Z. (ETZ) B, 1974, 26, No.2, 42-3. In German

NASAR, S.A., BOLDEA, I. and LAGUNA, N.: 'Performance of linear induction motors with dual windings', IEE Conference on Linear Electric Machines, London, 21-23 Oct. 1974, 191-6

OBERRETL, K.: 'Linear motors with extended and compact winding', Arch. Elektrotech. (Germany), 1974, 56, No.2, 55-8. In German

OBERRETL, K.: 'Linear motors with special double-layer windings', IEE Conference on Linear Electric Machines, London, 21-23 Oct. 1974, 15-20

SHUKELIS, A.Ch.V.: 'Some aspects of investigations and synthesis of linear stepper motors', ibid., 83-8

YOUNG, P.C.: 'Magnetic switches control linear motor (magneto-transistor IC)', Elect. Engr. (Australia), 1974, 51, No.3, 14-6

1975 BALCHIN, M.J. and EASTHAM, J.F.: 'Performance of linear induction motors with airgap windings', Proc. IEE, 1975, 122, No.12, 1382-90

BHATTACHARYYA, M. and MAHENDRA, S.N.: 'Self-oscillating single-phase linear motor', J. Instn. Engrs. (India) Elect. Engng. Div., 1975, Pt. EL3, 55, 134-8

DELEROI, W., GRUMBKOW, P.V. et al.: 'Short-stator linear motor - state of development', Elektrotech. Z. (ETZ) A, 1975, 96, No.9, 401-9. In German

EASTHAM, J.F. and BALCHIN, M.J.: 'Pole-change windings for linear induction motors', Proc. IEE, 1975, 122, No.2, 154-60

LAITHWAITE, E.R.: 'Linear electric machines - a personal view', Proc. IEEE, 1975, 63, No.2, 250-90

'Linear motor with coils and laminar core', Conz-Elektricitats GmbH, Patent UK 1382863, 5 April 1972; publ. 5 Feb. 1975; prior. 5 April 1971, W. Germany 116516

MELIK-SHAKHNAZAROV, A.M., SAVIN, V.V. and DADAYAN, Yu. A.: 'Linear phase digitizer for a selsyn', Izv. VUZ Priborostr. (USSR), 1975, 18, No.4, 55-60. In Russian

PAYEN, J.P.: 'Suspension and coupling device for the movable magnetic field unit of a linear induction motor', Soc. Dauphinoise Electrique, Patent USA 3874300, 6 March 1973; publ. 1 April 1975; prior. 15 March 1972, France 9749

SOKOLOV, M.M. and SOROKIN, L.K.: 'An experimental investigation of the mechanical characteristics of a linear induction motor', Elektrotekhnika (USSR), 1975, No.2, 43-6. In Russian

WATSON, D.B.: 'Speed and torque control of a single-phase linear induction motor', Proc. IEE, 1975, 122, No.2, 188-9

WEH, H.: 'Linear electric motors - state of the art', Naturwissenschaften (Germany), 1975, 62, No.3, 113-7. In German

INDUSTRIAL APPLICATIONS

1949 KOPECKI, E.S.: 'Induction stirring in electric furnace steelmaking', Iron Age, 1949, 164, No.12, 73-8

1953 BARNES, A.H.: 'D.C. electro-magnetic pumps', Nucleonics, 1953, 11, No.1, 16-21

ROBIN, M.: 'Electromagnetic flowmeter', J. Rech. Cent. Natn. Rech. Scient. (Paris), 1953, 5, 187-9

SHERCLIFF, J.A.: 'Steady motion of conducting fluids in pipes under transverse magnetic fields', Proc. Camb. Phil. Soc., 1953, 49, Pt.1, 136-44

WATT, D.A.: 'A.C. liquid metal pumps for laboratory use', A.E.R.E. (Harwell), Report No. CE/R 1089, 1953

1954 COTTRELL, W.B. and MANN, L.A.: 'Components for sodium or NaK systems (i.e. pumps)', Nucleonics, 1954, 12, No.12, 22-5

REMENIERAS, G. and HERMANT, C.: 'The electromagnetic measurement of velocities in liquids', Houille Blanche, 1954, 9, Special No.B, 732-46. In French

1955 GREENHILL, M.: 'Electromagnetic pumps and flowmeters', A.E.R.E. (Harwell), Inf. Bib. No.93, 1955

LOCK, R.C.: 'The stability of the flow of an electrically conducting fluid between parallel planes under a transverse magnetic field', <u>Proc. Royal Soc.</u>, 1955, 233A, 105-25

SHERCLIFF, J.A.: 'Experiments on the dependence of sensitivity and velocity profile in electromagnetic flow-meters', <u>J. Sci. Instrum.</u>, 1955, 32, No.11, 441-2

THÜRLEMANN, B.: 'On the electromagnetic measurement of the velocity of flow of liquids', <u>Helv. Phys. Acta</u>, 1955, 28, No.5-6, 483-5. In German

WATT, D.A.: 'A study in design of travelling field electromagnetic pumps for liquid metals', <u>A.E.R.E.</u> (Harwell), Report No. ED/R 1696, H.M.S.O., June 1955

WOOLLEN, W.B.: 'Electromagnetic pumping of liquid metals', <u>Fluid Handling</u>, 1955, No.62, 60-3 and No.63, 90-2

1956 DENISON, A.B. et al: 'Square wave electromagnetic flowmeter design', <u>Rev. Sci. Instrum.</u>, 1956, 27, No.9, 707-11

KOLIN, A.: 'Principle of electromagnetic flowmeter without external magnet', <u>J. Appl. Phys.</u> (USA), 1956, 27, No.8, 965-6

LIELPETER, Ya. Ya. and TYUTIN, I.A.: 'Berechnungs-Verfahren für Flüssigmetall-Induktions-Pumpen', <u>Prikladnaya Magneto-gidrodinamika</u>, 1956, No.8, 95-106

SHERCLIFF, J.A.: 'Entry of conducting and non-conducting fluids in pipes', <u>Proc. Camb. Phil. Soc.</u>, 1956, 52, 573-83

STRONG, G.H.: 'Electromagnetic ball conveyor', <u>Electrical Engineering</u> (USA), 1956, 75, No.12, 1064-9

TYUTIN, I.A. and YANKOP, E.K.: 'Elektromagnetische Pumpen für Flüssige Metalle', Prikladnaya Magnetogidrodinamika, 1956, No.8, 24-48

TYUTIN, I.A. and YANKOP, E.K.: 'Elektromagnetische Vorgänge in Induktions-Pumpen für Flüssige Metalle', ibid., 65-80

UL'MANIS, L. Ya.: 'Randeffekte in Linearen Induktions-Pumpen', ibid., 81-94

WATT, D.A.: 'Electromagnetic pumps for liquid metals, circulating reactor coolants and fuels', Engineering, 27 April 1956, 181, 264-8

1957 BLAKE, L.R.: 'Conduction and induction pumps for liquid metals', Proc. IEE, 1957, 104A, No.13, 49-63

FENEMORE, A.S.: 'Linear induction pumps for liquid metals', Engineer, 17 May 1957, 203, 752-5

HOLDAWAY, H.W.: 'A note on electromagnetic flowmeters of rectangular cross-section', Helv. Phys. Acta., 1957, 30, No.1, 85-8. In German

LAITHWAITE, E.R.: 'New developments in electromagnetic shuttle propulsion', Textile Weekly, 1957, 57(1), No.1506, 234-5

LAITHWAITE, E.R.: 'Recent developments in electromagnetic shuttle propulsion', Textile Manufacturer, March 1957, 131-3

LAITHWAITE, E.R. and DUXBURY, V.: 'Electromagnetic shuttle-propelling devices', Journal of the Textile Institute, 1957, 48, No.3, 214-24

LAITHWAITE, E.R. and LAWRENSON, P.J.: 'A self-oscillating induction motor for shuttle propulsion', Proc. IEE, 1957, 104A, No.14, 93-101

SHERCLIFF, J.A.: 'Electromagnetic flowmeter without external magnet', J. Appl. Phys. (USA), 1957, 28, No.1, 140

WATT, D.A., O'CONNOR, R.J. and HOLLAND, E.: 'Tests on an experimental D.C. pump for liquid metals', A.E.R.E. (Harwell), Report No. R/R 2274, 1957

1958 BROMKAMP, K.H., IDELBERGER, K. and SCHNEEFUß, P.: 'Drehstrom-Induktions-Pumpe', Diplomarbeit Technische Hochschule Aachen, 1958

CUSHING, V.: 'Induction flowmeter', Rev. Sci. Instrum., 1958, 29, No.8, 692-7

1959 BALLS, B.W. and BROWN, K.J.: 'The magnetic flowmeter', Trans. Soc. Instrum. Technol., 1959, 11, No.2, 119-30

BIRZVALKS, Yu. et al: 'Velocity distribution in electromagnetic pump channels with a rectangular cross section', Latv. PSR Zinat. Akad. Vestis (USSR), 1959, No.10 (147), 85-9. In Russian

HLASNIK, I.: 'Reaction of armature of an electromagnetic pump', Elektrotech. Obzor (Czechoslovakia), 1959, 48, No.3, 135-46. In Slovak

LIELPETER, Ya.: 'The thermal processes in electromagnetic induction pumps', Latv. PSR Zinat. Akad. Vestis (USSR), 1959, No.9 (146), 91-100. In Russian

WATT, D.A.: 'The design of electromagnetic pumps for liquid metals', Proc. IEE, 1959, 106A, No.26, 94-103

1960 CAMBILLARD, E. and SCHWAB, B.: 'Multiphase annular electromagnetic pumps - design and construction', Bull. Soc. Franc. Elect. (Ser. 8), 1960, 1, No.6, 417-24. In French

DEVIDAS, R.: 'Single phase AC electromagnetic pumps', ibid., 411-6. In French

HERMANT, C.: 'Improving the characteristics of the linear-type electromagnetic induction pump', ibid., No.3, 163-78. In French

LAITHWAITE, E.R. and NIX, G.F.: 'Further developments of the self-oscillating induction motor', Proc. IEE, 1960, 107A, No.35, 476-86

LAITHWAITE, E.R., TIPPING, D. and HESMONDHALGH, D.E.: 'The application of linear induction motors to conveyors', ibid., No.33, 284-94

OKHREMENKO, N.M.: 'Electromagnetic phenomenon in flat induction pumps for liquid metals', Elektrichestvo, 1960, No.3, 48-54. In Russian

OKHREMENKO, N.M.: 'Electromagnetic phenomenon in flat induction pumps for molten metals', Elect. Technol. USSR, 1960, 5, 132

ROLFF, J.J.P.: 'Magnetic flowmeters', Arch. Tech. Messen, 1960 No.297 (ref. V1249-2), 197-200. In German

SCHWAB, B.: 'Different types of electromagnetic pumps', Bull. Soc. Franc. Elect. (Ser. 8), 1960, 1, No.6, 404-10. In French

VAUTREY, L.: 'The use of electromagnetic pumps', ibid., 399-403. In French

VOL'DEK, A.I.: 'Electromagnetic pumps for liquid metals', Elektrichestvo, 1960, No.5, 22-7. In Russian

WILKINSON, K.: 'Linear motors applied to materials handling', Elec. Times, 1960, 154, No.3996, 31-6

1961 HANKA, L.: 'The geometry of induction pumps for liquid metals', Elektrotech. Obzor (Czechoslovakia), 1961, 50, No.1, 38-44. In Czech

LAITHWAITE, E.R.: 'Linear induction motors and their applications', Design and Components in Engineering, Dec. 1961, 589-94

LAITHWAITE, E.R.: 'Prospects for linear induction motors', New Scientist, 1961, 12, No.255, 42-5

LAITHWAITE, E.R., NIX, G.F., BRUNNSCHWEILER, D. and BINA, J.: 'The self-oscillating induction motor as a traverse mechanism for cone-winding machines', Journal of the Textile Institute, 1961, 52, No.11, 625-34

OKHREMENKO, N.M.: 'Optimal geometrical relations in the induction pumps for liquid metals', Elektrichestvo, 1961, No.9, 10-6. In Russian

'Pumps in atomic energy - survey', Nuclear Power (GB), 1961, 6, No.63, 68-73

1962 BROMKAMP, K.H.: 'Inductive stirring device for metallic melts', <u>Elektrotech. Z. (ETZ) A</u>, 1962, 83, No.24, 811-5. In German

BYESTE, L.A.: 'An electromagnetic chute for molten metal', <u>Elektrichestvo</u>, 1962, 82, No.5, 74-7. In Russian

KHABLENKO, A.I. and IL'YASHENKO, O.I.: 'Electromagnetic pump', <u>Pribory i Tekh. Eksper.</u> (USSR), 1962, No.2, 178-9. English trans. <u>Instrum. Exper. Tech.</u>(USA), 1962, No.2, 400

1963 GERMAIN, C.: 'Bibliographical review of the methods of measuring magnetic fields', <u>Nuclear Instrum. and Methods</u> (Netherlands), 1963, 21, No.1, 17-46 (280 refs)

LEBÉE, G., RICHARD, M. et al: 'Study and production of a rectangular field electromagnetic flowmeter', <u>Rev. Gen. Elec.</u>, 1963, 72, No.9, 439-43. In French

PANHOLZER, R.: 'Electromagnetic pumps', <u>Electrical Engineering</u> (USA), 1963, 82, No.2, 128-35

SLATER, R.A.C., JOHNSON, W. and LAITHWAITE, E.R.: 'An appraisal of the linear induction motor concept for high-energy-rate metal forming', <u>Sheet Metal Industries</u>, April 1963, 237-43

SLATER, R.A.C., JOHNSON, W. and LAITHWAITE, E.R.: 'An experimental investigation relating to the accelerated motion of various "translators" in the air gap of a linear induction motor', <u>Int. J. of Mach. Tool Des. and Res.</u>, 1963, 3, 111-135

VERTE, L.A.: 'Experimental induction pump for molten iron', <u>Elektrichestvo</u>, 1963, 82, No.12, 64. In Russian

1964 GUSEL'NIKOV, E.M. and ROTT, V.Zh.: 'A series of single-rod electrohydraulic linear actuators', Elektrotekhnika (USSR), 1964, No.8, 55. In Russian

'Impact extrusion by linear induction', Engineering, 4 Sept. 1964, 198, 302-3

JOHNSON, W., LAITHWAITE, E.R. and SLATER, R.A.C.: 'An experimental impact-extrusion machine driven by a linear induction motor', Proc. I. Mech. E., 1964-65, 179, Pt.I, No.1, 15-35

'Linear motor drives impact extruder', Elec. Times, 1964, 146, No.8, 259-60

1965 KIRSHTEIN, H.Kh. and RYBAKOV, E.K.: 'Method for measuring velocity of conducting medium using a pulsatory magnetic field', Latv. PSR Zinat. Akad. Vestis Fiz. Tehn. Ser. (USSR), 1965, No.2, 35-40. In Russian

MARIN, N.I., POVSTEN, V.A., DOKTOROVA, T.V. and AVILOVA, E.M.: 'Electromagnetic pumps for (liquid) alkali metals', Atomnaya Energiya (USSR), 1965, 18, No.3, 239-42. In Russian

OKHREMENKO, N.M.: Travelling magnetic wave induction pumps', Magn. Gidrodin (USSR), 1965, No.4, 3-23. In Russian

STEINER, D.: 'On the feasibility of electromagnetic pumps in space vehicles', Atomkernenergie (Germany), 1965, 10, No.9-10, 359-62. In German

TOURCHIN, N.M.: 'Magnetic flowmeters', Magn. Gidrodin (USSR), 1965, No.1, 147-53. In Russian

VASIL'EV, S.V. et al: 'Experimental investigation of an induction pump in a magnetic field', ibid., No.2, 111-22. In Russian

VOJ, P.: 'Electromagnetic flow meters for liquid metals', Arch. Tech. Messen, 1965, No.357 (ref. V1243-4), 217-8. In German

1966 DANCY, W.H. Jr. and TOWLER, W.R.: 'Three dimensional magnetically supported wind tunnel balance', Rev. Sci. Instrum., 1966, 37, No.12, 1643-8

DAVEY, A.W.: 'Linear motors for crane drives. Development and production', Elec. Rev., 1966, 179, No.26, 956-8

DOMRES, H.G.: 'The transport of molten metals by means of electromagnetic pumps', Elektrotech. Z. (ETZ) B, 1966, 18, No.26, 989-91. In German

LAITHWAITE, E.R.: 'Induction-type actuators', Design and Components in Engineering, 10 Dec. 1966, 16-24

NIX, G.F. and LAITHWAITE, E.R.: 'Linear induction motors for low-speed and standstill application', Proc. IEE, 1966, 113, No.6, 1044-56

ROSE, G.A.: 'Magnetic field measurements for electromagnetic flowmeters', Phys. Med. Biol. (GB), 1966, 11, No.1, 109-12

1967 ADAMS, W.J. and WHITE, B.A.: 'Applying linear induction motors', Automation, 1967, 14, No.6, 74-7

BERNARD, H.: 'The magnetic inductive measurement of flow', Elektrotech. Z. (ETZ) B, 1967, 19, No.1, 7-11. In German

BOWLES, L.F. and TAYLOR, D.: 'Prototype fast reactor sodium pumps', U.K.A.E.A. TRG Report No. 1538(r), 1967

COYNE, D.G. and MULLINS, J.H.: 'Critically damped oscillatory linear induction engine for driving a bubble chamber valve', Rev. Sci. Instrum., 1967, 38, No.5, 681-5

DORAIRAJ, K.R. and KRISHNAMURTHY, M.R.: 'Polyphase induction machine with a slitted ferromagnetic rotor', IEEE Trans., 1967, PAS-86, No.7, 835-43 ('I - Experimental investigations and a novel slipmeter'), ibid., 844-55 ('II - Analysis')

LAITHWAITE, E.R.: 'Applications of linear induction motors', Electrical and Electronics Technician Engineer, 1967, 1, No.3, 10-6

MARINESCU, M.: 'Automatic adjustment of the charge for the driving by linear electrical motor of a piston compressor with double effect', Rev. Roum. Sci. Tech. Electrotech. Energet., 1967, 12, No.4, 563-9. In French

WHITE, B.: 'Linear induction motor loads bobbins', Control Engineering, 1967, 14, No.7, 91

1968   DRAKE, J., GEE, R. and PIERRO, J.: 'Linear electric motors', Frontiers of Technology Study, 3, Sec. 10, North American Rockwell Corp., 1968

EZEKIEL, F.D.: 'Electromagnetic actuators', Instrum. Control Syst. (USA), 1968, 40, No.12, 90-6

GOLDHAMMER, A.B., MARKEY, P.J. and MATHUR, S.K.:
'Linear motors applied. Economic for special purposes
despite low efficiency', Elec. Times, 1968, 153, No.24,
967-71

KANT, M.: 'Contribution à l'étude du champ magnétique
dans un convertisseur magnétohydrodynamique à viene
liquide', Thèse de Doctorat d'Etat és Sciences Physiques,
Paris, 1968.

KVACHEV, G.S., MELNICHENKO, G.I. and STASENKO, P.:
'Linear motor drives for oil circuit breakers', Elek.
Sta. (USSR), 1968, No.2, 56-60

'Linear motor in crash-test rig', Electronics & Power,
May 1968, 14, 213

MADDON, D.: 'Linear induction motors: tomorrow's transit
power', Product Engineering, 20 May 1968, 39, 73-5

VICTORRI, M.: 'The rolling chute, a new solution for
the transport of bulk products', Manutention, Nov. 1968,
18, No.151. In French

WILKINSON, K.: 'Linear motors applied to materials
handling', Elec. Times, 1968, 154, No.3996, 31-6

1969 CHOUDHURY, J.K. and DAS, J.K.: 'Use of single-phase linear
induction motor for a self-balancing d.c. potentiometer',
J. Instn. Engrs. (India) Electronics Telecomm. Engng. Div.,
1969, Pt. ET3, 49, No.9, 94-6

DAVEY, A.W.: 'Linear motors thrust ahead in many roles',
Engineer, 1969, 229, No.5928, 42-5

GREEN, C.W. and PAUL, R.J.A.: 'Application of d.c. linear machines as short-stroke and static actuators', Proc. IEE, 1969, 116, No.4, 599-604

HERRNKIND, O.P.: 'Der Linearmotor und seine Anwendung', Elektro-Anz. (Germany), 1969, 22, No.17, 365-6.

HOPPE, L.: 'Linearmotoren-Richtungsweisende Elektroantriebe der Fördertechnik', Maschinenmarkt, 1969, 75, No.101, 2212-3

JONES, P.L.: 'D.C. linear motors for industrial applications', Elec. Times, 1969, 156, No.2, 48-51

KULP, P.G.: 'Operation mode and application possibilities of linear motors', Tech. Uberwach (Germany), 1969, 10, No.11, 395-7. In German

LAITHWAITE, E.R.: 'Moteurs linéaires: les applications futuristes d'une vieille idée', Atomes, 1969, 24, No.263, 179-81

'Linear induction motor', Herbert Morris Ltd, Patent UK 1152410, 15 Sept. 1967; publ. 21 May 1969; prior. UK 29 Sept. 1966, 50909/68

RÉMY, E.: 'A new linear motor conveyor for mining products', Annales des Mines, Feb. 1969, 73-81. In French

RÉMY, E. and VICTORRI, M.: 'The linear motor: II The application of the linear motor', Rev. Gen. Elec., 1969, 78, No.4, 362-70. In French

SUNDBERG, K.: 'Magnetic travelling fields for metallurgical processes', IEEE Spectrum, 1969, 6, No.5, 79-88

USAMI, Y., ISHIHARA, M., KOJIMA, N. and MITOMI, T.: 'Feasibility of linear motor for car-retarder or car-accelerator at marshalling yards', Bull. Internat. Railway Congress Assoc. (Belgium), 1969, 46, No.5, 311-20

VICTORRI, M.: 'The linear motor and its applications', Journées d'Informations Electro-industrielles, Grenoble, Oct. 1969, No.13. In French

1970 BEATSON, C.: 'Linear motors are poised to thrust into industry', Engineer, 1970, 230, No.5968, 38-9

BUDIG, P.-K.: 'Construction, mode of operation, operating behaviour and possibilities of application of linear motors', Elektrie, 1970, 24, No.10, 335-40. In German

KLOCKE, W.: 'Use of linear motors in the shipbuilding experimental station', ibid., 354-6. In German

KNOPEK, G. and KRAJEWSKI, J.: 'Design problems and the development of silent crane induction motors', Przeglad Elektrotech. (Poland), 1970, 46, No.7-8, 307-9. In Polish

KOCKISCH, K.-H.: 'Experiences with the design and construction of linear motors from the manufacturer's point of view', Elektrie, 1970, 24, No.10, 344-6. In German

LAITHWAITE, E.R. and NASAR, S.A.: 'Linear-motion electrical machines', Proc. IEEE, 1970, 58, No.4, 531-42

LINKE, H.: 'Betriebsverhalten und Steuerung eines speziellen Linearmotors', Industrieanzeiger, 1970, 92, No.25, 533-5

LORINET, J.P.: 'Handling by linear (induction) motor', Tech. Mod. (France), 1970, 62, No.9, 343-6. In French

'New handling system using the linear motor', Electricien (France), 1970, 83, No.2115, 52-4. In French

PEACH, N.: 'Linear motors have practical applications', Power (USA), 1970, 114, No.6, 96

RÉMY, E.: 'Possibilities of application of the linear motor', Elektrie, 1970, 24, No.10, 352-4. In German

ROSATI, S.: 'Linear induction motors and their possible use in industry', Alta Frequenza (Italy), 1970, 39, No.7 (supplement), 133-41. In Italian

SFAX, E.: 'Drive and positioning of vehicles for internal transport in workshops by means of linear motors', Elektrie, 1970, 24, No.10, 358-61. In German

STOLZEL, D.: 'Use of linear motors in mines', ibid., 361-5. In German

WIART, A.: 'Basic theory and applications of linear motors', Rev. Jeumont-Schneider, May 1970, No.8, 43-50. In French

WIART, A.: 'Theory and applications of linear motors', ibid., Aug. 1970, No.9, 39-46. In French

1971 ANCEL, J. and FAURE, A.: 'Design alternates for the linear motor according to its specific applications', Rev. Gen. Elec., 1971, 80, No.2, 135-7. In French

BIRZVALKS, Yu. A., BROKA, M.E. et al: 'An experimental investigation and generalization of channel characteristics in conduction pumps with independent excitation', Magn. Gidrodin. (USSR), 1971, No.4, 117-22. In Russian (English trans. in Magnetohydrodynamics (USA))

BUDIG, P.-K.: 'Application of linear motors', Elektrie, 1971, 26, No.6, 204-6. In German

BUDIG, P.-K.: 'Possible applications of linear motors', Automatisieningspraxis (Germany), 1971, 14, No.8, 149-53. In German

CONRAD, H. and NOPIRAKOWSKI, J.: 'Electromagnetic stirring of molten metal', Elektrie, 1971, 26, No.9, 375-7. In German

CYNOBER, S.: 'Applications of the linear motor to the handling of materials', Rev. Gen. Elec., 1971, 80, No.2, 114-20. In French

DAVEY, A.W.: 'Control systems for cranes', Herbert Morris Ltd, Patent USA 3596156, 15 July 1968; publ. 27 July 1971; prior. 21 July 1967, UK 336227/67

ERICSON, A.: 'Metallurgical aspects of the induction stirring of molten metal', Asea J. (Sweden), 1971, 44, No.4, 81-2

FARAH, O.G. and LAY, R.K.: 'Power conditioning unit for linear induction motor thrust and speed control', IEEE International Convention Digest, New York, 22-25 March 1971, 418-9

GUTT, H.-J.: 'Wanderfeldmotoren im Vergleich zu üblichen Drehfeldmaschinen', Elektrotech. Z. (ETZ) A, 1971, 92, 342-7

HANAS, B.: 'Design and plant considerations of induction stirrers', Asea J. (Sweden), 1971, 44, No.4, 87-92

KRAUYA, V.M.: 'A conduction pump with a circular channel', Magn. Gidrodin. (USSR), 1971, No.4, 133-7. In Russian (English trans. in Magnetohydrodynamics (USA))

POLOUJADOFF, M.: 'Linear induction machines: II Applications', IEEE Spectrum, 1971, 8, No.3, 79-86

ROUBECEK, O., PEJSEK, Z. and TUSLA, P.: 'The possibility of application of an electric translatory drive (with linear motion) of the mine drilling hammer', Elektrotech. Obzor (Czechoslovakia), 1971, 60, No.4, 202-7. In Czech

SADLER, G.V. and DAVEY, A.W.: 'Applications of linear induction motors in industry', Proc. IEE, 1971, 118, No.6, 765-76

SUNDBERG, Y.: 'Principles of the induction stirrer', Asea J. (Sweden), 1971, 44, No.4, 71-80

TORGERSEN, P.: 'The linear motor - simple and strong without moving parts. Possible applications in industry and the carriage of goods', Elektrotek. Tidsskr. (Norway), 1971, 84, No.21, 16-8. In Norwegian

VED'KALOV, I.R., KAPUSTA, A.B. et al: 'An electromagnetic pilot plant for the treatment of flowing molten metal', Magn. Gidrodin. (USSR), 1971, No.4, 127-32. In Russian (English trans. in Magnetohydrodynamics (USA))

WARNETT, K.: 'Linear electric motor', Electro-Lifts Ltd, Patent USA 3581127, 21 April 1969; publ. 25 May 1971; prior. 20 April 1968, UK 18777/68

1972 AHMAD, V.: 'Applications of linear induction motors in industry', Proc. IEE, 1972, 119, No.2, 233-4

ASTON, T.R.: 'Impact testing of motor vehicles: an application for the linear induction motor', Electrical and Electronics Technician Engineer, 1972, 6, No.4, 3-7

DATTA, S.K.: 'A static variable-frequency three-phase source using the cycloconverter principle for the speed control of induction motor', IEEE Trans., 1972, IA-8, No.5, 520-30

DAVEY, A.W.: 'Linear motors for industrial processes', Elec. Times, 1972, 161, No.19, 45-8

EADY, J.G.: 'Electromagnetic transport and metering of molten metals', Mining Technol. (GB), 1972, 54, No.615, 28-31

'Electric industrial furnaces', Mon. Tech. Rev. (Germany), 1972, 16, No.1, 5-7. In English

'Electromagnetic conveying trough', AEG Flotherm GmbH, Patent UK 1292576, 19 April 1971; publ. 11 Oct. 1972; prior. 30 Sept. 1970, Germany P2048026.6

GALE, K.: 'Shove a penny: linear motor sorts them out', Engineer, 1972, 23, No.6072-3, 64-5

GUTT, H.-J.: 'Applications of travelling-field motors of the sector and linear type', Siemens Rev. (Germany), 1972, 39, No.1, 32-6

HIYANE, M. INOUE, Y. and KURUSU, M.: 'Development of linear motion actuator', Fujitsu Sci. Tech. J. (Japan), 1972, 8, No.2, 59-92

KURZWEIL, F. Jr.: 'Linear eddy current actuator using moving conductive belt', IBM·Tech. Disclosure Bull. (USA), 1972, 15, No.1, 95-6

LANCASHIRE, R.: 'The application of linear thrust units to work transporters on metal finishing plant', Electroplating and Met. Finish. (GB), 1972, 25, No.7, 30-2

LENZKES, D.: 'Electric drives with linear motors', Electr. Ausreustung (Germany), 1972, 13, No.4, 13-8

RUMMICH, E.: 'Entwicklungsstand und Anwendungsmöglichkeiten von Linearmotoren', Bull. Assoc. Suisse Electr. (Switzerland), 1972, 63, No.19, 1093-9

RUMMICH, E.: 'Linear motors and their applications', Elektrotechnik und Maschinenbau (Austria), 1972, 89, No.2, 60-9. In German

'The linear motor: applications and perspectives', Achats & Entretien Mater. Ind. (France), 1972, No.237, 45-54. In French

1973 BAKANOV, Yu. A., DRONNIK, L.M. et al: 'An experimental investigation of a liquid metal induction machine in the pump regime', Magn. Gidrodin (USSR), 1973, No.1, 153-5. In Russian (English trans. in Magnetohydrodynamics (USA))

BAKANOV, Yu. A., VLASENKO, L.G. et al: 'An experimental study of a liquid metal alternating current conduction machine', ibid., No.2, 124-9. In Russian (English trans. in Magnetohydrodynamics (USA))

BERESFORD, J.E. and BARCLAY, R.: 'A cheap linear motor for teaching and research purposes', Elec. Rev., 1973, 192, No.21, 748-50

D'YAKOV, V.I. and OROLOV, A.N.: 'The plotting of the mechanical characteristics of a linear induction machine with a liquid metal working medium', Magn. Gidrodin. (USSR), 1973, No.2, 100-4. In Russian (English trans. in Magnetohydrodynamics (USA))

GIROUD, F.: 'Linear motor propelled catapult for automobile collision testing', Analyses (France), 1973, No.1. In French

HÜHNS, T. and KRATZ, G.: 'Steuerung von und mit Linearmotoren', Steuerungstechnik, 1973, 6, No.2, 23

LLOYD, R.G.: 'Linear motors for better performance and reliability in instrumentation', Elec. Rev., 1973, 193, No.19, 631-3

OLBRICH, O.E.: 'Layout and characteristics of electrodynamic linear motors as positioners for disc storages', Feinwerktech. & Micronic (Germany), 1973, 77, No.4, 151-7. In German

USAMI, Y., FUJIE, J., ISHIHARA, M. and NAKASHIMA, H.: 'Linear-motorised yard automation system', Quart. Rept. Railway Technical Research Institute, JNR, 1973, 14, No.4, 207-13

WITWICKI, A.: 'The development of linear motor applications', Wiad. Elektrotech. (Poland), 1973, 41, No.3, 72-8. In Polish

1974

BAMJI, P.J.F.: 'Some applications of a linear electric machine for handling molten aluminium', IEE Conference on Linear Electric Machines, London, 21-23 Oct. 1974, 68-76

BOLTON, H.: 'An electromagnetic bearing' ibid., 45-50

BROWN, M. and FERRY, G.A.: 'A high voltage thyristor regulator for control of a linear induction motor', IEE Conference on Power Semiconductors and their Applications, London, 3-5 Dec. 1974, 63-70

DAVEY, A.W.: 'How to assess the rating of a linear motor', Design Engineering, June 1974, 38-9

DAVEY, A.W.: 'Linear motor applications', IEE Conference on Linear Electric Machines, London, 21-23 Oct. 1974, 51-5

DAVEY, A.W.: 'Variable speeds for industry by versatile linear motor', Elec. Times, 1974, No.4286, 7-8

GUTT, H.-J.: 'Limitations to the application of linear motors', Elektrotechnik und Maschinenbau (Austria), 1974, 91, No.9, 4555-61. In German

JUFER, M. and CASSAT, A.: 'Application of linear stepping and induction motors', 3rd Annual Symposium on Incremental Motion Control Systems and Devices, University of Illinois, Urbana, 6-8 May 1974, L/1-13

JUFER, M. and MATTATIA, S.: 'The linear induction motor - characteristics and applications', <u>Bull. Assoc. Suisse Electr.</u> (Switzerland), 1974, 65, No.12, 880-91. In French

KHOZHAINOV, A.I.: 'Study of a conduction-type linear motor with liquid metal conductor', <u>Elektrichestvo</u> (USSR), 1974, No.10, 52-6. In Russian

LAZARUS, J.H., ENSLIN, N.C. and LIDDIARD, R.C.: 'Application of a linear motor to the hydraulic transportation of ore', IEE Conference on Linear Electric Machines, London, 21-23 Oct. 1974, 56-61

LIANDER, W.: 'Motors for servo and linear applications', <u>Eltek. Aktuell Elektron.</u> (Sweden), 1974, 17, No.4, 38-40. In Swedish

MARINESCU, M.: 'A linear electro-magnetic type of motor for operating compressors for cooling devices, by direct coupling', IEE Conference on Linear Electric Machines, London, 21-23 Oct. 1974, 118-23

RESIN, M.G., PIRUMIAN, N.M., SARAPULOV, F.N. and YASENEV, N.D.: 'Some aspects of the linear motor's design and application', <u>ibid.</u>, 21-4

SCHLEGEL, K.-D.: 'Linear drives for limited displacements - application perspectives of linear motors in the control of material flow', <u>Wiss. Z. Tech. Hochsch. Otto Von Guericke Magdeb.</u> (Germany), 1974, 18, No.5, 565-9. In German

SIKORA, R., LIPINSKI, W. and PURCZYNSKI, J.: 'Analysis of levitation force and power losses of a certain induction bearing', Arch. Elektrotech. (Poland), 1974, 23, No.2, 487-500. In Polish

TIISMUS, Kh.A.: 'Prospects for the development of an electrical drive with linear induction MHD motors', Elektrotekhnika (USSR), 1974, No.10, 6-10. In Russian

1975 CHIRIIOIU, N.: 'Linear motor used with magnetic-disk memory', Electroteh. Electron. & Autom. Electroteh. (Rumania), 1975, 23, No.3, 127-36. In Rumanian

'Drive for linear positioning of magnetic leads', Philips Electronic Ind., Patent UK 1381143, 3 May 1972; publ. 22 Jan. 1975; prior. 6 May 1971, Netherlands 6185

LARONZE, J.: 'Linear motor with multi-section armature and a drive device utilizing a linear motor of this kind', Brown Boveri & Co. Ltd, Patent USA 3860840, 17 July 1973; publ. 14 Jan. 1975; prior. 31 July 1972, France 72.27585

LUDA, G.: 'The range of application of the linear motor', Polytech. Tijdschr. Werktuigbouw (Netherlands), 1975, 30, No.14, 443-54. In Dutch

MARINESCU, M., MORARU, A. and BUNEA, V.: 'A linear electromagnetic type of motor for operating a cooling device compressor by direct coupling', Rev. Roum. Sci. Tech. Electrotech. Energet., 1975, 20, No.1, 57-63. In English

MORI, K.: 'Linear motor winding and method of fabricating the same', Patent USA 3903437, 1 April 1974; publ. 2 Sept. 1975; prior. 3 April 1973, Japan 4837420

OCHI, T., KUNITOMO, Y., TAKEI, K., MURAI, T. and ANDO, M.: 'Simulation of marshalling yard system using linear motor cars', Hitachi Rev. (Japan), 1975, 24, No.1, 34

PAUL, R.J.A. and THOMSON, A.F.: 'Dynamic performance of a linear helical reluctance stepping motor', 4th Annual Symposium on Incremental Motion Control Systems and Devices, University of Illinois, Urbana, 1-3 April 1975, F/1-16

LEVITATION

1912 'Foucault and eddy currents put to service', Engineer, 18 Oct. 1912, 114, 420-1

1923 MUCK, O.: 'Method of and devices for melting materials, in particular conductive materials, by means of induced electric currents', German Patent No.422004, 30 Oct. 1923. In German

1939 BEDFORD, B.D., PEER, H.B. and TONKS, I.: 'The electromagnetic levitator', Gen. Elec. Rev. (USA), 1939, 42, No.6, 246-7

1950 BEAMS, J.W.: 'Magnetic suspension for small rotors', Rev. Sci. Instrum., Feb. 1950, 21, 182-4

1951 LUNDQUIST, S.: 'On the stability of magneto-hydrostatic fields', Phys. Rev., 1951, 83, No.2, 307-11

1952 'Melting metals in mid air', *Life*, 16 June 1952, 49

OKRESS, E.C. and WROUGHTON, D.M.: 'Metals melted without crucibles', *Iron Age*, 1952, 170, No.5, 83-6

OKRESS, E.C., WROUGHTON, D.M., COMENETZ, G., BRACE, P.H. and KELLY, J.C.R.: 'Electromagnetic levitation of solid and molten metals', *J. Appl. Phys*. (USA), 1952, 23, No.5, 545-52

WROUGHTON, D.M., OKRESS, E.C., BRACE, P.H., COMENETZ, G. and KELLY, J.C.R.: 'A technique for eliminating crucibles in heating and melting metals', *J. Electrochem. Soc.*, 1952, 99, No.5, 205-11

1953 KEMPER, H.: 'Elektrisch Angetriebene Eisenbahnfahrzeuge mit elektromagnetischer Schwebefuhrung', *Elektrotech. Z. (ETZ) A*, 1953, 74, No.1, 11-4

SCHEIBE, W.: 'Some problems of high-vacuum metallurgy', *Metall*, 1953, 7, No.19-20, 751-4. In German (English Trans. TIL/T 4603, Ministry of Supply, Technical Information and Library Services, Sept. 1956)

WINKLER, O.: 'The melting of metals without crucible reaction', *Z. Metallkunde*, 1953, 44, No.8, 333-41. In German (English Trans. TIL/T 4701, Ministry of Supply, Technical Information and Library Services, Aug. 1957)

1954 KRUSKAL, M. and SCHWARZSCHILD, M.: 'Some instabilities of a completely ionized plasma', *Proc. Royal Soc.*, 1954, 223, 348-60

POLONIS, D.H., BUTTERS, R.G. and GORDON-PARR, J.:
'Levitation melting titanium and titanium alloys',
Research (London), 1954, 7, No.2, S10-2 (Correspondence
supplement)

POLONIS, D.H., BUTTERS, R.G. and GORDON-PARR, J.:
'Some techniques for melting reactive metals', ibid.,
No.7, 273-7

POLONIS, D.H. and GORDON-PARR, J.: 'Phase transformations
in titanium rich alloys of iron and titanium', J. Metals
(Met. Soc. AIME), 1954, 6, 1138-54

1955 LOVELL, W.V.: 'Electromagnet removes nonferrous metals',
Electronics, 1955, 28, No.9, 164-6

1956 BOERDIJK, A.H.: 'Technical aspects of levitation',
Philips Research Reports, 1956, 11, No.1, 45-56

1957 TAYLER, R.J.: 'Hydromagnetic instabilities of an ideally
conducting fluid', Proc. Phys. Soc., 1957, Section B,
70, 31-48

1958 BERNSTEIN, I.B., FRIEMAN, E.A., KRUSKAL, M.D. and
KULSRUD, R.M.: 'An energy principle for hydromagnetic
stability problems', Proc. Royal Soc., 1958, 244A,
No.1236, 17-40

COMENETZ, G. and SALATKA, J.W.: 'Ten-gram levitation
melted ingots', J. Electrochem. Soc., 1958, 105, No.11,
673-6

ZHEZHERIN, R.P.: 'The problems of the electromagnetic crucible', Conference on MHD, Riga, USSR, July 1958. In Russian (English Trans. US Atomic Energy Comm. No. AEC-tr-Physics, 242-55)

1959 BEGLEY, R.T., COMENETZ, G., FLINN, P.A. and SALATKA, J.W.: 'Vacuum levitation melting', Rev. Sci. Instrum., 1959, 30, No.1, 38

FOGEL, A.A.: 'Melting of laboratory specimens of metals in vacuum or an inert gas by levitation technique', Izv. Akad. Nauk SSSR, OTN, Metallurgiya i Toplivo, 1959, No.2, 24-34. In Russian (English trans. Henry Brutcher Technical Transl. No. 4702)

HARRIS, B. and JENKINS, A.E.: 'Controlled atmosphere levitation system', J. Sci. Instrum., 1959, 36, No.5, 238-40

VLADIMIRSKII, V.V. and KALEBIN, S.M.: 'The stability of rotating ferromagnetic bodies suspended in a magnetic field', Pribory i Tekh. Eksper. (USSR), 1959, No.2, 41-2. English trans. Instrum. Exper. (USA), 1959, No.2, 216-23

WEISBERG, L.R.: 'Levitation melting of Ga, In, Au and Sb', Rev. Sci. Instrum., 1959, 30, No.2, 135

1960 BIRDSAL, D.H., COLGATE, S.A., FURTH, H.P., HARTMAN, C.W. and SPOERLEIN, R.L.: 'Livermore pinch program', University of California, Report No.UCRL-9500, Sept.-Nov. 1960, 38-51

PORITSKY, H.: 'Conducting sphere in alternating magnetic fields', Trans. AIEE (Part 1 Communication and Electronics), 1960, 78, No.46, 937-42

SLATER, W.J., BARTON, J.W. and TAGGART, R.: 'Coil design in levitation melting', Sci. and Industry, 1960, 7, 89-96

1961 FURTH, H.P. and KILLEEN, J.: 'Instability of resistive sheet pinch', University of California, Report Nos. UCRL-9777, UCRL-9969, approx. 1961

JENKINS, A.E.: 'The physico-chemical applications of electromagnetic levitation', XVIII International Congress of Pure and Applied Chemistry, Montreal, Quebec, 1961

STERLING, H.F. and WARREN, R.W.: 'A cold crucible for high-temperature melting processes', Nature, 1961, 192, No.4804, 745

'Temperature control improves levitation', Chem. Eng. News, 21 Aug. 1961, 39, 38 & 40

1962 COLGATE, S.A., FURTH, H.P. and HALLIDAY, F.O.: 'Hydromagnetic equilibrium experiments with liquid and solid sodium', Rev. Mod. Phys., 1962, 32, No.4, 744-7

LEWIS, J.C., NEUMAYER, H.R.J. and WARD, R.G.: 'The stabilization of liquid metal during levitation melting', J. Sci. Instrum., 1962, 39, No.11, 569

REDLICH, R.: 'Stability of thin metal foils levitated by ac fields', J. Appl. Phys. (USA), 1962, 33, No.1, 231

VOLKOV, T.F.: 'Stability of a heavy conducting fluid contained by a rapidly varying magnetic field', Soviet Phys. - Tech. Phys., 1962, 7, No.1, 22-7

1963 BRISLEY, W. and THORNTON, B.S.: 'Electromagnetic levitation calculations for axially symmetric systems', Brit. J. Appl. Phys., 1963, 14, No.10, 682-6

HULSEY, W.J.: 'The design and performance of levitation melting coils', Union Carbide Nuclear Co., Report No. Y-1413, April 1963

JENKINS, A.E., HARRIS, B. and BAKER, L.: 'Electromagnetic levitation and its uses in physico-chemical studies at high temperature', Metallurgical Society Conference, Dallas, Texas, 1963, 22, 22-43

KIRKO, I.M. and MIKEL'SON, A.E.: 'Concerning the stability of free suspension of liquid metal in an alternating magnetic field', English Trans. in Problems of MHD and plasma dynamics II, Report No.FTD-TT-62-1301, 587-94, Wright Patterson AFB, Ohio, 1963

STERLING, H.F. and WARREN, R.W.: 'High temperature melting without contamination in cold crucibles', Metallurgia, 1963, LXVII, No.404, 301-7

WHITE, H.E. and WELTUM, H.: 'Electromagnetic levitator', Amer. J. Phys., 1963, 31, No.12, 925-9

1964 BUNSHAH, R.F. and JUNTZ, R.S.: 'Levitation melting of beryllium and aluminum', Amer. Vac. Met. Soc. Conference, Boston, Mass., 1964, 136-44

GEARY, P.J.: Magnetic and electric suspensions. A survey of their design, construction, and use (Chislehurst, Kent: British Scientific Instrument Research Association) 1964, Research Report R 314 (345 refs)

LAITHWAITE, E.R.: 'Electromagnetic levitation', Design and Components in Engineering, 10 Sept. 1964, 17-23

NIXON, J.D. and KENNEY, D.J.: 'Electronic damping for the magnetically suspended rotor', Rev. Sci. Instrum., 1964, 35, No.12, 1721-2

OLIVER, B.F.: 'The segregation of tantalum in iron in levitating zone melter', Trans. Metallurgical Society (AIME), 1964, 230, No.4, 1353-7

RONY, I.R.: 'The electromagnetic levitation of metals', Amer. Vac. Met. Soc. Conference, Boston, Mass., 1964, 145-53

SIDOROVA, T.A.: 'Power and energy relationships in an electromagnetic crucible', Collection: Industrial applications of high frequency currents, Izd-Vo. 'Mashinostroyeniye', 1964, 266-9

1965 BEAMS, J.W.: 'Magnetic support for non ferromagnetic bodies', Rev. Sci. Instrum., 1965, 36, No.12, 1892

FROMM, E. and JEHN, H.: 'Electromagnetic forces and power absorption in levitation melting', Brit. J. Appl. Phys., 1965, 16, 653-63

HATCH, A.J.: 'Potential-well description of electromagnetic levitation', J. Appl. Phys. (USA), 1965, 36, No.1, 44-52

LAITHWAITE, E.R.: 'Electromagnetic levitation', Proc. IEE, 1965, 112, No.12, 2361-75

LAITHWAITE, E.R.: 'Electromagnetic levitation', Electronics & Power, Dec. 1965, 11, 408-13

PEIFER, W.A.: 'Levitation melting, a survey of the state-of-the-art', J. Metals (Met. Soc. AIME), 1965, 17, No.5, 487-93

PONIZOVSKII, V.M.: 'A generator for magnetic suspension of ferromagnetic motors', Pribory i Tekh. Eksper. (USSR), 1965, No.5, 234-5. English trans. Instrum. Exper. Tech. (USA), 1965, No.5, 1269-70

SKOV, P.B.: 'A preliminary study of the problems of electromagnetic suspension and linear propulsion', Systems Technology, Westinghouse R & D, Report No.TDS-65-1D3-CSCON-R1, April 1965

SMITH, W.E.: 'Electromagnetic levitation forces and effective inductance in axially symmetric systems', Brit. J. Appl. Phys., 1965, 16, No.3, 377-83

STERLING, H.F.: 'Towards a perfect crucible', Discovery, 1965, 26, No.4, 30-3

VIGOUREUX, P.: 'Electromagnetic levitation forces', Brit. J. Appl. Phys., 1965, 16, No.5, 757

1966

ADAMS, J.B.: 'A review of nuclear fusion research' (1965 Guthrie Lecture), Proc. Phys. Soc., 1966, 89, 189-216

GETSELEV, Z.M.: Patent UK 1,157,977, 4 Oct. 1966

GRANEAU, P.: 'Electromagnetic levitation forces and self inductance', Brit. J. Appl. Phys., 1966, 17, No.1, 139-40

HAAS, F.A. and WESSON, J.A.: 'Stability of the theta-pinch', Phys. of Fluids, 1966, 9, No.12, 2472-7

MAGER, A.: 'Theory and experiments on levitation melting', Z. Metallkunde, 1966, 57, No.5, 358-63. In German

PIGGOTT, L.S. and NIX, G.F.: 'Electromagnetic levitation of a conducting cylinder', Proc. IEE, 1966, 113, No.7, 1229-35

POLGREEN, G.R.: New applications of modern magnets (London: Macdonald) 1966

ROSENWEIG, R.E.: 'Buoyancy and stable levitation of a magnetic body immersed in magnetizable fluid', Nature, 1966, 210, 613-4

ROSENWEIG, R.E.: 'Fluidmagnetic buoyancy', AIAA J. (USA), 1966, 4, No.10, 1751-8

STERLING, H.F.: 'Electromagnetic levitation', Proc. IEE, 1966, 113, No.8, 1395-6 (Discussion)

1967 CAMPBELL, I.E. and SHERWOOD, E.M. (Ed.): 'Levitation melting' in High temperature materials and technology (New York: Wiley) 1967, 600-6

HAAS, F.A. and WESSON, J.A.: 'Dynamic stabilization of the theta-pinch', Phys. Rev. Letters, 1967, 19, No.15, 833-5

HAAS, F.A. and WESSON, J.A.: 'Stability of the theta-pinch II', Phys. of Fluids, 1967, 10, No.10, 2245-52

HATCH, A.J. and SMITH W.E.: 'Inductance-variation method of measuring characteristics of electromagnetic levitation systems', J. Appl. Phys. (USA), 1967, 38, No.2, 742-4

JAYAWANT, B.V. and REA, D.P.: 'New electromagnetic suspension and its stabilisation', Electronics Letters, 1967, 13, No.9, 401-2

KAPLAN, B.Z.: 'Analysis of a method for magnetic levitation', Proc. IEE, 1967, 114, No.11, 1801-4

LAITHWAITE, E.R.: 'Electromagnetic levitation', Engineering, 6 Jan. 1967, 203, 37-40

LAITHWAITE, E.R.: 'Electromagnetic levitation', Student Technologist, Feb. 1967, 19-21

LAITHWAITE, E.R.: 'Research jumping ring experiment', Elec. Rev., 1967, 180, No.11, 406-7

LAITHWAITE, E.R.: 'The anti-magnetic effect', ibid., No.1, 18-9

'Trends in superconductivity related to electromagnetic suspension of HSGT vehicles', TRW Systems Group, Washington, DC, Report No.06818-6009-R000 for US Dept of Transportation under Contract C-353-66, 6 Oct. 1967

1968 FILE, J., MARTIN, G.D., MILLS, R.G. and UPHAM, J.L.: 'Stabilized, levitated superconducting rings', J. Appl. Phys. (USA), 1968, 39, 2623-6

FOGEL, A.A., SIDOROVA, T.A., KORKIN, I.V. and MEZDROGINA, M.M.: 'Obtaining a predetermined steady-state temperature during fusion of a metal in levitation', Russ. Met., 1968, No.1, 54-60

FOGEL, A.A., SIDOROVA, T.A., SMIRNOV, V.V., GUTZ, Z.A. and KORKIN, I.V.: 'Optimum parameters for apparatus for levitation melting of metals', ibid., No.2, 89-95

FREEMAN, E.M.: 'Levitation or attraction due to a travelling field', Proc. IEE, 1968, 115, No.6, 894 (Correspondence)

POLGREEN, G.R.: 'Railways with magnetic suspension', Engineer, 1968, 226, No.5883, 632-6

1969 GUDERJAHN, C.A.: 'Magnetic levitation and guidance of a rocket sled', Atomics International Report No. AI-68-149, Jan. 1969

LAITHWAITE, E.R.: 'Electromagnetic levitation', Encyclopaedic Dictionary of Physics, Supplementary Vol.3 (Pergamon Press), 1969, 82-7

'Magnetic suspension for the fastest rocket on earth', New Scientist, 1969, 42, No.646, 186

NICKL, J.C.: 'A new crucible-less induction melting process (the Stipp process)', Z. Metallkunde, 1969, 60, No.10, 800-2. In German

SAITO, T., SHIRAISHI, Y. and SAKUMA, Y.: 'Density measurement of molten metals by levitation technique at temperatures between 1800°C and 2200°C', Trans. Iron-Steel Inst. Japan, 1969, 9, No.2, 118-26

1970 WESSON, J.: 'Dynamic stabilization and the Rayleigh-Taylor instability', Phys. of Fluids, 1970, 13, No.3, 761-6

WOODS, L.C., COOPER, R.K., NEIL, V.K. and TAYLOR, C.E.: 'Stability analysis of a levitated superconducting current ring stabilized by feedback and eddy current', J. Appl. Phys. (USA), 1970, 41, No.8, 3295-305

1971 BARAN, W.: 'Optimizing a permanent magnet suspension system for high-speed ground transport', Z. Angew. Phys. (Germany), 1971, 32, No.3, 216-8. In German (Report of meeting of working group on magnetism, Salzburg, Austria, 29-31 March 1971)

BOCIAN, E.S. Jr. and YOUNG, F.J.: 'Some stability considerations in levitation melting', J. Electrochem. Soc., 1971, 118, No.12, 2021-6

BORCHERTS, R.H.: 'Mathematical analysis of "permanent" magnet suspension systems', J. Appl. Phys. (USA), 1971, 42, 1528

'Congress delegates see rival magnetic suspension cars', Railway Gaz. Int., 1971, 127, 434

DARDEL, Y.: 'The scopes of techniques of electromagnetic levitation on rectilinear network', Rev. Met., 1971, 69, No.1, 71-83

DAVIS, L.C. and WILKIE, D.F.: 'Analysis of the motion of magnetic levitation systems: implications for high speed vehicles', J. Appl. Phys. (USA), 1971, 42, No.12, 4779-93

DUKOWICZ, J.K.: 'Attraction/repulsion forces in a single-sided linear induction motor', Mitre Corp., McLean, Va., Final Report No.WP-7519 FRA-RT-71-78 for US Dept of Transportation under Contract OHSGT-7-35248, March 1971

FREEMAN, E.M.: 'The dolphin effect in linear induction motors', 6th Universities Power Engineering Conference, UMIST, Manchester, 4-6 Jan. 1971

FOGEL, A.A., SIDOROVA, T.A. and MEZDROGINA, M.M.: 'Characteristics of the "boat" inductor for retaining liquid metal in a state of levitation', Russ. Met., 1971, No.1, 48-52

GETSELEV, Z.M.: 'Casting in an electromagnetic field', J. Metals (Met. Soc. AIME), 1971, 23, No.10, 38-9

HIERONYMUS, H., MIERICKE, J. and BOGNER, C.: 'Preliminary results on an electrodynamically levitated superconducting coil', 2nd International Symposium on Electromagnetic Suspension, Southampton University, July 1971

JAYAWANT, B.V.: 'Electromagnetic suspensions using tuned LCR circuits', ibid.

JAYSINGHANI, N.D.: 'Power supply for a magnetic suspension system', ibid.

KNIGHT, C.F. and PERKINS, R.: 'Levitation melting of uranium mono-carbide', J. Nuclear Mater., 1971, 39, No.2, 224-5

ZAPATA, R.N.: 'Safety aspects of superconducting magnetic suspension systems', 2nd International Symposium on Electromagnetic Suspension, Southampton University, July 1971

1972   BOHN, G., ROMSTEDT, P., ROTHMAYER, W. and SCHWÄRZLER, P.: 'A contribution to magnetic levitation technology', 4th International Cryogenic Engineering Conference, Eindhoven, 24-26 May 1972 (IPC Science & Technology Press) 202-6

DAVIS, L.C.: 'Drag force on a magnet moving near a thin conductor', J. Appl. Phys. (USA), 1972, 43, 4256-7

JAYAWANT, B.V.: 'Magnetic and electrostatic suspension techniques', Elektrotechnik (Germany), 1972, 54, No.8, 22-8. In German

JAYAWANT, B.V.: 'Magnetic and electrostatic suspension techniques', New Zealand Elec. J., 1972, 45, No.3, 32-6

KOLM, H.H. and THORNTON, R.D.: 'The magneplane: guided electromagnetic flight', Applied Superconductivity Conference, Annapolis, Md., May 1972

REITZ, J.R. and DAVIS, L.C.: 'Force on a rectangular coil moving over a conducting slab', J. Appl. Phys. (USA), 1972, 43, 1547-53

1973   ATHERTON, D.L., CASTEL, B. and PUHACH, P.A.: 'Magnetic levitation forces - finite conductor size', J. Appl. Phys. (USA), 1973, 44, No.4, 1938-40

FREEMAN, E.M. and LOWTHER, D.A.: 'Normal force in single-sided linear induction motors', Proc. IEE, 1973, 120, No.12, 1499-506

LANGERHOLC, J.: 'Electrodynamics of magnetic levitation coil', J. Appl. Phys. (USA), 1973, 44, No.6, 2829-37

LICHTENBERG, A.: 'Research and development on electrodynamic levitation in West Germany', 2nd Intersociety Conference on Transportation, Denver, Colorado, 23-27 Sept. 1973

MIERICKE, J. and URANKAR, L.: 'Theory of electrodynamic levitation with a continuous sheet, I and II', Appl. Phys., 1973, 2, 202-11 and 1974, 3, 67-76

NASAR, S.A. and del CID, L. Jr.: 'Propulsion and levitation forces in a single-sided linear induction motor for high-speed ground transportation', Proc. IEEE, 1973, 61, No.5, 638-44

PUHACH, P.A., ATHERTON, D.L. and CASTEL, B.: 'Magnetic levitation forces - circular coils', Canadian J. Phys., 1973, 51, No.7, 731-6

RYBAR, J.: 'Theory of a linear induction A.C. motor with levitation effect', Elektrotech. Obzor (Czechoslovakia), 1973, 62, No.11, 670-3. In Czech

SAITO, R., TADA, N., KIMURA, I. and TAKAHASHI, T.: 'Superconducting magnet for magnetic suspension device', Hitachi Hyoron (Japan), 1973, 55, No.6, 31-5. In Japanese

STEKLY, Z.J.J., de WINTER, T.A., VITKEVICH, J.A., TARRH, J.M. and EMANUEL, A.E.: 'Design of superconducting magnetic levitation pads', Magnetic Corp. of America, Report No.MCA-TP124 for US Dept of Transportation under Sub Contract DOT-FRA-10026, April 1973

THORNTON, R.D.: 'Design principles for magnetic levitation', Proc. IEEE, 1973, 61, No.5, 586-98

THORNTON, R.D.: 'Flying low with Maglev', IEEE Spectrum, 1973, 10, No.4, 47-54

URANKAR, L. and MIERICKE, J.: 'Forces on null-flux magnetic levitation systems', J. Appl. Phys. (USA), 1973, 44, No.4, 1907-8

1974

APPLETON, A.D. and BOLSHAW, S.: 'Superconducting propulsion systems (d.c. machine)', 5th International Cryogenic Engineering Conference, Kyoto, Japan, May 1974 (IPC Science & Technology Press) 149-53

DANBY, G.T., JACKSON, J.W. and POWELL, J.R.: 'Force calculations for hybrid (ferro-nullflux) low-drag systems', IEEE Trans., 1974, MAG-10, No.3, 443-6

FREEMAN, E.M. and LOWTHER, D.A.: 'Normal force in single-sided linear induction motors', IEE Conference on Linear Electric Machines, London, 21-23 Oct. 1974, 210-5

IWASA, Y., HOENIG, M.O. and KOLM, H.H.: 'Design of a full-scale magneplane vehicle', IEEE Trans., 1974, MAG-10, No.3, 402-5

JACKSON, D.W.: 'Magneplane power supply costs', ibid., 406-9

LEE, S.W. and MENENDEZ, R.C.: 'Force on current coils moving over a conducting sheet with application to magnetic levitation', Proc. IEEE, 1974, 62, No.5, 567-77

MATSUMURA, F. and YAMADA, S.: 'A method to control the suspension system utilizing magnetic attractive force (railway)', Elec. Eng. Jap. (USA), 1974, 94, No.6, 50-7

PITTS, E.: 'The stability of pendent liquid drops, Part 2, Axial symmetry', J. Fluid Mech., 1974, 63, No.3, 487-508

SABNIS, A.V.: 'Analysis of forces in rectangular-pole geometries using numerical integration techniques', IEEE Trans., 1974, MAG-10, No.3, 447-50

SAITO, Y., TAKANO, I., MATSUDA, S. and OGIWARA, H.: 'Experimental studies on superconducting magnetic levitation for ultra-high-speed vehicles', Elec. Eng. Jap. (USA), 1974, 94, No.2, 92-105

SYKES, A. and WESSON, J.A.: 'Two-dimensional calculation of Tokamak stability', Nuclear Fusion, 1974, 14, 645-8

URANKAR, L.: 'Survey of basic magnetic levitation research in Erlangen', IEEE Trans., 1974, MAG-10, No.3, 421-4

URANKAR, L. and MIERICKE, J.: 'Forces on arbitrary plane multiple excitation current system used in magnetic levitation', Siemens Forsch. und Entwicklungsber., (Germany), May 1974. In English

USAMI, Y., KUZUU, T. and FUJIE, J.: Superconducting magnet levitation/propulsion test vehicle', Quart. Rept. Railway Technical Research Institute, JNR, 1974, 15, No.2, 62-8

WILKIE, D.F. and BORCHERTS, R.H.: 'Dynamic characteristics and control requirements of alternative magnetic levitation systems', American Society of Mechanical Engineers, Technical Paper No. ASME-73-ICT-17, publ. 1974

YAMAMURA, S., ABE, S. and HAYASHI, T.: 'Attractive electro-magnet levitation of vehicles', Elec. Eng. Jap. (USA), 1974, 94, No.3, 72-9

1975 ALBRECHT, C.: 'Electrodynamic supporting and guiding systems', Elektrotech. Z. (ETZ) A, 1975, 96, No.9, 383-90. In German

APPUN, P. and RITTER, G.R.: 'Calculation and optimization of the magnets for an electromagnetic levitation system', IEEE Trans., 1975, MAG-11, No.1, 39-44

GOODYER, M.J., HENDERSON, R.I. and JUDD, M.: 'The measurement of Magnus force and moment using a magnetically suspended wind tunnel model', ibid., No.5, 1514-6

GUNTHER, R., HEYM, Kl.D. and NAVE, P.M.W.: 'The Maglev support and guidance system as seen by the dynamicist', Elektrotech. Z. (ETZ) A, 1975, 96, No.9, 373-7. In German

HARRIS, M.R. and STEPHAN, S.Y.: 'Support of liquid metal surface by alternating magnetic field', IEEE Trans., 1975, MAG-11, No.5, 1508-10

HOCHHAUSLER, P.: 'Model of an electrodynamic levitation train', Elektrotech. Z. (ETZ) A, 1975, 96, No.9, 394-6. In German

KRATKI, N. and OBERRETL, K.: 'Transients and oscillations in the electrodynamic magnetic suspension system', Arch. Elektrotech. (Germany), 1975, 57, No.2, 59-64. In German

LANG, A., WEH, H. and MAY, H.: 'On the inductive levitation system with finite secondary width', ibid., No.5, 223-33. In German

LICHTENBERG, A.: 'Electrodynamic suspension in future long-distance traffic', Elektrotech. Z. (ETZ) A, 1975, 96, No.9, 378-83. In German

LOCKERBIE, N.A., SHERLOCK, R.A. and WYATT, A.F.G.: 'Linear motion at 0·1K', 14th International Conference on Low Temperature Physics, Otaniemi, Finland, 14-20 Aug. 1975, Pt. IV, 305-8

OBERRETL, K.: 'Comparison of electrodynamic suspension systems', Elektrotech. Z. (ETZ) A, 1975, 96, No.9, 391-4. In German

OBERRETL, K. and KRATKI, N.: 'Transients and oscillations in the repulsive magnetic levitation system', IEEE Trans., 1975, MAG-11, No.5, 1493-4

OOI, B.T.: 'Electromechanical dynamics in superconducting levitation systems, ibid., 1495-7

REITZ, J.R. and BORCHERTS, R.H.: 'US Department of Transportation program in magnetic suspension (repulsion concept)', ibid., No.2, 615-8

WEH, H.: 'The integration of magnetic levitation and electric propulsion', Elektrotech. Z. (ETZ) A, 1975, 96, No.3, 131-5. In German

WINKLE, G.: 'State of research and development of electromagnetic suspension engineering in the Federal Republic of Germany', ibid., No.9, 367-73. In German

WONG, J.Y., MULHALL, B.E. and RHODES, R.G.: 'The impedance modelling technique for investigating the characteristics of electrodynamic levitation systems', J. Phys. D, 1975, 8, No.16, 1948-55

TRANSPORT

1902   ZEHDEN, A.: 'New improvements in electric traction apparatus', Patent USA 88145, 4 June 1902

1905   ZEHDEN, A.: Patent USA 732312, 1905

1946   'A wound-rotor motor 1400 feet long', Westinghouse Engineer, Sept. 1946, 6, 160

1959   HUBER, J.: 'Electrodynamic force effects on a wheel set, movable on railway tracks', Elektrotechnik und Maschinenbau (Austria), 1959, 76, No.8, 169-74. In German

'Linear motor for rapid transit', Westinghouse Electric Corp., Rectifier and Traction Dept., Feasibility Study, Nov. 1959

1961   BEKKER, M.G.: 'Is the wheel the last thing in land locomotion?', New Scientist, 1961, 11, No.248, 406-10

1962   LAITHWAITE, E.R.: 'Electrical machines of the future', Proc. Royal Institution, 1962, 39, Pt.I, No.176, 119-35

LAITHWAITE, E.R.: 'Linear induction motor propulsion for high-speed railways', Engineers' Digest, 1962, 23, No.8, 61-8

LAITHWAITE, E.R.: 'Linear induction motors', Interavia, 1962, XVII, No.8, 1029-30

'Linear motor for traction', Elec. Times, 1962, 142, No.21, 759

'Rail transport traction of the future?', Engineering, 1962, 194, 731

1963   LAITHWAITE, E.R.: 'Linear induction motors for rail traction', Engineering, 1963, 195, No.5069, 784-5

'Linearbeschleuniger als Fahrzeugmotor', Deutsche Eisenbahntechnik (East Berlin), 1963, 11, No.8, 376

POWELL, J.R.: 'The magnetic railroad: a new form of transport', American Society of Mechanical Engineers Railroad Conference, 23-25 April 1963, Paper 63-RR4

'Versuchsfahrzeug mit linearem Elektromotor', Industrie-Elektrik-Elektrowelt, 1963, 8, No.B6, 113

1964 LAITHWAITE, E.R. and BARWELL, F.T.: 'Linear induction motors for high-speed railways', Electronics & Power, April 1964, 10, 100-3

'Linear motor may be used in channel tunnel', Elec. Rev., 1964, 174, No.21, 777-8

1965 KINGSLEY, C. Jr.: 'Linear traction motors', Project Report, MIT, Spring 1965

'Survey of technology for high speed ground transportation', MIT Report No.PB 168648 under US Dept of Transportation Contract C-85-65, 15 June 1965

1966 BARWELL, F.T.: 'Traction research', J. Instn. Locomotive Engineers, 1966-67, 36, No.2, 158-96

BARWELL, F.T. and LAITHWAITE, E.R.: 'Application of the linear induction motor to high-speed transport', Proc. I. Mech. E., 1966-67, 181, Pt.3G, 83-100 (Convention on Guided Land Transport, London, 27-28 Oct. 1966)

BAUMANN, D.M. and MEACHAM, G.B.K.: 'Preliminary design and test of linear induction traction motors and suspension systems', MIT Report DSR-6106-3 under US Dept of Transportation Contract C-85-65, 1 Nov. 1966

'British high speed transport system', <u>Engineering</u>, 11 March 1966, 201, 498

EBELT, R.: 'Der Linearmotor und sein Einsatz für den Bau elektrischer Triebfahrzeuge', <u>Deutsche Eisenbahntechnik</u> (East Berlin), 1966, 14, No.2, 66-70

'Electric rapid transport: a multi-faceted opportunity (Special Report)', <u>Elec. World</u>, 11 July 1966, 166, 71-90

POLGREEN, G.R.: 'Magnetic suspension', <u>Proc. I. Mech. E.</u>, 1966-67, 181, Pt.3G, 145-50 (Convention on Guided Land Transport, London, 27-28 Oct. 1966)

WHITE, D.C., THORNTON, R.D., KINGSLEY, C. Jr., NAVON, D.H. and NONAKA, S.: 'Some problems related to electric propulsion', MIT Report DSR-76104 under US DOT Contract C-85-65, 1 Nov. 1966

1967 ARMSTRONG, D.S.: 'Application of the linear motor to transport', <u>Railway Gaz. Int.</u>, 1967, 123, 145-50

CHAPOUTHIER, F.: 'The French aerotrain', <u>High Speed Ground Transportation J.</u>, 1967, 1, No.1, 100-4

'Der französische Luftkissenfahrzug Aérotrain', <u>Die Bundesbahn</u>, 1967, 41, No.10, 346-7

FRIEDLANDER, C.D.: 'Railway versus highway', <u>IEEE Spectrum</u>, 1967, 4, No.9, 62-76

'Green light for the hovertrain', <u>Engineering</u>, 15 Sept. 1967, 224, 406-7

'Hovertrain goes ahead - in stages', Air Cush. Veh., 1967, 10, No.63, 30

'Kommt der 500 km/h-Zug?', Die Bundesbahn, 1967, 41, No. 13/14, 494

KURIHARA, S.: 'Wheel-axle traverser with linear induction motor. An application of a linear induction motor for a car', Toshiba Rev. (Internat. Ed.), July-Aug. 1967, 20-3

LAITHWAITE, E.R.: 'Machines with open magnetic circuits', Elec. Rev., 1967, 181, No.26, 942-3

LAITHWAITE, E.R.: 'Propulsion without wheels', The Advancement of Science (British Association), Sept. 1967, 24, 119-28

LEARY, F.: 'High-speed ground transportation', Space Aeronaut., 1967, 48, No.4, 85-101.

LEE, C.H.: 'Study of linear induction motor, its feasibility for high-speed ground transportation', Garrett Corp. AiResearch Mfg. Div., Los Angeles, Final Report No.67-1948 for US Dept of Transportation under Contract C-145-66(Neg), June 1967

'NRDC gives hovertrain go-ahead', Air Cush. Veh., 1967, 9, No.59, 48

'Plan for rapid transit in Manchester', Railway Gaz. Int., 1967, 123, 865-9

BIBLIOGRAPHY

POWELL, J.R. and DANBY, G.T.: 'High speed transport by magnetically suspended trains', American Society of Mechanical Engineers Winter Annual Meeting (Railroad Division), New York, Nov. 1966, Technical Paper No. 66-WA/RR-5, published 1967

'Proposals for tracked hovercraft', Engineer, 1967, 223, No.5809, 784-5

RÉMY, E.: 'Moteur linéaire et aérotrain', Cahiers de l'Alpe, Oct. 1967, No.34

RODWELL, R.R.: 'Aerotrain advancing on two fronts', Air Cush. Veh., 1967, 9, No.59, 66-7

ROSS, H.R.: 'High speed ground transport in 1980', Interavia, April 1967, 22, 492-4

'Schiene oder Luftkissen? Untersuchungen über den Städteschnellverkehr des Jahres 1980', Moderne Eisenbahn, 1967, 5, No.29, 20-1

'Technical extracts from tracked air cushion vehicle system study and analysis report', TRW Report No.06818-6008-R000 for US Dept of Transportation, Oct. 1967

THORNTON, R.D., NAVON, D.H., LICHTENBERGER, J., ERDELYI, C. and MILLER, E.: 'Application of high power solid-state electronics to electric propulsion', MIT Report No.PB-176 920 for US Dept of Transportation under Contract C-85-65, Oct. 1967

'Track section chosen for UK Hovertrain', Air Cush. Veh., 1967, 10, No.65, 71-2

'Versuche mit dem Aérotrain bei 345 km/h', L'Usine Nouvelle, 1967, No.51, 21

1968   AUTRUFFE, H.: 'The linear motor', Revue Générale des Chemins de Fer, April 1968. In French

BARBEE, T.W., BYCROFT, G.N., CHILTON, E.G., CHILTON, F.M. and COFFEY, H.T.: 'The hypervelocity rocket sled - a design analysis', SRI Project PMU-7014, July 1968

BARTHALON, M.: 'Urba, a new transport system for tomorrow's cities', Sciences et Techniques, 1968, No.10, 21-5. In French

BARWELL, F.T.: 'Problems of support, guidance and propulsion involved in high speed transport systems', IRCA-UIC 'High Speeds' Symposium, Vienna, July 1968

BUDIG, P.-K., TIMMEL, H. and DITTRICH, W.: 'Design of linear motors as propulsion means of railway vehicles', Elektrie, 1968, 22, No.10, 405-8. In German

CHIRGWIN, K.M.: 'Linear induction motor research in the USA', IRCA-UIC 'High Speeds' Symposium, Vienna, July 1968

FREEMAN, E.M. and LAITHWAITE, E.R.: 'Unbalanced magnetic push', Proc. IEE, 1968, 115, No.4, 538 (Correspondence)

'Future transport by "airborne bus"', Ind. Handling, May 1968, 59-61

GIRANE, F.L.: 'Tracked air cushion vehicles for ground transportation systems', Proc. IEEE, 1968, 56, No.4, 646-53

GIRAUD, F.L.: 'The Aérotrain system: air cushion guided ground transportation: description and performance of the experimental vehicle', Aeroglide Systems Inc., New York, Report No. PB-178961, 1968

'High-speed surface transportation', UK Scientific Miss., Jan. 1968, 47, 1-2

LEWIS, H.E.: 'Tracked hovercraft: a future challenge to civil engineers', Consulting Engineer, Aug. 1968, 32, 54-7

LILEN, H.: 'Public transport and the linear electric motor', Electronique Indust., 1968, 112, No.4, 284-7. In French

'Linear induction motor propulsion system', Garrett Corp. Progress Report No.67-2439(5), 19 June 1968, Supplementary Report, 'Performance of 2500 horsepower LIM', No.68-4027, 22 July 1968, Technical Progress Report No.67-2439(6), 3 Oct. 1968, Progress Report, 27 Dec. 1968

'Linear induction motors: tomorrow's transit power', Product Engineering, 20 May 1968, 39, 73-4

MILLAR, J. and DEAN, J.: 'Practical considerations of rapid transit: summary of Manchester study', High Speed Ground Transportation J., 1968, 2, No.3, 409-22

NOGI, T. and KUROIWA, M.: 'Electric traction in Japanese National Railways', IEE Conference on Performance of Electrified Railways, London, Oct. 1968

'Nonfrictional power collection for guided high-speed ground vehicles', General Electric Co., Schenectady, New York, Final Report (Pt.2) No.S-68-1056 for US Dept of Transportation under Contract C-7-35121, 12 April 1968

'NRDC and tracked hovercraft', Inv. for Industry, Jan. 1968, 31, 21-2

PELENC, Y.: 'Nouvelle méthode de propulsion électrique', Firmenschrift Merlin Gerin, 1968, No.D/P 186

'Putting the hovertrain upside down', Design, July 1968, 235, 61

'Rapid transit on rubber tyres', Engineering, 1968, 205, No.5312, 210

RÉMY, E.: 'Le moteur linéaire au service d'un nouveau moyen de transport: l'Urba', Cahiers de l'Alpe, 1968, No.37

RÉMY, E.: 'Why a new means of transportation?', Flux (Revue des Anciens Élèves de l'École Supérieure d'Électricité), 1968, No.52, 5-8. In French

SOOKAWA, H.: 'Standing test of trial linear induction motor', Quart. Rept. Railway Technical Research Institute, JNR, 1968, 9, No.3, 175-7

'Specifications for linear induction motor P/N 546230, etc.', Garrett Corp. Report No.FRA 68-01 TACRV for US Dept of Transportation, 1968

STONE, P.: 'French hovertrain streaks ahead', Br. Ind. Week, 1968, 2, No.17, 30-1

'Technical discussion of LIM capabilities', Garrett Corp. Progress Report No.68-4185-1 for US Dept of Transportation, 15 Aug. 1968

'The design of linear motors for vehicle drives', Wiss. Z. Tech. Hochsch. Karl-Marx-Stadt (Germany), 1968, No.3, 321-4. In German

'The "Urba" monorail transport system', Engineers' Digest, 1968, 29, No.4, 71-3

'Transportation', Proc. IEEE Special Issue, 1968, 56, No.4, 377-786

USAMI, Y., ISHIHARA, M., KOJIMA, N. and MITOMI, T.: 'Feasibility of linear motor for car-retarder or car-accelerator at marshalling yards', Quart. Rept. Railway Technical Research Institute, JNR, 1968, 9, No.2, 91-6

'Vacuum-supported suspension railway', Engineer, 1968, 225, No.5852, 478-9

VICTORRI, M. and REYX, P.: 'For Urba, a linear induction motor', Science et Techniques, 1968, No.10, 26-9. In French

WANG, T.C.: 'A preliminary study of the linear induction motor for high speed ground transportation', TRW Systems, Report No.06818-W454-R0-12 for US Dept of Transportation under Contract C-353-66, Jan. 1968

1969 BUCHANAN, Colin (and Partners): 'Tracked Hovercraft study', Hovering Craft & Hydrofoil, 1969, 9, No.2, 11-2

CARPETIS, C., GANN, A. and PESCHKA, W.: 'A contribution of the liquid-metal induction-MHD-converter-theory to the calculations of the linear motor for very high speed ground vehicles', 4th Intersociety Energy Conversion Engineering Conference, Washington, D.C., 22-26 Sept. 1969, 818-25

CHAPA, J.: 'Control and instrumentation for a high speed rail vehicle propelled by a linear induction motor', Proc. 24th Annual ISA Conference, 1969, 24

CHIRGWIN, K.M., LEE, C.H. and LARSEN, P.J.: 'Linear induction motor for high-speed tracked vehicles', 4th Intersociety Energy Conversion Engineering Conference, Washington, D.C., 22-26 Sept. 1969, 795-806

COFFEY, H.T., BARBEE, T.W. Jr. and CHILTON, F.: 'Magnetic suspension and guidance of high speed vehicles', Int. Inst. of Refrigeration Conference on Low Temperatures and Electric Power, London, 24-28 March 1969, 311-7

COFFEY, H.T., CHILTON, F. and BARBEE, T.W. Jr.: 'Suspension and guidance of vehicles by superconducting magnets', J. Appl. Phys. (USA), 1969, 40, No.5, 2161

BIBLIOGRAPHY

'Electric power systems for high speed ground transportation', Westinghouse Electric Corp., Pittsburgh, Pa., Final Report for US Dept of Transportation under Contract DOT-9-0025, 25 Aug. 1969

GUDERJAHN, C.A., WIPF, S.L. et al: 'Magnetic suspension and guidance for high speed rockets by superconducting magnets', J. Appl. Phys. (USA), 1969, 40, No.5, 2133-40

'HSGT systems engineering study, tracked air cushion vehicles', TRW Systems Group, Washington, D.C., Final Report No. NECPT-219 for US Dept of Transportation under Contract DOT-C-353-66, Dec. 1969

KALMAN, G.P. and HAFELE, B.W.: 'Feasibility study of linear induction motor thrust boosters for diesel-electric locomotives', Garrett Corp. AiResearch Mfg. Div., Los Angeles, Report No. 69-4862 for US Dept of Transportation under Contract DOT-FR-9-0014, 21 March 1969

KALMAN, G.P., IRANI, D. and SIMPSON, A.U.: 'Electric propulsion system for linear induction motor test vehicle', 4th Intersociety Energy Conversion Engineering Conference, Washington, D.C., 22-26 Sept. 1969, 807-17

KETTERER, H.: '"Urba" - eine Schwebebahn, die wirklich Schwebt', Elektrotech. Z. (ETZ) B, 1969, 21, No.1, 14-5. In German

LAITHWAITE, E.R.: 'Linear induction motors for high-speed vehicles', Electronics & Power, July 1969, 15, 230-3

LAITHWAITE, E.R.: 'The application of linear induction motors to high speed vehicles', Design and Components in Engineering, 7 May 1969, 36-42

LAITHWAITE, E.R. and BARWELL, F.T.: 'Applications of linear induction motors to high speed ground transport systems', Proc. IEE, 1969, 116, No.5, 713-24

LAMBERMONT, P.: '80 passenger Aerotrain nears completion', Hovercraft World, 1969, 3, No.2, 41-2

'Linear induction motor', Zavod Elektrotransporta Jmeni F.E. Dzerzhinskogo, Patent UK 1175487, 22 Feb. 1968; publ. 23 Dec. 1969, UK 8589/68

MORTIMER, J.: 'Konkurrenz für den Aérotrain: hovercraft', VDI-Nachrichten, 17 Dec. 1969, No.51, 2

PEARCE, T.G. and MAY, B.J.: 'A study of the stability and dynamic response of the linear induction motor test vehicle', British Railways Board Research Dept., Derby, Final Report No. FRA-RT-70-25 for US Dept of Transportation under Contract DOT-FR-3-0261, Sept. 1969

POWELL, J.R. and DANBY, G.T.: 'Magnetically suspended trains for very high speed transport', 4th Intersociety Energy Conversion Engineering Conference, Washington, D.C., 22-26 Sept. 1969, 953-63

POWELL, J.R. and DANBY, G.T.: 'Magnetically suspended trains: the application of superconductors to high speed transport', Cryogenics Industr. Gases (USA), 1969, 4, No.10, 19-24

'Preliminary design study of a linear induction motor with aluminium bonded to a steel reaction rail', Westinghouse Electric Corp., Pittsburgh, Pa., Final Report No. PB 186231 for US Dept of Transportation under Contract DOT-9-0025, 25 Aug. 1969

RECHOU, J., du MERLE, G., ETIENNE, M. and VICTORRI, M.: 'New technical means of mass land transport, Sciences et Techniques, 1969, No.16, 3-24. In French

ROUSSIAUX, P.: 'Aerotrains', Rev. Alumin., Jan. 1969, 370, 51-5

SOOKAWA, H., YASUKAWA, S. and YUDA, S.: 'Standing and running tests of the trial linear induction motor M-300', Quart. Rept. Railway Technical Research Institute, JNR, 1969, 10, No.2, 94-100

STOCKFORD, M.A.: 'Tracked hovercraft transport systems', Engineering, 1969, 207, No.5363, 249-52

'Tracked hovercraft and the third London Airport', Hovering Craft & Hydrofoil, 1969, 9, No.2, 12

'URBA, a new short-distance transport vehicle', ITA Bull., 12 May 1969, 19, 443-6

WANG, T.C.: 'Linear induction motor for high speed ground transportation', 4th Intersociety Energy Conversion Engineering Conference, Washington, D.C., 22-26 Sept. 1969, 964-74

1970

'A cost comparison of three tracked air cushion vehicle configurations', Tracked Hovercraft Ltd, London, Final Report No. FRA-RT-71-68 for US Dept of Transportation under Contract DOT-FR-9-0032, July 1970

BLISS, D.S.: 'The evolution of tracked air cushion vehicles, Parts I and II', Hovering Craft & Hydrofoil, 1970, 9, No.11, 17-47 and No.12, 6-39

DAVIES, K.J.: 'Tracked hovercraft', Chart. Mech. Engr., 1970, 17, No.7, 330

'Electromagnetic traction systems', Rotax Ltd, Patent UK 1217514, 10 July 1968; publ. 31 Dec. 1970; prior. 17 July 1967, UK 32729/67

'Experimental linear motor vehicles for high speed transport', Elec. Rev., 1970, 187, No.22, 761-2

FORD, T.: 'Tracked Hovercraft Ltd', Hovering Craft & Hydrofoil, 1970, 9, No.7, 4-6

'High-speed ground transportation systems engineering study. Tracked air cushion vehicle systems', TRW Systems Group, Redondo Beach, Calif., Final Report No. 06818-6039-RO-00 FRA-RT-59 for US Dept of Transportation under Contract DOT-C-353-66, May 1970

'Hovertrains for Japan?', Nature, 1970, 225, No.5229, 214

'Japanese 300 mph linear motor magnetic suspension project', Modern Railways, 1970, 26, No.262, 321

KANTER, H.: 'The importance of linear motors to metropolitan railway services', Elektrie, 1970, 24, No.10, 347-51. In German

KLAPPER, C.F.: 'Tracked hovercraft: the Buchanan Report', Modern Railways, 1970, 26, No.257, 64-8

LAWES, G.: 'Hovertrain comes on apace', New Scientist, 1970, 47, No.709, 68-70

MUHLENBERG, J.D.: 'Control systems for linear induction motor guidance in high speed vehicles', American Society of Mechanical Engineers Joint Transportation Engineering Conference, New York, Oct. 1970

MUHLENBERG, J.D.: 'LIM guidance control systems', Mitre Corp. McLean, Va., Report No. MTR-4136-Rev-1, FRA-RT-71-46 for US Dept of Transportation under Contract DOT-7-35248, June 1970

PETERS, D.: 'The hovertrain: takeoff point for new technique', Engineer, 1970, 231, No.5978, 29-31

POWELL, J.R. and DANBY, G.T.: 'Dynamically stable cryogenic magnetic suspension for vehicles in very high velocity transport systems' in Recent advances in engineering science, 5 (Gordon and Breach) 1970

SANHUEZA HARDY, H. and STRAUGHEN, A.: 'Track-powered linear induction motors as a rapid-transit drive', Engineering J. (Canada), 1970, 53, No.4, 26-32

SHINRYO, Y.: 'Test facility for linear induction motor for high speed train', <u>Railway Electric Rolling Stocks</u>, 1970, 23, No.5, 20-5

'Steps towards practical linear motor propulsion in France', <u>Elec. Rev</u>., 1970, 186, No.1, 11-5

'Technical background on linear induction motors in transportation', Garrett Corp., Torrance, Calif., Report prepared for the US Dept of Transportation, June 1970

'The flying railroad', <u>Time</u>, 24 Aug. 1970, 39

'UK tracked hovercraft to be tested this summer', <u>Science Journal</u>, 1970, 6, No.5, 17

USAMI, Y., FUJIE, J. and NAKASHIMA, H.: 'Linear-motorized flat yard system', <u>Quart. Rept. Railway Technical Research Institute, JNR</u>, 1970, 11, No.2, 101-4

1971 ASAGOE, Y., SINRYO, Y. and OHNO, E.: 'Fundamental study of high-speed ground transportation', <u>Mitsubishi Denki Eng</u>. (Japan), 1971, No.30, 10-9

BOLTON, H.R., FELLOWS, T.G., LAITHWAITE, E.R., EASTHAM, J.F. and NEEDHAM, E.F.: 'Linear induction motor', Tracked Hovercraft Ltd, Patent USA 3585423, 1 May 1970; publ. 15 June 1971; prior. 2 May 1969, UK 225427/69

BORCHERTS, R.H. and REITZ, J.R.: 'High speed transportation via magnetically supported vehicles: a study of the magnetic forces', <u>Transportation Research</u> (Pergamon Press), 1971, 5, 197-209

CARPETIS, C.: 'On the use of linear induction motors and magnetic suspension for high speed ground vehicles and trains', Tech. Chron. (Greece), 1971, No.3, 187-96. In Greek

CHILTON, F. and COFFEY, H.T.: 'Magnetic levitation. Tomorrow's transportation', The Helium Society, Washington, D.C., 1971, 288

COLLING, N.W. and QUAYLE, G.P.: 'Linear induction motor stator assembly', Tracked Hovercraft Ltd, Patent USA 3626858, 10 July 1969; publ. 14 Dec. 1971; prior. 11 July 1968, UK 33182/68

DANNAN, J. and D'SENA, G.O.: 'Linear induction motor research. Vol.I. Introduction and background', Garrett Corp. AiResearch Mfg. Co., Torrance, Calif., Final Report No. 71-7094-Vol-1 FRA-RT-73-4 for US Dept of Transportation under Contract OHSGT-7-35399, Oct. 1971

'Development and manufacture of a linear induction motor propulsion system for the tracked air cushion research vehicle', Garrett Corp. AiResearch Mfg. Co., Torrance, Calif., Final Report on Task 1, No. 71-7289 FRA-RT-72-35 for US Dept of Transportation under Contract DOT-FR-00029, April 1971

EASTHAM, A.R. and RHODES, R.G.: 'The levitation of high speed trains using superconducting magnets', 2nd International Symposium on Electromagnetic Suspension, Southampton University, July 1971

'Electromagnetic transport system', North American Rockwell Corp., Patent UK 1240574, 21 Oct. 1969; publ. 28 July 1971; prior. 25 Oct. 1968, USA 770.695

ERLER, K.: 'Application of linear motors for transport systems', Elektrie, 1971, 26, No.6, 211-3. In German

GUDERJAHN, C.A. and WIPF, S.L.: 'Magnetically levitated transportation', Cryogenics, June 1971, 171

HARDING, J.T.: 'Progress in magnetic suspension applied to high speed ground transportation', 17th Annual Conference on Magnetism and Magnetic Materials, Chicago, Ill., 16-19 Nov. 1971, AIP Conf. Proc., 1971, 5, Pt.2, 938-48

HOCHBRUCK, H., BOPP, K. and MARTEN, F.: 'Developments in unconventional guideway, carrying and propulsion systems in the Federal Republic of Germany', International Congress on Electric Railways, Munich, Germany, 11-15 Oct. 1971, Paper 7.2 (Berlin: VDE-Verlag). In German

HOLDEN, W.H.T.: 'Electrical systems for rapid transit railroads', IEEE Trans., 1971, IGA-7, No.5, 580-7

JEFFS, E.: 'Tracked air-cushion vehicles offer speed and high frequency', Des. Eng. Mater. & Compon., July 1971, 50-4

KAHLEN, H., RUNGE, W. and WEIGEL, W.D.: 'Research on propulsion systems of electrically powered urban vehicles', Elektrotech. Z. (ETZ) A, 1971, 92, No.10, 571-5. In German

KAYE, D.: 'Linear motor for mass transport improved by pulsed-DC technique', Electron. Des. (USA), 1971, 19, No.10, 30

KIKUCHI, T.: 'Application of linear motor at Toyama marshalling yard', J. Soc. Instrum. & Contr. Eng. (Japan), 1971, 10, No.8, 611-6. In Japanese

KURZ, K.: 'Unconventional means of transport in public rapid transit systems', International Congress of Electric Railways, Munich, Germany, 11-15 Oct. 1971, Paper 7.3 (Berlin: VDE-Verlag). In German

LAITHWAITE, E.R. and BLISS, D.S.: 'Linear induction motor for vehicle propulsion', Tracked Hovercraft Ltd, Patent USA 3585939, 25 Sept. 1969; publ. 22 June 1971; prior. 26 Sept. 1968, UK 45277/68

LAITHWAITE, E.R., BOLTON, H.R. and EASTHAM, J.F.: 'High speed linear motor testing at Imperial College', 6th Universities Power Engineering Conference, UMIST, Manchester, 4-6 Jan. 1971, Paper 3.1

LAITHWAITE, E.R., EASTHAM, J.F., BOLTON, H.R. and FELLOWS, T.G.: 'Linear motors with transverse flux', Proc. IEE, 1971, 118, No.12, 1761-7

LEHMANN, H.: 'Problems of integrating transportation systems with conventional and unconventional media', International Congress on Electric Railways, Munich, Germany, 11-15 Oct. 1971, Paper 7.4 (Berlin: VDE Verlag). In German

'Linear induction motor', Linerail Manutention Moteur Linéaire, Patent UK 1247257, 19 Dec. 1969; publ. 22 Sept. 1971; prior. 26 Dec. 1968, France 180268

'Linear induction motor research: Linear induction motor and test vehicle design and fabrication. Vol.II, Books 1 and 2', Garrett Corp. AiResearch Mfg. Co., Torrance, Calif., Final Report Nos. 71-7094-Vol-2-Bk-1 FRA-RT-73-2 and 71-7094-Vol-2-Bk-2 FRA-RT-73-5 for US Dept of Transportation under Contract OHSGT-7-35399, Oct. 1971

'Linear induction motor research: Data acquisition, data analyses, and low-speed testing. Vol.III', Garrett Corp. AiResearch Mfg. Co., Torrance, Calif., Final Report No. 71-7094-Vol-3 FRA-RT-73-3 for US Dept of Transportation under Contract OHSGT-7-35399, Oct. 1971

MACHEFERT-TASSIN, Y.: 'Application of the linear motor in transportation', Rev. Gen. Elec., 1971, 80, No.2, 121-34. In French

'Magnetic support for HSB test car', Railway Gaz. Int., 1971, 127, No.6, 233

MEISENHOLDER, S.G., GRAHAM, H.R. and BIRCHILL, J.: 'Dynamic response tests of an air cushion suspension system for the linear induction motor (LIM) of the tracked air cushion research vehicle (TACRV)', TRW Systems Group, Redondo Beach, Calif., Report No. 17617-6003-RO-00 FRA-RT-72-24 for US Dept of Transportation under Contract DOT-FR-00044, July 1971

MOYSE, S.A.: 'Vehicle linear induction motor', Patent UK 1233250, 30 Dec. 1968; publ. 26 May 1971; prior. 9 Jan. 1968, France 235266

NAVE, P.M.W.: 'The Messerschmitt-Bölkow-Blohm magnetically suspended vehicle', 2nd International Symposium on Electromagnetic Suspension, Southampton University, July 1971

POWELL, J.R. and DANBY, G.T.: 'Magnetic suspension for levitated tracked vehicle', *Cryogenics,* June 1971, 11, 192-204

POWELL, J.R. and DANBY, G.T.: 'The linear synchronous motor and high speed ground transport', Intersociety Energy Conversion Engineering Conference, Boston, Mass., 3-5 Aug. 1971, 118-31

RICHARDS, P.L.: 'Magnetic suspension and propulsion systems for high-speed transportation', 17th Annual Conference on Magnetism and Magnetic Materials, Chicago, Ill., 16-19 Nov. 1971, *AIP Conf. Proc.*, 1971, 5, Pt.2, 935-7

ROBERT, G.: 'FS moves towards 200 km/h', *Railway Gaz. Int.*, 1971, 127, No.11, 425-6

SCHLOSSER, A.: 'Tracked air cushion research vehicle. Subsystems analysis', Grumman Aerospace Corp., Bethpage, N.Y., Design Report No. FRA-RT-72-33 for US Dept of Transportation under Contract DOT-FR-00005, March 1971

'Study backs train at 1000 mph', *Christian Science Monitor*, 23 March 1971

TAKAHASHI, N., KAWAI, S. and AKIHASHI, K.: 'Analysis of rail eddy-current brake for high-speed railroad vehicles', *Elec. Eng. Jap.* (USA), 1971, 90, No.1-2, 95-104

TAKEI, K., AKIHAMA, Y., TAKAYAMA, M., TAKAHASHI, H and ANDO, M.: 'Type L_4 linear motor system wagon booster-retarder', Hitachi Rev. (Japan), 1971, 20, No.11, 457-64

TAMAKI, M. and USAMI, Y.: 'Linear-motorized yard automation system', Rail Int. (Belgium), 1971, 2, No.6, 534-40

'The way ahead for tracked hovercraft', Railway Gaz. Int., 1971, 127, No.6, 229-32

WANG, T.C.: 'Linear induction motor for high-speed ground transportation', IEEE Trans., 1971, IGA-7, No.5, 632-42

WEH, H.: 'Asynchronous linear motors for traction duties', Energ. & Tech. (Germany), 1971, 23, No.12, 431-4. In German

WEH, H. and APPUN, P.: 'Eddy currents in the stationary part of magnets for magnetic suspension of vehicles', Elektrotech. Z. (ETZ) A, 1971, 92, No.11, 623-7. In German

'West Germans unveil new hovertrain', Int. Ry. J. (Netherlands), 1971, 11, No.7, 42

WILKIE, D.F.: 'A study of the control of magnetically levitated vehicles', Ford Motor Co., Dearborn, Mich., Report No. SR 71-104, 3 Aug. 1971

YAMADA, T. and IWAMOTO, M.: 'Theoretical analysis of lift and drag forces on magnetically suspended high speed trains', Mitsubishi Electric Corp., Japan, 1971 (unpublished)

1972   ALSTON, I.A. and HAYDEN, J.T.: 'A preliminary technical assessment of magnetically suspended trains', 4th International Cryogenic Engineering Conference, Eindhoven, Netherlands, 24-26 May 1972 (IPC Science & Technology Press) 198-201

ATHERTON, D.L., LOVE, L.E.G. and PRENTISS, P.O.: 'Magnetic levitation: linear synchronous motor efficiency', Canadian J. Phys., 1972, 50, No.24, 3143-6

BARAN, W.: 'The present position and further development in the area of permanent magnet suspension systems for track-bound high speed transport', Int. J. Magn. (GB), 1972, 3, No.1-3, 103-11. In German

BOPP, K., HOCHBRUCK, H. and MARTEN, F.: 'Developments in the area of unconventional vehicle, drive and guidance components in the German Federal Republic', Elek. Bahnen (Germany), 1972, 43, No.5, 98-106. In German

BORCHERTS, R.H. and DAVIS, L.C.: 'Force on a coil moving over a conducting surface including edge and channel effects', J. Appl. Phys. (USA), 1972, 43, No.5, 2418-27

BORCHERTS, R.H., DAVIS, L.C., REITZ, J.R. and WILKIE, D.F.: 'High-speed ground transportation using superconducting magnetic suspension', 4th International Cryogenic Engineering Conference, Eindhoven, 24-26 May 1972 (IPC Science & Technology Press) 185-7

BUNTING, P.M.: 'Magnetic suspension for guided transport vehicles', Transp. Plann. & Technol., 1972, 1, No.1, 49-74

COFFEY, H.T., CHILTON, F. and HOPPIE, L.O.: 'The feasibility of magnetically levitating high speed ground vehicles. Final Report on Task 1', Stanford Research Institute, Menlo Park, Calif., No. SRI-FR-1080 FRA-RI-72-39 for US Dept of Transportation under Contract DOT-FR-10001, Feb. 1972

COFFEY, H.T., HOPPIE, L.O. and CHILTON, F.: 'Performance of a magnetically levitated vehicle', 7th Annual Meeting of IEEE Industry Applications Society, Philadelphia, Pa., 9-12 Oct. 1972, 331

DANBY, G.T. and POWELL, J.R.: 'Integrated systems for magnetic suspension and propulsion of vehicles', Applied Superconductivity Conference, Annapolis, Md., May 1972

DUKOWICZ, J.K.: 'The single-sided LIM with saturated back iron', Mitre Corp., McLean, Va., Final Report No. MTR-6094, FRA-RT-72-28 for US Dept of Transportation under Contract DOT-FR-7-35248, Jan. 1972

DUKOWICZ, J.K.: 'The single-sided LIM with saturated guideway back iron', 7th Annual Meeting of IEEE Industry Applications Society, Philadelphia, Pa., 9-12 Oct. 1972, 321

ENGLISH, C.D.: 'A linear induction motor propulsive system, I and II', Modern Railways, 1972, 29, No.287, 292-5 and No.290, 432-4

FALLSIDE, F., JACKSON, R.D., MISKIN, D.J. and PORRELLI, C.N.: 'A cyclo-converter-linear motor drive for air-cushioned vehicles', IEE Conference on Electrical Variable Speed Drives, London, 10-12 Oct. 1972, 243-8

FOGG, J.C.: 'Tracks into the future', Electrical and Electronics Technician Engineer, 1972, 7, No.3, 3-10

GAVIN, J.G. Jr.: 'Tracked air cushion research vehicle - today's test bed for tomorrow's high speed ground transportation', 2nd Urban Technology Conference and Display, San Francisco, 24-26 July 1972 (American Inst. Aeronautics and Astronautics) 72-614/1-7

HAIGHT, E.C. and HUTCHENS, W.A.: 'Lower boundary condition effects on LIM (linear induction motors) reaction rail mechanical behavior: analysis and experiments', Mitre Corp., McLean, Va., Technical Report No. MTR-6038 FRA-RT-73-11 for US Dept of Transportation under Contract DOT-FR-7-35248, 18 July 1972

HAIGHT, E.C., HUTCHENS, W.A. and WILLIAMS, J.G.: 'Experimental determination of the compressive behavior of a linear induction motor reaction rail', Mitre Corp., McLean, VA., Report No. MTR-374 for US Dept of Transportation under Contract DOT-FR-7-35248, 1 Nov. 1972

'High-speed railway train', Voith Getriebe KG, Patent UK 1292683, 6 April 1970; publ. 11 Oct. 1972; prior. 5 April 1969, West Germany P1917 640.0

'High-speed traction line-up', Railway Gaz. Int., 1972, 128, No.8, 296-7

JOHNSON, R.A.: 'Aerodynamic drag and stability characteristics of a magnetically levitated vehicle', AVCO Report No. AVSD-0391-72-CR, 1 Nov. 1972

KYOTANI, Y.: 'Magnetic levitation research vehicle', Japanese Railway Engineering (Japan Railway Engineers' Assoc.), 1972, 13, No.4, 6-9

LEVI, E.: 'Design of the electric propulsion system for an 820 km/h train', Ingegneria Ferroviaria, 1972, No.2, 1-8. In Italian

'Linear induction motor reaction rail', Tracked Hovercraft Ltd, Patent UK 1292801, 17 Oct. 1969; publ. 11 Oct. 1972; prior. 17 July 1968, UK 34096

MEISENHOLDER, S.G. and WANG, T.C.: 'Dynamic analysis of an electromagnetic suspension system for a suspended vehicle system', TRW Systems, Redondo Beach, Calif., Final Report No. FRA-RT-73-1 for US Dept of Transportation, Jan. 1972

MOON, F.C. and MARKIEWICZ, W.D.: 'Calculations for the design of magnetically suspended vehicles', Report to Boeing-Vertol Corp., Philadelphia, Pa., Dec. 1972

MUHLENBERG, J.D. et al: 'Comparative analysis of non-contacting suspensions for high speed ground vehicles', Mitre Corp., McLean, Va., Report No. MTR-6223, Dec. 1972

OGIWARA, H. and TAKANO, N.: 'Large scale superconducting magnets for suspended high speed trains', Toshiba Rev., 1972, 27, 631

OGIWARA, H., TAKANO, N., MATSUKUMA, M. and YAMAYA, T.: 'Magnetically suspended superhigh-speed train', ibid., 625

OGIWARA, H., TAKANO, N., SAITO, Y., NAKAYAMA, Y. and YONEMITSU, H.: 'Magnetic levitation and power transmission', 4th International Cryogenic Engineering Conference, Eindhoven, 24-26 May 1972 (IPC Science & Technology Press) 194-7

POLGREEN, G.R.: 'Controlled permanent magnets for tracked transport', Elec. Rev., 1972, 191, No.2, 41-4

POWELL, J.R. and DANBY, G.T.: 'Integrated magnetic suspension and propulsion systems', 7th Annual Meeting of IEEE Industry Applications Society, Philadelphia, Pa., 9-12 Oct. 1972, 517-26

PRIEBE, E.P.: 'Propulsion motor requirements for mass transportation', IEEE Trans., 1972, IA-8, No.3, 310-15

REITZ, J.R., BORCHERTS, R.H., DAVIS, L.C. and WILKIE, D.F.: 'Technical feasibility of magnetic levitation as a suspension system for high speed ground transportation vehicles', Ford Motor Co., Technical Report No. FRA-RT-72-40 for US Dept of Transportation under Contract DOT-FR-10026, Feb. 1972

'Report on a study of the magneplane high-speed transportation system', Jackson & Moreland, Division of United Engineers and Constructors Inc. for Raytheon Company Advanced System Engineering Dept., P.O. 50-5073-AS-8386, R.N. 5633-00, 10 Jan. 1972

RICHARDS, P.L. and TINKHAM, M.: 'Magnetic suspension and propulsion systems for high-speed transportation', J. Appl. Phys. (USA), 1972, 43, No.6, 2680-91

'Single-sided linear motor key to hovertrain design', Elec. Rev., 1972,190, No.12, 402-4

'Statistical analysis of LIMRV reaction rail survey data', Mitre Corp. Working Paper No. WP-10125, 30 Nov. 1972

'Study of magnetic levitation and linear synchronous motor propulsion', Canadian Institute of Ground Transport, Queen's University at Kingston, Ontario, Annual Report, 1972

'Superconducting solenoids for magnetically levitated high speed trains', VDI Z. (Germany), 1972, 114, No.17, 1276-7. In German

TAKANO, N., OGIWARA, H., SAITO, Y., NAKAYAMA, Y. and YONEMITSU, H.: 'Experimental studies on large superconducting magnets for magnetically suspended trains', 4th International Cryogenic Engineering Conference, Eindhoven, 24-26 May 1972 (IPC Science & Technology Press) 191-3

TAKANO, N., SAITO, Y. and OGIWARA, H.: 'Characteristics of magnetic levitation for high-speed trains, ibid., 188-9

'Thyristor control of high-speed trains', Railway Gaz. Int., 1972, 128, No.8, 289-93

VIERLING, B.J.: 'Personal rapid transit – applying electronics to urban transportation', IEEE Electronics and Aerospace Systems Convention, Washington, D.C., 16-18 Oct. 1972, 130-5

WAGNER, A.: 'Bibliography on the linear induction motor in high speed ground transportation', IEEE Summer Power Meeting, San Francisco, Calif., 1972

WARD, E.J.: 'Linear electric motors for high speed ground transport', IEEE Electronics and Aerospace Systems Convention, Washington, D.C., 16-18 Oct. 1972, 169-73

WILKIE, D.F.: 'Dynamics, control and ride quality of a magnetically levitated high speed ground vehicle', Transportation Research (Pergamon Press), 1972, 6, 343-64

WILKIE, D.F., DAVIS, L.C. and BORCHERTS, R.H.: 'A design for a magnetically levitated high speed vehicle', 7th Annual Meeting of IEEE Industry Applications Society, Philadelphia, Pa., 9-12 Oct. 1972, 333-42

YAMADA, T. and IWAMOTO, M.: 'Theoretical analysis of lift and drag forces on magnetically suspended high-speed train', Elec. Eng. Jap. (USA), 1972, 92, No.1, 53-62

YAMADA, T., IWAMOTO, M. and ITO, T.: 'Levitation performance of magnetically suspended high speed trains', IEEE Trans., 1972, MAG-8, No.3, 634-5

YAMAMURA, S., ISHIKAWA, Y. and HAYASHI, T.: 'Magnetic levitation of trains by means of normal conductive electromagnets', ibid., 629

1973

ARP, V.D., CLARK, A.F. and FLYNN, T.M.: 'Some applications of cryogenics to high speed ground transportation', Nat. Bur. Stand., Washington, D.C., Report No. NBS-TN-635, Feb. 1973

ATHERTON, D.L. and EASTHAM, A.R.: 'Magnetic levitation of high speed guided ground transport', IEEE International Electrical, Electronics Conference and Exposition Digest, Toronto, 1-3 Oct. 1973, 80-1

ATHERTON, D.L. and EASTHAM, A.R.: 'The Canadian magnetic levitation program', Annual Conference, Roads and Transportation Association of Canada, Halifax, 9-12 Oct. 1973

BORCHERTS, R.H., DAVIS, L.C., REITZ, J.R. and WILKIE, D.F.: 'Baseline specifications for a magnetically suspended high-speed vehicle', Proc. IEEE, 1973, 61, No.5, 569-78

COFFEY, H.T., CHILTON, F. and HOPPIE, L.O.: 'Magnetic levitation for tomorrow's transportation' in Jane's Surface Skimmers (London: Sampson Low, Marston & Co.) 1973-74

COFFEY, H.T., COLTON, J.D. and MAHRER, K.D.: 'Study of a magnetically levitated vehicle', Stanford Research Institute, Menlo Park, Calif., Final Report No. DOT-FR-73-24 for US Dept of Transportation under Contract DOT-FR-10001, Feb. 1973

COOPER, B.K.: 'Thyristors for high-speed traction', Modern Railways, 1973, 30, No.292, 34-5

DANNAN, J.H., DAY, R.N. and KALMAN, G.P.: 'A linear-induction-motor propulsion system for high-speed ground vehicles', <u>Proc. IEEE</u>, 1973, 61, No.5, 621-30

FAHLENBRACH, H.: 'Magnetic suspension for tracked transport systems', <u>Elektrotech. Z. (ETZ) B</u> , 1973, 25, No.3, 48-9. In German

FORGACS, R.L.: 'Evacuated tube vehicles versus jet aircraft for high-speed transportation', <u>Proc. IEEE</u>, 1973, 61, No.5, 604-16

FRIEDRICH, R.: 'Guideways for repulsion MAGLEV tests', 2nd Status Seminar of BMFT, Berlin, Feb. 1973

GOTTZEIN, E. and LANGE, B.: 'Magnetic suspension control systems for the German high speed train', 5th IFAC Symposium on Automatic Control in Space, Genoa, Italy, June 1973

GREENE, A.H., HARROLD, W.J., KASEVICH, R.S., TANG, C.H. and WEISS, E.: 'Final Report on Magneplane synchronous motor study', for MIT National Magnet Laboratory, June 1973

'Ground transportation for the eighties', <u>Proc. IEEE</u> Special Issue, 1973, 61, No.5, 513-688

GUDERJAHN, C.: 'Hybrid magnetically levitated bus', Greater Los Angeles Area Transportation Symposia, 1972-73 (North Hollywood: Western Periodicals) 1973, 209-14

HANLON, J.: 'Magnets boost high speed trains', New Scientist, 1973, 57, No.833, 360-2

HARROLD, W.J., KASEVICH, R.S., TANG, C.H. and VIENS, N.P.: 'Electromagnetic propulsion for magnetically levitated vehicles', 19th Conference on Magnetism and Magnetic Materials, Boston, Mass. 16 Nov. 1973, AIP Conf. Proc., 1973, No.18, Pt.2, 1240-4

HARROLD, W.J. and TANG, C.H.: 'Optimization of magnet configuration for the Magneplane model', IEEE Trans., 1973, MAG-9, No.3, 248-52

'Herringbone windings key to high speed linear motor', Elec. Rev., 1973, 192, No.12, 435-6

HOCHHAUSLER, P.: 'New data on the operation of the magnetic suspended railway', Elektrotech. Z. (ETZ) B, 1973, 25, No.3, 45-7. In German

HOLLINGBERY, P.L.: 'High-speed transport: are we on the right lines?', Electronics & Power, 1973, 19, No.17, 420-2

IWASA, Y.: 'Electromagnetic flight stability by model impedance simulation', J. Appl. Phys. (USA), 1973, 44, No.2, 858-62

IWASA, Y.: 'Magnetic shielding for magnetically levitated vehicles', Proc. IEEE, 1973, 61, No.5, 598-603

KATZ, R.M., NENE, V.D. et al: 'Performance of magnetic suspensions for high speed vehicle operating over flexible guideways', Intersociety Conference on Transportation, Denver, Colorado, Sept. 1973, Paper No. 73-ICT-89

KILL, E.: 'Electrical engineering in guided transport systems of the future', Elektrotech. Z. (ETZ) A, 1973, 94, No.12, 717-25. In German

KOLM, H.H. and THORNTON, R.D.: 'Electromagnetic flight', Scientific American, 1973, 229, No.4, 17-25

KONOVALOV, V.S.: 'Linear drive motors for lifting and transporting machinery', Vestn. Mashinostr. (USSR), 1973 No.9, 21-2. In Russian.   English trans. Russ. Eng. J. (GB), 1973, 53, No.9, 16-8

KRINGS, B.J.: 'Alternative systems for rapid-transit propulsion and electrical braking', Westinghouse Engineer, 1973, 33, No.2, 34-41

KUSTER, W., JORAN, H. and KOEWIUS, A.: 'Aluminium for linear motor reaction rails', Elektrotech. Z. (ETZ) A, 1973, 94, No.8, 494-8. In German

LAITHWAITE, E.R.: 'Linear motors for high speed vehicles', New Scientist, 1973, 58, No.852, 802-5

LAITHWAITE, E.R.: 'Linear motors with transverse flux', Indian & Eastern Engineer, 1973, 115, No.10, 529-31

LANGERHOLC, J.: 'Torques and forces on a moving coil due to eddy-currents (transportation systems)', J. Appl. Phys. (USA), 1973, 44, No.4, 1587-94

LEVI, E.: 'Linear synchronous motors for high-speed ground transportation', <u>IEEE Trans</u>., 1973, MAG-9, No.3, 242-8

MACHEFERT-TASSIN, Y. and WIART, A.: 'Linear induction-type rail brakes', <u>Rev.Jeumont-Schneider</u>, 1973, No.16, 32-42. In French

MELVILLE, P.H.: 'Magnetic propulsion for magnetically levitated trains', <u>Cryogenics</u>, 1973, 13, No.12, 716-7

NAGAHIRO, T., TERADA, K., KASAI, Y. and MOTONAGA, M.: 'Magnetically suspended experimental vehicle - strength of structure and dynamic analysis', <u>Hitachi Hyoron</u> (Japan), 1973, 55, No.6, 37-42. In Japanese

NONAKA, S., HAYASHI, K. and YOSHIDA, K.: 'Thrust characteristics of high-speed linear motors driven by the three-phase invertor', <u>Technol. Rep. Kyushu Univ</u>. (Japan), 1973, 46, No.2, 102-8. In Japanese

OHNO, E., IWAMOTO, M., OGINO, O., KAWAMURA, T. and SHINOBU, M.: 'Studies on magnetic levitation for high-speed trains', <u>Mitsubishi Electr. Eng</u>. (Japan), 1973 No.37, 23-9

OHNO, E., IWAMOTO, M. and YAMADA, T.: 'Characteristics of superconductive magnetic suspension and propulsion for high-speed trains', <u>Proc. IEEE</u>, 1973, 61, No.5, 579-86

RAPOSA, F.I., KNUTRUD, T. and WAWZONEK, J.J.: 'Power conditioning for high speed tracked vehicles', Transportation Systems Centre, Cambridge, Mass., Report No. DOT-TSC-FR-72-13, DOT-TSC-FR-71-1A for US Dept of Transportation under Contract DOT-RR-205, Jan. 1973

REITZ, J.R., BORCHERTS, R.H., DAVIS, L.C., HUNT, T.K. and WILKIE, D.F.: 'Preliminary design studies of magnetic suspensions for high speed ground transportation', Ford Motor Company, Report No. FRA-RT-73-27 for US Dept of Transportation under Contract DOT-FR-10026, 1973

ROSS, J.A.: 'ROMAG transportation system', Proc. IEEE, 1973, 61, No.5, 617-20

TINKHAM, M.: 'AC losses in superconducting magnet suspensions for high-speed transportation', J. Appl. Phys. (USA), 1973, 44, No.5, 2385-90

WEH, H.: 'Magnetische Schwebetechnik für Schnellbahnen', Bull. Assoc. Suisse Electr. (Switzerland), 1973, 64, No.9, 564-71

1974   ABEL, E., CORBETT, A.E., MULHALL, B.E. and RHODES, R.G.: 'Levitation and propulsion of guided vehicles using superconducting magnets', IEE Conference on Linear Electric Machines, London, 21-23 Oct. 1974, 223-9

ALBRECHT, C., ELSEL, W., FRANKSEN, H., PARSCH, C.P. and WILHELM, K.: 'Superconducting levitated systems: first results with the experimental facility at Erlangen', 5th International Cryogenic Engineering Conference, Kyoto, Japan, May 1974 (IPC Science & Technology Press) 28-34

ANDREI, R. and SORAN, I.: 'Study of a configuration for an asynchronous linear motor used in electrical traction', Stud. Cercet. Energ. Electrotehn. (Rumania), 1974, 24, No.1, 179-91. In Rumanian

ARIMA, K., NAKASHIMA, H. and KUZUU, T.: 'Refrigeration system for magnetically levitated trains', 5th International Cryogenic Engineering Conference, Kyoto, Japan, May 1974 (IPC Science & Technology Press) 99-101

ASZTALOS, S., BALDUS, W., KNEUER, R. and STEPHAN, A.: 'On-board cryogenic system for magnetic levitation of trains: cryogenic system of EET', ibid., 37-41

ATHERTON, D.L. and EASTHAM, A.R.: 'Flat guidance systems for magnetically levitated high speed guided ground transport', J. Appl. Phys. (USA), 1974, 45, No.3, 1398-1405

ATHERTON, D.L. and EASTHAM, A.R.: 'Guidance of a high speed vehicle with electrodynamic suspension', IEEE Trans., 1974, MAG-10, No.3, 413-6

ATHERTON, D.L. and EASTHAM, A.R.: 'High-speed Maglev studies in Canada', 5th International Cryogenic Engineering Conference, Kyoto, Japan, May 1974 (IPC Science and Technology Press) 46-50

ATHERTON, D.L. and EASTHAM, A.R.: 'Limitations of levitation by iron-cored electromagnets', IEEE Trans., 1974, MAG-10, No.3, 410-2

ATHERTON, D.L. and EASTHAM, A.R.: 'Propulsion requirements for high speed vehicles with electrodynamic suspension', Proc. IEEE Industry and Applications Society Meeting, Pittsburgh, Pa., 1974

BIBLIOGRAPHY

AUGENREICH, K.: 'Control of a bulk-goods transport medium with linear motor drive', <u>Foerdern & Heben</u> (Germany), 1974, 24, No.6, 578-83. In German

BAHMANYAR, H. and ELLISON, A.J.: 'The application of permanent magnets to the suspension of surface-guided vehicles', IEE Conference on Linear Electric Machines, London, 21-23 Oct. 1974, 203-9

BARTHALON, M.: 'A new linear electromagnetic motor TRAKLEC opens a wide field of application to linear machines', <u>ibid</u>., 62-7

BOBINEAU, B. and TEYSSANDIER, C.: 'A test station for high-speed linear motors', High Speed Electric Propulsion Meeting (Société des Électriciens, des Électroniciens et des Radioélectriciens), Grenoble, 16 May 1974. In French

BOHM, H.: 'Magnetically suspended vehicles', <u>Elektrotech. Z. (ETZ) B</u>, 1974, 26, No.16, 411. In German

BOLDEA, I. and NASAR, S.A.: 'Some baseline specifications for an optimal design of a high-speed linear induction motor', International Symposium on Linear Electric Motors, Lyon, France, May 1974

BOMPA, L.: 'The linear motor and its future in high speed land transport', <u>Rail Int</u>. (Belgium), 1974, 5, No.6, 411-3

BOPP, K.: 'Electrical engineering problems of future high speed tracked land transport', <u>Elek. Bahnen</u> (Germany), 1974, 45, No.10, 222-30. In German

BORCHERTS, R.H. and DAVIS, L.C.: 'Lift and drag forces for the attractive electromagnetic suspension system', *IEEE Trans.*, 1974, MAG-10, No.3, 425-8

BORCHERTS, R.H., REITZ, J.R. and WAN, C.C.: 'Superconducting magnetic suspensions using slotted or ladder tracks', Digests of the IEEE Intermag Conference, Toronto, 14-17 May 1974

BURKE, P.E., TURTON, R.A. and SLEMON, G.R.: 'The calculation of eddy current losses in guideway conductors and structural members of high-speed vehicles', *IEEE Trans.*, 1974, MAG-10, No.3, 462-5

CHAHAL, J.S.: 'Some aspects of a transverse flux linear induction motor design, suitable for high speed applications', IEE Conference on Linear Electric Motors, London, 21-23 Oct. 1974, 161-6

CHIRGWIN, K.M.: 'Test results from the US linear induction motor research vehicle program', *ibid.*, 236-43

COFFEY, H.T.: 'Magnetic suspensions for high speed vehicles', *Advances in Cryogenic Engineering*, 1974, 19, 137-53

COFFEY, H.T., COLTON, J.D., SOLINSKY, J.C. and WOODBURY, J.R.: 'An evaluation of the dynamics of a magnetically levitated vehicle', Report No. FRA-ORD D-74-41 for US Dept of Transportation, March 1974

EASTHAM, J.F., BOLTON, H., LAITHWAITE, E.R. and FREEMAN, E.M.: 'Transverse flux self-levitated linear induction motors', International Symposium on Linear Electric Motors, Lyon, France, May 1974

EASTHAM, J.F. and LAITHWAITE, E.R.: 'Linear induction motors as electromagnetic rivers', Proc. IEE, 1974, 121, No.10, 1099-108

EASTHAM, J.F. and WILLIAMSON, S.: 'Experiments on the lateral stabilization and levitation of linear induction motors', IEEE Trans., 1974, MAG-10, No.3, 470-3

ELLISON, A.J. and BAHMANYAR, H.: 'Surface-guided transport systems of the future', Proc. IEE Reviews, 1974, 121, No.11R, 1224-8

ELLISON, A.J. and BAHMANYAR, H.: 'The application of linear electric motors to the proposed surface-guided transport systems of the future', IEE Conference on Linear Electric Machines, 21-23 Oct. 1974, 124-30

FARRER, W.: 'Inverter system vital to rail traction by linear motor', Elec. Times, 20 June 1974, No.4286, 10-11

FELLOWS, T.G.: 'High speed surface transport', Railway Eng. J., 1974, 3, No.2, 4-13

GREENE, A.H., HARROLD, W.J., KASEVICH, R.S., MORRISON, F.P. and TANG, C.H.: 'LSM control of maglev vehicle ride quality', IEEE Trans.,1974, MAG-10, No.3, 431-4

GUARINO, M. Jr.: 'Integrated linear electric motor propulsion systems for high speed transportation', International Symposium on Linear Electric Motors, Lyon, France, May 1974

GUTBERLET, H.G.: 'The German magnetic transportation program', IEEE Trans., 1974, MAG-10, No.3, 417-20

HIERONYMUS, H., MIERICKE, J., PAWLITSCHEK, F. and RUDEL, M.: 'Experimental study of magnetic forces on normal and null flux coil arrangements in the inductive levitation system', Appl. Phys., 1974, 3, 359-66

HOCHHAUSLER, P.: 'A catamaran as a magnetically levitated vehicle', IEE Conference on Linear Electric Machines, London, 21-23 Oct. 1974, 216-22

HOCHHAUSLER, P.: 'The catamaran as a magnetic suspension vehicle', Elektrotech. Z. (ETZ) B, 1974, 26, No.16, 12-3. In German

ICHIKAWA, H. and OGIWARA, H.: 'Design considerations for superconducting magnets as a maglev pad', 5th International Cryogenic Engineering Conference, Kyoto, Japan, May 1974 (IPC Science & Technology Press) 86-9

ICHIKAWA, H. and OGIWARA, H.: 'Design considerations of superconducting magnets as a maglev pad', IEEE Trans., 1974, MAG-10, No.4, 1099-103

ISHIZAKI, Y., KURODA, T. and OHTSUKA, T.: 'Sealed cryostat system for magnetically levitated vehicles', 5th International Cryogenic Engineering Conference, Kyoto, Japan, May 1974 (IPC Science & Technology Press) 102-5

IWAHANA, T. and KUZUU, T.: 'Characteristics of the ride quality of superconducting magnetic levitation test vehicle', ibid., 106-7

IWAMOTO, M., YAMADA, T. and OHNO, E.: 'Magnetic damping force in electrodynamically suspended trains', IEEE Trans., 1974, MAG-10, No.3, 458-61

JAYAWANT, B.V., HODKINSON, R.L., WHEELER, A.R. and WARTON, R.J.: 'Transducers and their influence in the design of magnetically suspended vehicles', IEE Conference Publication No.117, 1974, 200-6

KALMAN, G.P.: 'Linear motors to power DoT's high-speed research vehicles', Railway Gaz. Int., 1974, 130, No.10, 378-83

KARSTEN, P.: 'Survey and assessment of possible magnetic suspension and guidance systems for high-speed tracked transportation', Elektrotechnick (Netherlands), 1974, 52, No.1, 26-31, 34-5. In Dutch

KASAHARA, T., SAITO, R. et al: 'A superconducting magnet for ML-100', 5th Cryogenic Engineering Conference, Kyoto, Japan, May 1974 (IPC Science & Technology Press), 82-5

KOLM, H.H.: 'Electromagnetic flight', IEEE Trans., 1974, MAG-10, No.3, 397-401

LAITHWAITE, E.R.: 'Inverter system is not vital to traction by linear motor', Elec. Times, 17 Oct. 1974, No.4302, 6-7

LAMB, C.St.J.: 'Electric linear motor urban transportation system', Elect. Engr. (Australia), 1974, 51, No.1, 6-8

LAMB, C. St.J: 'Parallel-connected linear motor for high speed transportation and rapid transit systems', IEE Conference on Linear Electric Machines, London, 21-23 Oct. 1974, 137-42

LAMB, C. St.J.: 'Using a parallel-connected linear motor for urban transportation', Annual Engineering Conference, Newcastle, N.S.W., 20-25 May 1974 (Sydney: Inst. Engrs. Australia) 15-20

LEHMANN, H.: 'Latest developments affecting magnetic levitation vehicles', Rail Int. (Belgium), 1974, 5, No.10, 629-37

LEVI, E.: 'High-speed, iron-cored synchronously operating linear motors', IEE Conference on Linear Electric Machines, London, 21-23 Oct. 1974, 155-60

'Levitation lineup - 1974', Railway Gaz. Int., Oct. 1974, 377

MANDT, K.: 'Measuring and control system for testing new components for guided transportation systems employing magnetic levitation', Siemens Rev. (Germany), 1974, 41, No.12, 527-32

MATSUI, K., UMEMORI, T., TAKETSUNA, Y. and HOSODA, Y.: 'D.C. linear motor controlled by thyristors and the testing equipment for its high speed characteristics', IEE Conference on Linear Electric Machines, London, 21-23 Oct. 1974, 149-53

MOON, F.C.: 'Laboratory studies of magnetic levitation in the thin track limit', IEEE Trans., 1974, MAG-10, No.3, 439-42

MOON, F.C. and DOWELL, E.H.: 'Dynamic interaction between a linear induction motor and elastic reaction rail', IEE Conference on Linear Electric Machines, London, 21-23 Oct. 1974, 173-8

NAKASHIMA, H. and ARIMA, K.: 'Vertical cryostat for guidance and propulsion of superconducting magnetic levitation vehicle', 5th International Cryogenic Engineering Conference, Kyoto, Japan, May 1974 (IPC Science & Technology Press) 97-8

OGIWARA, H. and TAKANO, N.: 'Development of superconducting magnets for a magnetically suspended high-speed train in Toshiba', ibid., 94-6

OHNO, E., IWAMOTO, M., OGINO, O. and KAWAMURA, T.: 'Development of superconducting magnets for magnetically levitated trains', ibid., 90-3

PRAST, G.: 'On-board refrigeration for high-speed trains', ibid., 35-6

RHODES, R.G., MULHALL, B.E., HOWELL, J.P. and ABEL, E.: 'The Wolfson Maglev project', IEEE Trans., 1974, MAG-10, No.3, 398-401

ROUBICEK, O.: 'A controllable low-frequency oscillator linear electric drive system', Elektrichestvo, 1974, No.10, 86-8. In Russian

SAITO, Y., TAKANO, I., MATSUDA, S. and OGIWARA, H.: 'Experimental studies on superconducting magnetic levitation for ultrahigh-speed vehicles', Elec. Eng. Jap. (USA), 1974, 94, No.2, 92-105

SHIRASHOJI, A., MUTO, T., OTA, S. and WORKS, I.: 'A control system for magnetically levitated vehicles', Mitsubishi Electr. Eng. (Japan), 1974, No.42, 6-9

SLEMON, G.R., TURTON, R.A. and BURKE, P.E.: 'A linear synchronous motor for high-speed ground transport', IEEE Trans., 1974, MAG-10, No.3, 435-8

TAKAHASHI, T., MAKI, N. and MIYASHITA, T.: 'Combined system for propulsion and guidance of magnetically suspended vehicles', 5th International Cryogenic Engineering Conference, Kyoto, Japan, May 1974 (IPC Science & Technology Press), 78-81

TANG, C.H. and CHU, R.S.: 'Magneplane magnetic levitation study evaluation of guideway edge effects', Final Report No. NSF-RANN under Contract NSF-C670 to M.I.T., Dec. 1974

THORNTON, R.D.: 'The magneplane linear synchronous motor', Digests of IEEE Intermag Conference, Toronto, 14-17 May 1974

THORNTON, R.D.: 'The magneplane linear synchronous motor propulsion system', IEE Conference on Linear Electric Machines, London, 21-23 Oct. 1974, 230-5

THORNTON, R.D., IWASA, Y. and KOLM, H.H.: 'The magneplane system', 5th International Cryogenic Engineering Conference, Kyoto, Japan, May 1974 (IPC Science & Technology Press) 42-5

USAMI, Y., FUJIE, J. and FUJIWARA, S.: 'Studies on linear motor in the Institute of JNR', IEE Conference on Linear Electric Machines, London, 21-23 Oct. 1974, 131-6

VORONKOV, V.S., POZDEEV, O.D. and SANDALOV, V.M.: 'Dynamics of magnetic suspension', <u>Izv. VUZ Elektromekh</u>. (USSR), 1974, No.10, 1082-9. In Russian

WIART, A.: 'Aerotrains: propulsion systems with linear motors and electronic convertors', <u>Rev. Jeumont-Schneider</u>, 1974, No.17, 23-32. In French

YAMADA, T., IWAMOTO, M. and ITO, T.: 'Magnetic damping force in inductive magnetic levitation system for high speed trains', <u>Elec. Eng. Jap</u>. (USA), 1974, 94, No.1, 80-4

YASUMOCHI, R. and MORI, H.: 'Electromagnetic forces of the linear synchronous motor for magnetically levitated vehicle', International Symposium on Linear Electric Motors, Lyon, France, May 1974

1975   APPUN, P. and von THUN, H.J.: 'An electro-magnetic suspension and guidance system for tracked, high-speed transportation', <u>Elek. Bahnen</u> (Germany), 1975, 46, No.4, 86-94. In German

ATHERTON, D.L. and EASTHAM, A.R.: 'Superconducting Maglev and LSM development in Canada', <u>IEEE Trans</u>., 1975, MAG-11, No.2, 627-32

BERTLING, A.: 'Switch arrangement for a magnetic suspension railroad', Siemens AG, Patent USA 3869990, 6 Sept. 1973, publ. 11 March 1975; prior. 29 Sept. 1972, Germany 2247858

BOBINEAU, B. and TEYSSANDIER, C.: 'A test station for high-speed motors', Rev. Gen. Elec., 1975, 84, No.2, 137-40. In French

BOLDEA, I.: 'On the performance of linear electric motors for the propulsion of fast vehicles (300-500 km/h)', Electroteh. Electron. & Autom. Electroteh. (Rumania), 1975, 23, No.3, 103-13. In Rumanian

BORCHERTS, R.H.: 'Comparison of two HSGT magnetic suspension systems (attraction)', Ford Motor Company, Technical Report No. FRA-ORD-73-27A for US Dept of Transportation under Contract No. DOT-FR-10026, Feb. 1975

BORCHERTS, R.H.: 'Repulsion magnetic suspension research - U.S. progress to date', Cryogenics, 1975, 15, No.7, 385-93

BROWN, W.S.: 'The effect of long magnets on inductive maglev ride quality', IEEE Trans., 1975, MAG-11, No.5, 1498-500

BUCHBERGER, H. and LEITGEB, W.: 'Traction drives with linear synchronous motors', Elek. Bahnen (Germany), 1975, 46, No.4, 82-5. In German

BURKE, P.E.: 'The use of stranded conductors to reduce eddy losses in guideway conductors of high speed vehicles', IEEE Trans., 1975, MAG-11, No.5, 1501-3

COHO, O.C., KLIMAN, G.B. and ROBINSON, J.I.: 'Experimental evaluation of a high speed double sided linear induction motor', ibid., PAS-94, No.1, 10-7

'Conceptual design and analysis of the tracked magnetically levitated vehicle technology program (TMLV) - repulsion scheme; Volume I - Technical Studies', Ford Motor Co., Dearborn, Mich., Report No. FRA-OR&D-75-21 for US Dept of Transportation under Contract No. DOT-FR-40024, Feb. 1975

'Control system for magnetically levitated vehicle', Messerschmitt-Bölkow-Blohm, Patent UK 1404987, 11 Oct. 1972; publ. 3 Sept. 1975; prior. 16 Oct. 1971, West Germany 151608

'Cooling system for rail-rotor of a linear motor', Conz Elektricit, Patent UK 1401592, 16 Jan. 1973; publ. 30 July 1975; prior. 19 Jan. 1972, West Germany 202426

EASTHAM, A.R. (Editor): 'Superconducting magnetic levitation and linear synchronous motor propulsion for high speed ground transportation', Canadian Institute of Guided Ground Transport, Queen's University, Kingston, Report CIGGT 75-5, March 1975

GOTTZEIN, E., LANGE, B. and OSSENBERG-FRANZES, F.: 'Control system concept for a passenger carrying maglev vehicle', International Conference on High Speed Ground Transportation, Arizona State Univ., Jan. 1975 (publ. High Speed Ground Transportation Journal) 435-45

'High-speed railway conductor rail', Vahle Paul KG, Patent UK 1405039, 1 Aug. 1973; publ. 3 Sept. 1975; prior. 11 Nov. 1972, West Germany 244492

HOGAN, J.R. and FINK, H.J.: 'Comparison and optimization on lift and drag forces on vehicles levitated by eddy current repulsion for various null and normal flux magnets with one or two tracks', IEEE Trans., 1975, MAG-11, No.2, 604-7

HOWELL, J.P., WONG, J.Y., RHODES, R.G. and MULHALL, B.E.: 'Stability of magnetically levitated vehicles over a split guideway', ibid., No.5, 1487-9

IWAHANA, T.: 'Study of superconducting magnetic suspension and guidance characteristics on loop tracks', ibid., No.6, 1704-11

IWASA, Y., BROWN, W.S. and WALLACE, C.B.: 'An operational 1/25-scale Magneplane system with superconducting coils', ibid., No.5, 1490-2

KIMURA, H., OGATA, H., SATO, S., SAITO, R. and TADA, N.: 'Superconducting magnet with tube-type cryostat for magnetically suspended train', ibid., No.2, 619-22

KING, S.Y.: 'Inter urban rapid transit systems', Elect. Engr. (Australia), 1975, 52, No.6, 19

KOLM, H.H., THORNTON, R.D., IWASA, Y. and BROWN, W.S.: 'The magneplane system', Cryogenics, 1975, 15, No.7, 377-84

KYOTANI, Y.: 'Development of superconducting levitated trains in Japan', ibid., 372-6

LAITHWAITE, E.R.: 'Electromagnetic rivers', Polytech. Tijdschr. Elektrotech. Elektron. (Netherlands), 1975, 30, No.7, 226-33. In Dutch

LAMB, C. St.J.: 'A rapid transit system for inner-city transportation', Elect. Engr. (Australia), 1975, 52, No.2, 11-3

LEE, S.W. and MENENDEZ, R.: 'Forces at low- and high-speed limits in magnetic levitation systems', J. Appl. Phys. (USA), 1975, 46, No.1, 422-5

LEVI, E.: 'A preliminary evaluation of electrical propulsion by means of iron-cored synchronously operating linear motors', Report for US Dept of Transportation under Contract DOT-FR-30030, Jan. 1975

LICHTENBERG, A.: 'Electromagnetic track guidance arrangement for a vehicle', Siemens AG, Patent USA 3861320, 5 April 1974; publ. 21 Jan. 1975; prior. 13 April 1973, Germany 2318756

'Linear induction motor', Zavod Elektrotransporta, Patent UK 1382904, 22 Feb. 1972; publ. 5 Feb. 1975; prior. 22 Feb. 1972, UK 8020

'Linear motor driven and magnetically levitated railway carriage', Augsburg Nurnberg AG, Patent UK 1404975, 26 Oct. 1972; publ. 3 Sept. 1975; prior. 29 Oct. 1971, West Germany 153928

MACHEFERT-TASSIN, Y.: 'Concerning a few myths linked to the use of linear motors at high speed', Rev. Gen. Elec., 1975, 84, No.2, 91-101. In French

'Magnetic vehicle drive/suspension system', Rohr Ind. Inc., Patent UK 1385212, 24 March 1972; publ. 26 Feb. 1975; prior. 19 April 1971, US 131041

MAKI, N.: 'Linear motor for high speed railroads', Hitachi Ltd, Patent USA 3858522, 11 June 1973; publ. 7 Jan. 1975; prior. 15 June 1972, Japan 47-59689

MANDT, K.: 'Measurement and guidance techniques in the Erlangen test track for magnetic floating trains', Elek. Bahnen (Germany), 1975, 46, No.4, 94-8. In German

MELVILLE, P.H., MULHALL, B.E. and WILSON, M.N.: 'Propulsion of magnetically levitated vehicles', Cryogenics, 1975, 15, No.5, 295

MUHLENBERG, J.D. and NENE, V.D.: 'Analysis of a combined attraction-maglev-propulsion system for a high speed vehicle', Technical Report No. FRA-ORD-75-61 for US Dept of Transportation, 1975

MURATO, K., TAKEI, K., AKIHAMA, Y., TAKAHASHI, H. and AKABANE, H.: 'Type L4 linear motor system wagon booster/retarder for Shiohama shunting yard of the J.N.R. (Japanese National Railways)', Hitachi Rev. (Japan), 1975, 24, No.6, 269-76

NAVE, P.M.: 'Maglev test facilities at Messerschmitt-Bölkow-Blohm', International Conference on High Speed Ground Transportation, Arizona State Univ., Jan. 1975 (publ. High Speed Ground Transportation Journal), 255-60

OGIWARA, H., TAKANO, N., OKAMATO, H. and HAYASHI, K.: 'Development of superconducting magnets for magnetically suspended high-speed trains', Toshiba Rev. (Internat. Ed.), July-Aug. 1975, No.98, 7-11

OHTSUKA, T. and KYOTANI, Y.: 'Superconducting levitated high speed ground transportation project in Japan', IEEE Trans., 1975, MAG-11, No.2, 608-14

OOI, B.-T.: 'Levitation, drag and transverse forces in finite width sheet guideways for repulsive magnetic suspension', International Conference on High Speed Ground Transportation, Arizona State Univ., Jan. 1975, (publ. High Speed Ground Transportation Journal), 369-73

OOI, B.-T.: 'Transverse force in magnetic levitation with finite width sheet guideways', IEEE Trans., 1975, PAS-94, No.3, 994-1002

'Polyphase linear induction motor', Tracked Hovercraft Ltd, Patent UK 1394998, 25 Aug. 1972; publ. 21 May 1975; prior. 10 Sept. 1971, UK 42378

'Regulator system for hover railway vehicle', Messerschmitt-Bölkow-Blohm, Patent UK 1401313, 29 Aug. 1972; publ. 30 July 1975; prior. 17 Sept. 1971, West Germany 146499

REITZ, J.R. and BORCHERTS, R.H.: 'US Department of Transportation program in magnetic suspension (repulsion concept)', IEEE Trans., 1975, MAG-11, No.2, 615-8

ROSS, J.A. and HARRIS, B.C.: 'Railway truck magnetic suspension', Rohr Ind. Inc., Patent USA 3886871, 8 May 1972; publ. 3 June 1975, US 251118

SEN, P.C.: 'On linear synchronous motor (LSM) for high speed propulsion', IEEE Trans., 1975, MAG-11, No.5, 1484-6

'Short stator induction motor circulating current reduction', Tracked Hovercraft Ltd, Patent UK 1379676, 7 June 1973; publ. 8 June 1975

SLEMON, G.R.: 'The Canadian Maglev project on high-speed interurban transportation', IEEE Trans., 1975, MAG-11, No.5, 1478-83

SLEMON, G.R.: 'The Canadian Maglev project on high-speed interurban transportation', Canadian Electric Power Symposium, Poland, 1975

STOSCHEK, J.: '5-MVA frequency converter for linear motor propulsion', BBC Nachr. (Germany), 1975, 57, No.4, 192-7. In German

TANG, C.H., HARROLD, W.J. and CHU, R.S.: 'A review of the Magneplane project', IEEE Trans., 1975, MAG-11, No.2, 623-6

THORNTON, R.D.: 'Magnetic levitation and propulsion, 1975', ibid., No.4, 981-95 (124 refs)

'Transport system with electromagnetically-suspended vehicle', Augsburg Nurnberg AG, Patent UK 1381241, 15 Jan. 1972; publ. 22 Jan. 1975; prior. 15 Jan. 1972, West Germany 201923

UHER, R.A.: 'Electric power systems and high-speed ground transportation', IEEE Trans., 1975, IA-11, No.1, 100-13

URANKAR, L.: 'Basic magnetic levitation systems with a continuous sheet track', <u>Siemens Forsch. und Entwicklungsber</u>. (Germany), 1975, 4, No.1, 25-32. In English

WEH, H.: 'Synchronous long-stator drive with controlled normal attractive forces', <u>Elektrotech. Z. (ETZ) A</u>, 1975, 96, No.9, 409-13. In German

WIART, A.: 'Propelling set with linear motor and electronic convertor', <u>Rev. Gen. Elec</u>., 1975, 84, No.2, 112-20. In French

WIEGNER, G.: 'The drive of the experimental vehicle for the evaluation of the electro-dynamic suspension technique - an example of linear motor drive', <u>Elek. Bahnen</u> (Germany), 1975, 46, No.5, 118-24. In German

YAMAMURA, S. and ITO, T.: 'Analysis of speed characteristics of attracting magnet for magnetic levitation of vehicles', <u>IEEE Trans</u>., 1975, MAG-11, No.5, 1504-7

THEORY

1946 SHTURMAN, G.I.: 'Induction machines with open magnetic circuits', <u>Elektrichestvo</u>, 1946, No.10, 43-50. In Russian

1947 SHTURMAN, G.I. and ARANOV, R.L.: 'Edge effect in induction motors with open magnetic field', <u>Elektrichestvo</u>, 1947, No.2, 54-9. In Russian

1950 REZIN, M.G.: 'Armature reaction effects and the mechanical characteristics of a motor with an arc-shaped stator', <u>Elektrichestvo</u>, 1950, No.2, 51-2. In Russian

1951 REZIN, M.G.: 'Some peculiarities of the electromagnetic phenomena in a motor with an arc-shaped stator', Elektrichestvo, 1951, No.6, 25-9. In Russian

SHTURMAN, G.I.: 'Open squirrel-cages in short-circuited induction motors', ibid., No.9, 36-43. In Russian

1956 MURGATROYD, W.: 'Theory of the ideal A.C. conduction pump', A.E.R.E. (Harwell), Report No. ED/R 1566, 1956

1958 CULLEN, A.L. and BARTON, T.H.: 'A simplified electromagnetic theory of the induction motor using the concept of wave impedance', Proc. IEE, 1958, 105C, No.8, 331-6

1959 SCHILDER, J.: 'Electromagnetic field moving in a conducting medium', Elektrotech. Obzor (Czechoslovakia), 1959, 48, No.3, 146-53. In Czech

SZULKIN, P.: 'Matrix form of the equations of propagation in an anisotropic medium with constant magnetic field', Bull. Acad. Polon. Sci. Ser. Sci. Tech. (Poland), 1959, 8, No.10, 581-7

WILLIAMS, F.C., LAITHWAITE, E.R. and EASTHAM, J.F.: 'Development and design of spherical induction motors', Proc. IEE, 1959, 106A, No.30, 471-84

1960 BIRZVALKS, Yu.: 'Analysis of optimal criteria being applied in calculations of pumps of constant current', Latv. PSR Zinat. Akad. Vestis (USSR), 1960, No.11 (160), 91-6. In Russian

BIRZVALKS, Yu.: 'Current distribution in the region of the longitudinal axis of the channel of a constant-current pump', ibid., No.2 (151), 63-72. In Russian

CARPENTER, C.J.: 'Surface-integral methods of calculating forces on magnetised iron parts', Proc. IEE, 1960, 107C, No.11, 19-28

LAITHWAITE, E.R.: 'Design of spherical motors', Elec. Times, 1960, 137, No.23, 921-5

STIER, E.: 'The determination of the energy of a plane magnetic field by considering the equivalent electric current field', Arch. Elektrotech. (Germany), 1960, 45, No.5, 343-6. In German

1961 CIAMPOLINI, F.: 'Introduzione allo studio dei motori a induzione lineari e sferici', Istuto di Elettrotenica del Universita di Bologna, Tipografia Gamma, 1961

SHARIKADZE, D.V.: 'Motion of a medium with finite conductivity in the presence of a uniform magnetic field', Dokl. Akad. Nauk SSSR, 1961, 138, No.4, 817-9. English trans. Soviet Physics, 1961, 6, No.6, 470-1

1962 DAMBIER, Th.: 'Zugkräfte eines Linearmotors im Stillstand', Archiv für Eisenbahntechnik, 1962, No.14, 2-10

HARRIS, M.R.: 'A modified Poynting vector for local energy loss in a ferromagnetic lamination', J. Sci. Instrum., 1962, 39, No.8, 441-2

RICHMOND, J.H.: 'Transmission through inhomogeneous plane layers', IRE Trans. Antennas & Propagation (USA), 1962, AP-10, No.3, 300-5

VATAZHIN, A.B. and REGIRER, S.A.: 'Approximate calculation of the current distribution when a conducting fluid flows along a channel within a magnetic field', Priklad. Mat. Mekh. (USSR), 1962, 26, No.3, 548-56. English trans. Appl. Math. Mech. (GB), 1962, 26, No.3, 816-28

WEST, J.C. and HESMONDHALGH, D.E.: 'The analysis of thick-cylinder induction machines', Proc. IEE, 1962, 109C, 172-81

1963 CHEKMAREV, I.B.: 'The influence of anisotropy of electrical conductivity on the development of flow of a viscous fluid in the initial section of a plane channel', Priklad. Mat. Mekh. (USSR), 1963, 27, No.3, 532. English trans. Appl. Math. Mech. (GB), 1963, 27, No.3, 789-94

KALISKI, S.: 'Attenuation of surface waves between perfectly conducting fluid and solid in a magnetic field normal to the contact surface', Proc. Vibration Problems (Poland), 1963, 4, No.4, 375-85

KANER, E.A.: 'Anomalous penetration of metals by an electromagnetic field', Zh. Eksper. Teor. Fiz. (USSR), 1963, 44, No.3, 1036-49. (English trans. in Soviet Physics - JETP (USA))

SHERCLIFF, J.A.: The theory of electromagnetic flow measurement (Cambridge University Press) 1963

SUDAN, R.N.: 'Interaction of a conducting fluid stream with a travelling wave of magnetic field of finite extension', J. Appl. Phys. (USA), 1963, 34, No.3, 641-50

1964   BURMAN, R. and GOULD, R.N.: 'On certain exact wave functions for electromagnetic fields in planar stratified media', J. Atmos. Terrest. Phys., 1964, 26, No.11, 1127-9

OKHREMENKO, N.M.: 'Magnetic field of a plane induction pump', Elektrichestvo, 1964, No.8, 18-26. In Russian

RATCLIFF, G. and GRIFFITHS, J.: 'A linear D.C. motor', J. Sci. Instrum., 1964, 41, No.5, 267-8

SUCHY, K.: 'Coupled wave equations for inhomogeneous anisotropic media', Z. Naturforsch, 1964, 9a, No.7-8, 630-6. In German

TAI, C.T.: 'A study of electrodynamics of moving media', Proc. IEEE, 1964, 52, No.6, 685-9

TIPPING, D.: 'The analysis of some special-purpose electrical machines', Ph.D. Thesis, University of Manchester, 1964

1965   BARLOW, H.E.M.: 'Travelling-field theory of induction-type instruments and motors', Proc. IEE, 1965, 112, No.6, 1208-14

BERTINOV, A.I. et al: 'Some problems of 3-phase induction pump calculation', Magn. Gidrodin. (USSR), 1965, No.3, 103-10. In Russian

BROWN, P.M. and TAI, C.T.: 'A study of electrodynamics of moving media', Proc. IEEE, 1965, 52, No.11, 1362-3

HESMONDHALGH, D.E. and TIPPING, D.: 'General method for prediction of the characteristics of induction motors with discontinuous exciting windings', Proc. IEE, 1965, 112, No.9, 1721-35

LAITHWAITE, E.R.: 'Differences between series and parallel connection in machines with asymmetric magnetic circuits', ibid., No.11, 2074-82

LAITHWAITE, E.R.: 'The goodness of a machine', ibid., No.3, 538-41

LAITHWAITE, E.R.: 'The goodness of a machine', Electronics & Power, March 1965, 11, 101-3

MARINESCU, M.: 'On the production of alternating (attractive and repulsive) electromagnetic forces. New means for linear motor development', Rev. Roum. Sci. Tech. Electrotech. Energet, 1965, 10, No.3, 543-52. In French

OKHREMENKO, N.M.: 'Transverse fringe effect in flat linear induction pumps', Magn. Gidrodin. (USSR), 1965, No.3, 87-95. In Russian

POSTNIKOV, J.M. et al: 'Calculating an electromagnetic travelling field in a laminar three-layer conducting medium', Elektrichestvo, 1965, No.9, 1-7. In Russian

REGIRER, S.A.: 'Laminar flow of conducting fluid pipes and channels in the presence of transverse magnetic field', Magn. Gidrodin. (USSR), 1965, No.1, 5-17. In Russian

SATO, G.: 'Propagation of the electromagnetic waves in layered inhomogeneous media', J. Phys. Soc. Japan, 1965, 20, No.8, 1463-75

SMITH, W.E.: 'An electromagnetic force theorem for quasi-stationary currents', Austral. J. Physics, 1965, 18, No.3, 195-204

VALDMANIS, Ya. Ya.: 'Electrodynamic forces acting upon a conducting layer in a one-sided 3-phase inductor field', Latv. PSR Zinat. Akad. Vestis Fiz. Tehn. Ser. (USSR), 1965, No.1, 51-4. In Russian

VALDMANIS, Ya. Ya., KUNIN, P.F., MIKEL'SON, Ya. Ya. and TAKSAR, I.M.: 'Conducting strip in the travelling electromagnetic field of a plane inductor', Magnetohydrodynamics (USA), 1965, 1, No.2, 75-82

VESKE, T.A.: 'Solution of the electromagnetic field equations for a plane linear induction machine with secondary boundary effects', ibid., No.1, 64-71

VEZE, A.K. and KRUMIN, Yu. K.: 'Electromagnetic force acting on an infinitely wide conducting sheet in the travelling magnetic field of plane inductors', ibid., No.4, 62-8

VOL'DEK, A.I.: 'Compensation of pulsating magnetic fields in induction machines and pumps with an open magnetic circuit', Elektrichestvo, 1965, No.4, 50-3. In Russian

YANTOVSKII, E.I.: 'Determination of magnetic Reynolds number', Magn. Gidrodin. (USSR), 1965, No.4, 153-4. In Russian

1966 BETH, R.A.: 'Complex representation and computation of two-dimensional magnetic fields', J. Appl. Phys. (USA), 1966, 37, No.7, 2568-71

BRUNELLI, J.B.: 'Study of an induction motor with an arc shaped stator', L'Energia Elettrica, 1966, 43, No.10, 605-17. In Italian

ERASLAN, A.H.: 'Duct flow of conducting fluids under arbitrary oriented applied magnetic fields', AIAA J. (USA), 1966, 4, No.4, 620-6

ERMILOV, M.A. et al: 'Conductivity of a cylindrical PM-EM induction pump rotor', Magn. Gidrodin. (USSR), 1966, No.4, 121-5. In Russian

GUREVICH, B.Ya. et al: 'Flow of a plane sheet of electrically conducting liquid in a transverse magnetic field', ibid., 69-77. In Russian

KRUMINS, Yu.K.: 'Influence of conducting body with finite length on ponderomotive forces due to travelling magnetic field', ibid., No.3, 115-24. In Russian

OOI, B.-T.: 'Fringe end effects of short stator', S.M. Thesis, M.I.T., Jan. 1966

PARTS, I.R.: 'Current distribution in the liquid metal of plane induction pumps with shorting strips', Magn. Gidrodin. (USSR), 1966, No.4, 108-12. In Russian

PELENC, Y., PILLET, E., POLOUJADOFF, M., RÉMY, E. and REYX, P.: 'Influence of air-gap length on the performance of polyphase linear induction motors', Rev. Gen. Elec., 1966, 75, No.11, 1300-5. In French

BIBLIOGRAPHY

POLOUJADOFF, M.: 'Calcul des caractéristiques des moteurs asynchrones polyphases à entrefer plan', ibid., No.6, 859-63

POLOUJADOFF, M.: 'Improvement in the theory of linear induction motors for traction', C.R. Acad. Sci. (Paris), 1966, 263, Ser.B, No.9, 605-7. In French

POLOUJADOFF, M. and REYX, P.: 'Étude theorique et experimentalle de l'influence des sections de retour sur le fonctionnement de moteurs d'induction linéaires à inducteur court', ibid., No.14, 799-802

RASHCHEPKIN, A.P.: 'Longitudinal end effects of linear induction machines', Magn. Gidrodin. (USSR), 1966, No.3, 106-10. In Russian

VALDMANIS, J.J. et al: 'On the theory of longitudinal edge effect in linear MHD machines', ibid., 101-5. In Russian

VESE, A.K.: 'On the determination of electromagnetic forces in finite width conducting sheet in the presence of travelling magnetic field of one stator', ibid., 111-4. In Russian

VIGLIN, A.S.: 'Electrodynamics of an homogeneous anisotropic and dispersive medium', Zh. Eksper. Teor. Fiz. (USSR), 1966, 50, No.1, 85-92. English trans. Soviet Physics - JETP (USA), 1966, 23, No.1

VILNITIS, A. Va.: 'Transverse edge effect in flat induction magnetodynamic machines' in *Movement of conducting bodies in magnetic fields*, Latvian Academy of Sciences Institute of Physics, Monograph, 1966, 63-94

1967 GREIG, J. and FREEMAN, E.M.: 'Travelling-wave problem in electrical machines', *Proc. IEE*, 1967, 114, No.11, 1681-3

HARRIS, M.R.: 'Analysis of power factor in linear induction generators', *ibid.*, No.8, 1179. (Correspondence)

KRUMINS, J.K.: 'On the maximum ponderomotive force at the in-phase connection of two flat stators', *Magn. Gidrodin.* (USSR), 1967, No.2, 154-5. In Russian

LAITHWAITE, E.R.: 'Magnetic equivalent circuits for electrical machines', *Proc. IEE*, 1967, 114, No.11, 1805-9

LOPUKHINA, E.M. et al: 'Comparing asynchronous actuator motors with hollow non-magnetic and with squirrel-cage rotors', *Elektrotekhnika* (USSR), 1967, No.6, 21-3. (English trans. in *Soviet Elect. Engng.* (USA))

NARAYANAN, M.S.T.: 'Design, development and construction of a linear oscillating electric machine', *J. Indian Inst. Sci.*, 1967, 49, No.1, 28-35

POIRIER, Y.: 'Dragging motion of a liquid by plane walls in the presence of an induction magnetic field and at a low electromagnetic Reynolds number. Theoretical aspects', C.R. Acad. Sci. (Paris), 1967, 265, No.25, 889-91. In French

SAAL, C.: 'Study of an oscillating biphase synchronous linear motor', Bull. Inst. Politeh. Iasi (Rumania), 1967, 13, No.1-2, 345-62. In English

SAVY, C.: 'On the theory of linear induction motors', Elettrotecnica, 1967, 42, No.2, A.1.9 (Electrical Rotating Machines, 68th Annual Meeting of Italian Electrotechnical and Electrical Society, San Remo). In Italian

VALDMANIS, Ya. Ya. and LIELPETER, Y.Y.: 'The structure of a magnetic field in the airgap of a linear MHD machine', Magn. Gidrodin. (USSR), 1967, No.1, 115-21. In Russian

YANTOVSKY, Ye. I.: 'On the determination of magnetic Reynolds number', ibid., No.3, 154-5. In Russian

1968 BRADFORD, M.: 'Unbalanced magnetic pull in a 6-pole, 10 kW, induction motor with a series-connected stator winding', Electrical Research Assoc., Leatherhead, Surrey, Report No.5216, 1968

BYOLER, G.A. and LUKOVSKAYA, T.A.: 'Eddy currents in a metallic plate, moving in a constant magnetic field', Izv. VUZ Fiz. (USSR), 1968, No.2, 151-3. (English trans. in Soviet Phys. J. (USA))

FREEMAN, E.M.: 'Travelling waves in induction machines: input impedance and equivalent circuits', Proc. IEE, 1968, 115, No.12, 1772-6

IVANOV-SMOLENSKII, A.V. et al: 'A method for calculating currents and losses in conducting layers located in travelling or rotating magnetic fields, Elektrichestvo, 1968, No.8, 49-54. In Russian

KANT, M.: 'General equations of the sliding magnetic field of a magnetohydrodynamics machine', C.R. Acad. Sci. (Paris), 1968, 266, Ser.A, No.3, 164-7. In French

KANT, M.: 'The generation of a travelling magnetic field in a linear machine of finite dimensions', ibid., Ser.B, No.24, 1455-8. In French

KANT, M. and BONNEFILLE, R.: 'Determination of an ac liquid metal MHD generator' in Electricity from MHD, Vol.3 (Vienna: International Atomic Energy Agency) 1968, 1717-29

KANT, M. and EFTHYMIATOS, D.: 'Contribution to the study of optimized efficiency of a linear induction machine', C.R. Acad. Sci. (Paris), 1968, 267, Ser.B, No.25, 1328-31. In French

KESAVAMURTHY, N. et al: 'A general method of solving the field equations inside ferromagnetic sheets subjected to alternating magnetization', Proc. Royal Soc., 1968, 303A, 103-25

LAITHWAITE, E.R.: 'Shaded pole techniques applied to unusual electric motors', Electronics & Power, May 1968, 14, 206-8

BIBLIOGRAPHY

LAITHWAITE, E.R.: 'Some aspects of electrical machines with open magnetic circuits', Proc. IEE, 1968, 115, No.9, 1275-83

'Longitudinal end effect of a linear induction motor', TRW Systems Group, Report No. 06816-6030-R000, Oct. 1968

MATSUMIYA, T. and TAKAGI, K.: 'Magnetic field of linear induction machines', Elec. Eng. Jap. (USA), 1968, 88, No.4, 74-82

MEYEROVICH, E.A. et al: 'Using the duality principle for modelling a three-dimensional field', Elektrichestvo, 1968, No.8, 55-60. In Russian

MOFFATT, H.K.: 'On the suppression of turbulence by a uniform magnetic field', J. Fluid Mech., 1968, 28, Pt.3, 571-92

OHNO, E. and KISHIMOTO, T.: 'Basic characteristics of linear motors', Mitsubishi Denki Giho, Mitsubishi Electric Corporation, Marunouti, Tokyo, 1968, 42, No.12, 1581-8

POLOUJADOFF, M. and SABONNADIÈRE, J.-C.: 'An equivalent circuit for linear induction machines and M.H.D. induction generators', C.R. Acad. Sci. (Paris), 1968, 267, Ser.B, 1412-15. In French

POLOUJADOFF, M. and SABONNADIÈRE, J.-C.: 'Determination of lines of current in the secondary of a short primary linear induction motor with a balanced current supply', ibid., 266, Ser.B, 272-5. In French

POLOUJADOFF, M. and SABONNADIÈRE, J.-C.: 'Utilisation d'une méthode de partition dans la résolution de certaines équations aux derivées partielles dont le domaine comporte une bande infinie', ibid., 230-3

VOL'DEK, A.I. and LAZARENKO, L.F.: 'The effects of a finite length on linear induction machines', Proc. 6th Symposium on Magnetohydrodynamics, Institute of Physics, USSR Academy of Sciences, Riga, 1968, 2, 54. In Russian

WOODSON, H.H. and MELCHER, J.R.: Electromechanical dynamics, Part I: Discrete systems, Part II: Fields, forces and motion (New York: Wiley) 1968

1969 BOLTON, H.: 'Transverse edge effect in sheet-rotor induction motors', Proc. IEE, 1969, 116, No.5, 725-31

BRUNELLI, B.: 'Calculation of corrective coefficients for linear induction motors', L'Energia Elettrica, 1969, 46, No.1, 16-22. In Italian

MATSUMURA, Y.: 'Analysis of flux distribution and end effect in linear motor', Quart. Rept. Railway Technical Research Institute, JNR, 1969, 10, No.4, 253-4

MORARU, A.: 'Contributions to the theory of the linear electric motor: characteristic operating equations', Stud. Cercet Energ. Electrotehn. (Rumania), 1969, 19, No.2, 315-35. In Rumanian

NASAR, S.A.: 'Electromagnetic fields and forces in a linear induction motor, taking into account edge effects', Proc. IEE, 1969, 116, No.4, 605-8

POLOUJADOFF, M.: 'Influence of the form of the magnetic circuit on the edge effects in linear induction motors and on the braking by a magnet', C.R. Acad. Sci. (Paris), 1969, 269, Ser.B, No.24, 1215-8. In French

POLOUJADOFF, M.: 'Simplified theory of the linear induction motor', Elektrotech. Z. (ETZ) A, 1969, 90, No.21, 545-8. In German

PRESTON, T.W. and REECE, A.B.J.: 'Transverse edge effects in linear induction motors', Proc. IEE, 1969, 116, No.6, 973-9

RANADE, S.B.: 'Eddy currents and forces in thin sheets', M.Eng. Thesis, Nova Scotia Technical College, Halifax, Canada, 1969

SAAL, C.: 'Operational equations of the two phase reactive oscillomotor with linear movement', Bul. Inst. Politehn. Brasov (Rumania), 1969, 11, 49-62. In Rumanian

SABONNADIÈRE, J.-C.: 'Contribution to the study of induction motors', Doctoral Thesis, Grenoble, 27 Feb. 1969. In French

VALDMANIS, Ya.Ya., KALNIN, T.K., PETROVICHA, R.A. and POLMANIS, Ya.E.: 'The problem of an induction pump with a linearly varying magnetic field across the duct', Magnetodydrodynamics (USA), 1969, No.1, 86-91

YAMAMURA, S., ITO, H. and AHMED, F.I.: 'End-effect of induction-type linear motor', J. Fac. Engng. Univ. Tokyo A, 1969, No.7, 32-3. In Japanese

1970

ANDREI, R.: 'Determining the distribution of magnetic field in the air gap of a linear motor with inductor in short circuit', Electrotehnica (Rumania), 1970, 18, No.11, S417-25. In Rumanian

BOLTON, H.: 'Forces in induction motors with laterally asymmetric sheet secondaries', Proc. IEE, 1970, 117, No.12, 2241-8

BONNEFILLE, R. and KANT, M.: 'Application of electromagnetic field theory to linear induction machines', Rev. Phys. Appl. (France), 1970, 5, No.5, 743-57. In French

BONNEFILLE, R. and KANT, M.: 'Contribution to the study of the linear induction machine', Rev. Gen. Elec., 1970 79, No.10, 846-52. In French

de FLEURY, B.: 'Contribution to the study of end effects in the linear motor', Thèse de Maîtrise, Université de Laval, Quebec, June 1970

de FLEURY, B., POLOUJADOFF, M. and ROBERT, J.: 'A contribution to the study of end effect in linear induction machines and the calculation of braking due to a magnet', Symposium on Linear Induction Motors, Grenoble, 15 April 1970

DIMBOIU, E.: 'The study of the electromagnetic field in the air gap and the stator core of the linear induction motor', Electrotehnica (Rumania), 1970, 18, No.3, 77-83. In Rumanian

BIBLIOGRAPHY

DIMBOIU, E. and PESTEANU, O.: 'Present position of power calculation for travelling field linear motors', Wiss. Z. Tech. Hochsch. Karl-Marx-Stadt (Germany), 1970, No.3, 333-43. In German

KALNIN, T.K., PETRAVICHA, R.A. and PRIEDNICK, E.V.: 'The skin-effect in non-salient rotor induction pumps', Magn. Gidrodin. (USSR), 1970, No.1, 121-6. In Russian (English trans. in Magnetohydrodynamics (USA))

KANT, M. and BONNEFILLE, R.: 'Contribution to the study of the diffusion of travelling magnetic fields in a linear machine of many different zones', C.R. Acad. Sci. (Paris), 1970, 270, Ser.B, No.9, 592-5. In French

KANT, M., MOUILLET, A. and SCHEUER, J.-M.: 'Theoretical and experimental study of LIM windings', Symposium on Linear Induction Motors, Grenoble, 15 April 1970

MATSUMURA, Y.: 'Magnetic flux density distribution of the unloaded linear motor', Quart. Rept. Railway Technical Research Institute, JNR, 1970, 11, No.4, 223-8

NONAKA, S. and YOSHIDA, K.: 'Analysis of double-sided linear motors', Elec. Eng. Jap. (USA), 1970, 90, No.3, 21-31

NONAKA, S. and YOSHIDA, K.: 'Equivalent circuits for double-sided linear motors', ibid., 32-41

NORTH, G.G.: 'An analysis of a linear induction machine', Ph.D. Dissertation, University of California, Irvine, 1970

OOI, B.T. and WHITE, D.C.: 'Traction and normal forces in the linear induction motor', IEEE Trans., 1970, PAS-89, No.4, 638-45

PARTS, I.R. and PARTS, R.R.: 'Parameters of the equivalent circuit for cylindrical linear induction machines', Magn. Gidrodin. (USSR), 1970, No.3, 109-16. In Russian (English trans. in Magnetohydrodynamics (USA))

RADULET, R. and ILFRIM, A.: 'The calculus of the force acting on the thin narrow plate rotor of linear motors', Rev. Roum. Sci. Tech. Electrotech. Energet., 1970, 15, No.1, 3-15. In German

STIER, E.: 'Digital measurement and control of speed of a trailing car driven by linear motors', Elektrie, 1970, 24, No.10, 356-8. In German

TIMMEL, H.: 'Contribution to the determination of the stationary operation of linear short-stator motors', ibid., 341-3. In German

YAMADA, H.: 'Equivalent circuit of the induction motor with a double-layer solid rotor', Elec. Eng. Jap. (USA), 1970, 90, No.3, 92-101

YAMAMURA, S., ITO, H. and AHMED, F.J.: 'End effects of linear induction motor', ibid., No.2, 20-30

1971 ABRITSKA, M.Yu., BUGROV, N.S., MIKEL'SON, A.E. and CHEMODUROV, N.P.: 'The influence of channel shape on the parameters of a cylindrical pump', Magn. Gidrodin. (USSR), 1971, No.4, 99-104. In Russian (English trans. in Magnetohydrodynamics (USA))

BIBLIOGRAPHY

BRUNELLI, B.: 'Corrective coefficients and interference torques for an induction motor with a discontinuous stator', <u>Rev. Gen. Elec.</u>, 1971, 80, No.2, 90-4. In French

BURNHAM, D.C.: 'Asymptotic lift-to-drag ratios for magnetic suspension systems', <u>J. Appl. Phys.</u> (USA), 1971, 42, No.9, 3455-7

de FLEURY, B., POLOUJADOFF, M. and ROBERT, J.: 'The linear motor: contribution to the end effect in linear induction machines and to the calculation of magnet braking', <u>Rev. Gen. Elec.</u>, 1971, 80, No.2, 83-9. In French

DIMBOIU, E.: 'Contributions to the calculation of force for flat linear, travelling field motors', <u>Elektrie</u>, 1971, 26, No.6, 207-8. In German

FINK, H.J. and HOBRECHT, C.E.: 'Instability of vehicles levitated by eddy current repulsion - case of an infinitely long current loop', <u>J. Appl. Phys.</u> (USA), 1971, 42, No.9, 3446-50

FREEMAN, E.M. and LOWTHER, D.A.: 'Transverse edge effects in linear induction motors', <u>Proc. IEE</u>, 1971, 118, No.12, 1820-1

HAGEMANN, W.: 'Computerised calculation of the working performance of linear motors with short stators', <u>Elektrie</u>, 1971, 26, No.6, 209-10. In German

JUFER, M.: 'Determination of the specific characteristics of the linear motor', <u>Rev. Gen. Elec.</u>, 1971, 80, No.2, 105-13. In French

KANT, M., MOUILLET, A. and SCHEUER, J.-M.: 'Theoretical and experimental survey on the windings of linear induction motors', *ibid*., No.1, 13-9. In French

LIPKIS, R.S. and WANG, T.C.: 'Single-sided linear induction motor (SLIM). A study of thrust and lateral forces', TRW Systems Group, McLean, Va., Final Report No. 06818-W032-R0-00, FRA-RT-72-25 for US Dept of Transportation under Contract DOT-C-353-66, June 1971

MATSUMIYA, T. and TAKAGI, K.: 'End effect and equivalent circuit of linear induction machines', *Elec. Eng. Jap*. (USA), 1971, 91, No.1, 117-36

NONAKA, S. and YOSHIDA, K.: 'Characteristics of double-sided linear motors with secondary conductors vertically displaced from a symmetrical position', *Technol. Rep. Kyushu Univ*. (Japan), 1971, 44, No.6, 769-75. In Japanese

NONAKA, S. and YOSHIDA, K.: 'Characteristics of linear motors with sandwich compound conducting plates', *Elec. Eng. Jap*. (USA), 1971, 91, No.1, 183-93

ONUKI, T. and LAITHWAITE, E.R.: 'Optimised design of linear-induction-motor accelerators', *Proc. IEE*, 1971, 118, No.2, 349-55

POLOUJADOFF, M. and REYX, P.: 'Intermediate method for the analysis of a linear induction motor', *Rev. Gen. Elec*., 1971, 80, No.2, 99-104. In French

POLOUJADOFF, M., SABONNADIÈRE, J.-C., PELENC, Y. and REYX, P.: 'Design hypotheses for linear induction motors', *ibid*., No.1, 29-33. In French

SABONNADIÈRE, J.-C. and POLOUJADOFF, M.: 'Determination of the lines of current and characterization of the edge effect', ibid., 34-8. In French

VALDMANIS, Ya.Ya., VEZE, A.K. and ZUSMAN, I.M.: 'The force density distribution in an approximate model for an open core electromagnetic pump', Magn. Gidrodin. (USSR), 1971, No.4, 87-93. In Russian (English trans. in Magnetohydrodynamics (USA))

WANG, T.C., SMYLIE, J.W. and PEI, R.Y.: 'Single-sided linear induction motor', IEEE 6th Annual Meeting of Industry and General Applications Group, Cleveland, Ohio, USA, 18-21 Oct. 1971, 1-10

WEH, H., BRAESS, H. and MOSEBACH, H.: 'The analytical treatment of asynchronous linear machines', Energy Conversion, 1971, 11, 25-37. In German

WIART, A.: 'Separation of the variables in the study of linear or rotary motors with eddy currents', Rev. Jeumont-Schneider, 1971, No.12, 52-62. In French

WIART, A.: 'Separation of variables in the study of currents of linear or rotary motors with eddy currents', Rev. Gen. Elec., 1971, 80, No.1, 20-8. In French

YAMAMURA, S., ITO, H. and ISHIKAWA, Y.: 'Influence of end effect on characteristics of linear induction motors', Elec. Eng. Jap. (USA), 1971, 91, No.1, 136-47

1972

ALEKSANDROV, A.K. and BALTADZHIEV, G.M.: 'Magnetic field in the window of magnetic systems with rectilinear movement of the armature', Izv. Vmei 'Lenin' (Bulgaria), 1972, 30, No.2, 49-57. In Russian

ALLIN, G., CREIGHTON, G.K. and HALL, J.K.: 'Operation and analysis of an invertor-fed-linear-motor system', Proc. IEE, 1972, 119, No.11, 1587-94

BERZIN, Ya.Ya.: 'Effect of departure from airgap cylindricality on the balance of electromagnetic forces on the rotor of an electric motor', Izv. VUZ Priborostr. (USSR), 1972, 15, No.3, 53-8. In Russian

BLAZHKO, Yu.M. and OVCHARENKO, T.I.: 'An experimental investigation of linear systems with travelling magnetic fields', Elektrotekhnika (USSR), 1972, No.2, 32-3. In Russian (English trans. in Soviet Elect. Engng.)

BOLDEA, I.: 'Method of electromagnetic calculation of low velocity linear induction motors', Electrotehnica (Rumania), 1972, 20, No.1, 16-23. In Rumanian

BUGENIS, S., CESONIS, V., SMILGEVICIUS, A. and RUMMICH, E.: 'The effect of leakage flux in the coupling yoke of a linear induction motor', Elektrotechnik und Maschinenbau (Austria), 1972, 89, No.11, 458-60. In German

CHEMERIS, V.T. and PIS'MENNYI, A.S.: 'The electromagnetic field in the air gap of an asymmetrically fed linear induction machine', Probl. Tekh. Elektr. (USSR), 1972, No.38, 33-43. In Russian

DAL HO IM: 'Study on the travelling magnetic field in the linear induction motor with its end effect taken into consideration', J. Korean Inst. Elect. Engrs., 1972, 21, No.4, 7-14. In Korean

DIMBOIU, E.: 'An equivalent circuit of the asynchronous motor applied to the travelling field linear motor', Bul. Univ. Brasov A (Rumania), 1972, 14, 151-61. In Rumanian

DONCHEV, D., BOZHINOV, Ya. and BOZHINOVA, M.: 'Single inductor linear induction motor with ferromagnetic armature', Tekh. Misul (Bulgaria), 1972, 9, No.5, 93-6. In Bulgarian

D'YAKOV, V.I. and FROLOV, A.N.: 'Speed control of induction motors with an open stator by a thyristor voltage regulator', Izv. VUZ Elektromekh. (USSR), 1972, No.2, 198-201. In Russian

D'YAKOV, V.I., FROLOV, A.N. and D'YAKOVA, N.V.: 'Determination of the mechanical characteristics of linear induction motors from the equivalent circuit diagram', Izv. VUZ Energ. (USSR), 1972, No.7, 54-8. In Russian

ELLIOTT, D.G.: 'Numerical analysis method for linear induction machines', 12th Symposium on Engineering Aspects of Magnetohydrodynamics, Argonne, Illinois, 27-29 March 1972

HAYASHI, N.: 'Analysis of induction motors controlled by symmetrically triggered delta-connected thyristors, Elec. Eng. Jap. (USA), 1972, 92, No.5, 105-15

HUBNER, Kl.-D., MOSEBACH, H. and WEH, H.: 'A contribution to the calculation of the air gap in asynchronous linear motors', Elektrotech. Z. (ETZ) A, 1972, 93, No.11, 644-6. In German

IWAMOTO, M., YAMAMURA, S. and IWANTO, M.: 'End effect of high-speed linear induction motors', Elec. Eng. Jap. (USA), 1972, 92, No.3, 94-101

JUFER, M. and WAVRE, N.: 'The linear motor development theory and applications', Bull. Assoc. Suisse Electr. (Switzerland), 1972, 63, No.15, 844-56. In French

KAPCSOS, P. and TEVAN, G.: 'Application and calculation of linear eddy-current motors with ferromagnetic lining', Elektrotechnika (Hungary), 1972, 65, No.12, 458-65. In Hungarian

MOSEBACH, H.: 'The effects of finite length and width in linear induction motor, for both short primary and short secondary types', Ph.D. Dissertation, T.U. Braunschweig, 1972. In German

NITKA, S.: 'Dynamic processes of linear induction motors', Przeglad Elektrotech. (Poland), 1972, 48, No.4, 153-6. In Polish

NONAKA, S. and YOSHIDA, K.: 'Characteristics of double-sided linear motors with sandwich compound secondary conductors vertically displaced from a symmetrical position', Technol. Rep. Kyushu Univ. (Japan), 1972, 45, No.1, 72-9. In Japanese

NONAKA, S. and YOSHIDA, K.: 'Qualitative studies of vertical forces of double-sided linear induction motors', ibid., No.4, 503-5. In Japanese

NONAKA, S. and YOSHIDA, K.: 'Quantitative studies of vertical forces of double-sided linear induction motors', ibid., 506-13. In Japanese

OOKA, H.: 'Effect of wide air gap part in shaded pole motor', Hitachi Hyoron (Japan), 1972, 54, No.3, 203-6. In Japanese

PARTS, R.R.: 'Equivalent circuit of an induction motor with a large air gap between the stator and the rotor', Izv. VUZ. Elektromekh. (USSR), 1972, No.4, 385-7. In Russian

PELENC, Y.: 'Two slot linear induction motor', Merlin Gerin SA, Patent USA 3679952, 19 Jan. 1971; publ. 25 July 1972; prior. 25 Feb. 1970, France 7006858

POLOUJADOFF, M.: 'The circle diagram of linear motors with large air gap and very resistant rotor', C.R. Acad. Sci. (Paris), 1972, 275, Ser.B, No.22, 813-6. In French

RUMMICH, E.: 'A contribution on the calculation of suspension and driving properties of asynchronous linear motors', Elek. Bahnen (Germany), 1972, 43, No.11, 242-5. In German

STEPINA, J.: 'Theory of an induction motor with elliptical stator-bore', Elektrotech. Z. (ETZ) A, 1972, 93, No.4, 187-9. In German

SYUSYUKIN, A.I.: 'An approach to the calculation of an active type linear induction machine', Probl. Tekh. Elektr. (USSR), 1972, No.38, 88-93. In Russian

SYUSYUKIN, A.I.: 'Exciting field of a single phase linear inductor with varying magnetic field wave phase velocity', ibid., 43-7. In Russian

TIMMEL, H.: 'Contribution to the prediction of the stationary operating performance of short-stator linear motors', Wiss. Z. Tech. Hochsch. Karl-Marx-Stadt (Germany), 1972, No.6, 751-74. In German

TIMMEL, H.: 'Current density distribution in the moving plate of a travelling field linear motor', ibid., 775-88. In German

TIMMEL, H.: 'Velocity variation in travelling-field linear motors', Elektrie, 1972, 26, No.8, 228-31. In German

VAKLEV, I.I.: 'Allowing for the influence of the secondary system in a linear induction machine with an armature broader than the induction coil', Izv. Vmei 'Lenin' (Bulgaria), 1972, 29, No.2, 65-85. In Russian

VASIL'EV, S.V. and KOZLOV, A.I.: 'Electromagnetic effects in cylindrical linear induction pumps', Magn. Gidrodin. (USSR), 1972, No.4, 87-94. In Russian (English trans. in Magnetohydrodynamics (USA))

WEH, H., von GRUMBKOW, V. and MOSEBACH, H.: 'Kraftwirkungen orthogonal zur Bewegungsrichtung beim asynchronen Linearmotor', Elektrotech. Z.(ETZ) A, 1972, 93, No.1, 1-7

YAMAMURA, S.:   The theory of linear induction motors
(New York: John Wiley, Halsted Press) 1972

YAMAMURA, S., ITO, H. and ISHIKAWA, Y.:   'Theories of the linear induction motor and compensated linear induction motor', IEEE Trans., 1972, PAS-91, No.4, 1700-10

1973   AGARWAL, P.D. and WANG, T.C.:   'Evaluation of fixed and moving primary linear induction motor systems', Proc. IEEE, 1973, 61, No.5, 631-7

ALVES, M.F. and BURKE, P.E.:   'Single-sided linear induction motor with magnetic material in the secondary', IEEE 8th Annual Meeting of Industry Applications Society, Milwaukee, Wisconsin, USA, 8-11 Oct. 1973, 321-9

ANDREI, R.:   'Normal and tangential forces developing in high speed motors with reciprocating movement', Electrotehnica (Rumania), 1973, 21, No.2, 49-55. In Rumanian

BUDIG, P.-K., HAGEMANN, W. and TIMMEL, H.:   'Design of three-phase linear motors', Elektrie, 1973, 27, No.5, 253-6. In German

BUGYANIS, S.A. and CHESONIS, V.I.:   'End effects in three phase linear induction machines taking account of the shunting fluxes on the reverse side of the stators', Magn. Gidrodin. (USSR), 1973, No.3, 140-2.   In Russian (English trans. in Magnetohydrodynamics (USA))

CARRER, A.:   'Linear induction motor contribution to the study of the magnetic field in the gap', Atti Accad. Sci. Torino I (Italy), 1973, 107, No.6, 774-51. In Italian

CHESONIS, V.I.: 'The magnetic field structure and the number of poles in linear induction machines', *Magn. Gidrodin.* (USSR), 1973, No.3, 76-80. In Russian (English trans. in *Magnetohydrodynamics* (USA))

DUKOWICZ, J.: 'Analysis of linear induction machines with discrete windings and finite iron length', IEEE 8th Annual Meeting of Industry Applications Society, Milwaukee, Wisconsin, USA, 8-11 Oct. 1973, 311-9

FIENNES, J.: 'New approach to general theory of electrical machines using magnetic equivalent circuits', *Proc. IEE*, 1973, 120, No.1, 94-104

FOGGIA, A., GRELLET, G. and SABONNADIÈRE, J.-C.: 'Numerical and experimental study of the two-dimensional distribution of the inductance of linear electromagnetic device', *C.R. Acad. Sci.* (Paris), 1973, 276, Ser.B, No.14, 591-4. In French

HESMONDHALGH, D.E.: 'High-torque low-speed motor using magnetic attraction to produce rotation', *Proc. IEE*, 1973 120, No.1, 61-6

HOLLEY, H.J., NASAR, S.A. and del CID, L. Jr.: 'Computations of field and forces in a two-sided linear induction motor', *IEEE Trans.*, 1973, PAS-92, No.4, 1310-5

IWAMOTO, M., OHNO, E., ITOH, T. and SHINRYO, Y.: 'End effect of high-speed linear induction motor', *ibid.*, IA-9, No.6, 632-9

LAITHWAITE, E.R.: 'How an experiment with iron filings provided new motor design information', *Elec. Rev.*, 1973, 192, No.3, 93-5

BIBLIOGRAPHY

LAITHWAITE, E.R.: 'Magnetic or electromagnetic? The great divide', Electronics & Power, 1973, 19, No.14, 310-12

LANG, A.: 'The effect of stray fields on the design and performance of linear induction motors', Ph.D. Dissertation, T.U. Braunschweig, 1973. In German

NASAR, S.A. and del CID, L. Jr.: 'Certain approaches to the analysis of single-sided linear induction motors', Proc. IEE, 1973, 120, No.4, 477-83

NICOLAS, A. and SABONNADIÈRE, J.-C.: 'Étude des caractéristiques électromécaniques d'un moteur en tenant compte des effets de bord et des effets d'extrêmités', Symposium on Linear Motors, Capri, Italy, 19-21 June 1973

NONAKA, S. and YOSHIDA, K.: 'Characteristics of double-sided linear induction motors with secondary conductor vertically displaced from a symmetrical position', Elec. Eng. Jap. (USA), 1973, 93, No.6, 38-46

NONAKA, S. and YOSHIDA, K.: 'Space harmonic analysis of linear induction motors', ibid., No.2, 42-50

NONAKA, S. and YOSHIDA, K.: 'Studies of vertical forces of double-sided linear induction motors', JIEE (Japan), 1973, 93-B, No.10, 471-8

NORTH, G.G.: 'Harmonic analysis of a short stator linear induction machine using a transformation technique', IEEE Trans., 1973, PAS-92, No.5, 1733-43

OBERRETL, K.: 'Dreidimensionale Berechnung des Linearmotors mit Berücksichtigung des Endeffekte und der Wicklungsverteilung', Symposium on Linear Motors, Capri, Italy, 19-21 June 1973

OBERRETL, K.: 'Three-dimensional calculation of linear motor taking edge effects and winding distribution into account', Arch. Elektrotech. (Germany), 1973, 55, No.4, 181-90. In German

OOI, B.-T.: 'A generalized machine theory of the linear induction motor', IEEE Trans., 1973, PAS-92, No.4, 1252-9

PASCALE, D. and PUFLEA, I.: 'The study of linear motors', Stud. Cercet. Energ. Electrotehn. (Rumania), 1973, 23, No.1, 177-99. In Rumanian

POLOUJADOFF, M.: 'A study of the form of the field of force normal to the air-gap in linear induction motors', Rev. E. (Belgium), 1973, 7, No.5, 114-8. In French

SKOBELEV, V.E.: 'Limiting factors in operation in linear motors at superhigh speeds', Rail Int. (Belgium), 1973, 4, No.2, 269-76

TIMMEL, H.: 'Contribution to the transverse effect of short-stator linear motors', Elektrie, 1973, 27, No.5, 257-9. In German

TUROWSKI, J.: 'Methods of the calculation of field and secondary parameters of linear induction motors', Razpr. Electrotech. (Poland), 1973, 19, No.2, 371-96. In Polish

VOL'DEK, A.I., KARASEV, A.V. and KIENKO, A.I.: 'Engineering methods of calculating secondary medium parameters in the equivalent networks for flat linear induction machines with side bars', Magn. Gidrodin. (USSR), 1973, No.1, 99-104. In Russian (English trans. in Magnetohydrodynamics (USA))

VOL'DEK, A.I., MIKIRTICHEV, A.A., SOLDATENKOVA, N.A. and TOLVINSKAYA, E.V.: 'The influence of end effects on the operation of a linear induction machine without complex elements', ibid., No.2, 82-8. In Russian (English trans. in Magnetohydrodynamics (USA))

1974   ALDEN, R.T.H. and NOLAN, P.J.: 'Transfer-matrix analysis of linear induction machines with finite width and depth', Proc. IEE, 1974, 121, No.11, 1393-8

BOGDANOV, V.I.: 'Structural stability of linear electromechanical systems', Izv. VUZ Elektromekh. (USSR), 1974, No.12, 1291-4. In Russian

BOLDEA, I. and NASAR, S.A.: 'Simulation of high-speed linear-induction motor end effects in low-speed tests', Proc. IEE, 1974, 121, No.9, 961-4

BROUGH, J.J.: 'Anomalies in the design of direct current linear motors', Elec. Times, 20 June 1974, No.4286, 8-9

EDWARDS, J.D.: 'Scaling laws for electromagnets on magnetically levitated vehicles', Elec. Rev., 1974, 195, No.21, 745

FAWZI, T.H. and BURKE, P.E.: 'Edge effects in induction problems', IEEE Trans., 1974, MAG-10, No.3, 429-30

KLIMAN, G.B. and ELLIOTT, D.G.: 'Linear induction motor experiments in comparison with mesh/matrix analysis', IEEE Trans., 1974, PAS-93, No.5, 1624-33

LIN, S.-C.: 'A Fourier series analysis of linear induction machines', Thesis, University of Illinois, Chicago Circle, 1974 (available from Univ. Microfilms, Ann Arbor, Mich., Order No.75-1796)

LOWTHER, D.A. and FREEMAN, E.M.: 'Electromagnetic scale models of linear induction motors', IEE Conference on Linear Electric Machines, London, 21-23 Oct. 1974, 167-72

MIDDLEMISS, J.J.: 'Current pulsation of induction motor driving a reciprocating compressor', Proc. IEE, 1974, 121, No.11, 1399-403

NICOLAS, A. and SABONNADIÈRE, J.-C.: 'Computation of constant voltage operation characteristics of linear induction motors', IEE Conference on Linear Electric Machines, London, 21-23 Oct. 1974, 185-90

NONAKA, S. and YOSHIDA, K.: 'The characteristics of high-speed linear induction motors analysed using a space harmonic technique', ibid., 179-84

OBERRETL, K.: 'Single sided linear motor with cage in the secondary', Arch. Elektrotech. (Germany), 1974, 56, No.6, 305-19. In German

OLLENDORFF, F.: 'Relativistic electrodynamics of the linear induction machine', ibid., No.5, 278-83. In German

PESTEANU, O., DIMBOIU, E. and BIDIAN, D.: 'The effect of the longitudinal static end of an asynchronous linear motor', Stud. Cercet. Energ. Electrotehn. (Rumania), 1974, 24, No.1, 139-53. In Rumanian

RADHAKRISHNA, C. and RAO, B.C.: 'Energy transients in controlling the speed of linear induction motors', *J. Instn. Engrs. (India) Elect. Engng. Div.*, 1974, 55, Pt.EL1, 11-14

RYASHENTSEV, N.P., MALOV, A.T., UGAROV, G.G. and FEDONIN, V.N.: 'Research into electromagnetic motors of impact action with linear travelling of armature', IEE Conference on Linear Electric Machines, London, 21-23 Oct. 1974, 95-100

SCHNEIDER, J.: 'Calculation of eddy currents in linear motor geometries. A two-dimensional integral equation approach', *IEEE Trans.*, 1974, MAG-10, No.4, 1097-9

SKALSKI, C.A.: 'Application of a general analysis for single-sided linear induction motors', IEE Conference on Linear Electric Machines, London, 21-23 Oct. 1974, 197-202

SKOBELEV, V.E.: 'Influence of longitudinal fringe effect on the operation of high speed traction induction linear motors', *Rail Int.* (Belgium), 1974, 5, No.12, 767-81

SKOBELOW, V., SOLOWJEW, H. and EPIFASOW, A.: 'Edge effects in the end zones of linear railway motors and their influence on performance', *Wiss. Z. Tech. Univ. Dres.* (Germany), 1974, 23, No.2, 375-83. In German

SLEMON, G.R., TURTON, R.A., BURKE, P.E. and DEWAN, S.B.: 'Analysis and control of a linear synchronous motor for high-speed ground transport', IEE Conference on Linear Electric Machines, London, 21-23 Oct. 1974, 143-8

TAKIZAWA, M., OTSUKI, M. and SUZUKI, T.: 'Theoretical analysis of a passive magnetic suspension system with an eight-pole stator - study on a magnetic suspension system for floated inertial sensors (1st Report)', J. Jap. Soc. Precis. Eng. (Japan), 1974, 40, No.7, 557-63. In Japanese

VASILEV, I.N. and POPOVA-CHURANOVA NIPKIEP, G.B.: 'On calculating the parameters of three-phase linear induction motors', Elektro. Prom.-st & Priborostr. (Bulgaria), 1974, 9, No.5, 135-8. In Bulgarian

WEH, H., VOLLSTEDT, W. and MEINS, J.: 'Modell eines integrierten Trag- und Vortriebsaggregates auf elektromagnetischer Grundlage', Elektrotech. Z. (ETZ) A, 1974, 95, No.12, 684-5

YUN JONG LEE, DAL HO IM and SOO HYUN BAEK: 'A study on transverse edge effect in linear induction motor with sheet rotor', J. Korean Inst. Elect. Engrs., 1974, 23, No.4, 39-45. In Korean

1975 BOHN, G.H. and LANGERHOLC, J.: 'Theoretical and experimental investigation of eddy current effects', International Conference on High Speed Ground Transportation, Arizona State University, Jan. 1975 (publ. by High Speed Ground Transportation Journal)

BOLDEA, I. and NASAR, S.A.: 'Quasi-one-dimensional theory of linear induction motors with half-filled primary end-slots', Proc. IEE, 1975, 122, No.1, 61-6

BOLDEA, I. and NASAR, S.A.: 'Thrust and normal forces in a segmented-secondary linear reluctance motor', ibid., No.9, 922-4

ELLIOTT, D.G.: 'Matrix analysis of linear induction machines', Jet Propulsion Laboratory, Pasadena, Calif., Final Report No.FRA-OR&D-75-77 for US Dept of Transportation under Contract No.RD 152, Sept. 1975

FOGGIA, A., SABONNADIÈRE, J.-C. and SILVESTER, P.: 'Finite element solution of saturated travelling magnetic field problems', IEEE Trans., 1975, PAS-94, No.3, 866-71

FREEMAN, E.M.: 'Equivalent circuit for the transverse-flux tubular induction motor', Proc. IEE, 1975, 122, No.7, 744-5

FREEMAN, E.M., LOWTHER, D.A. and LAITHWAITE, E.R.: 'Scale-model linear induction motors', ibid., 721-6

HOLTZ, J.: 'Linear asynchronous motor without iron, force components and their control', Elektrotech. Z. (ETZ) A, 1975, 96, No.9, 396-400. In German

MENENDEZ, R.C. and LEE, S.-W.: 'Side force in coil-sheet magnetic levitation systems', Proc. IEEE, 1975, 63, No.5, 768-76

OLLENDORFF, F.: 'Relativistic electrodynamics of the linear synchronous machine', Arch. Electrotech. (Germany), 1975, 57, No.1, 27-30. In German

OOI, B.-T. and EASTHAM, A.R.: 'Transverse edge effects of sheet guideways in magnetic levitation', IEEE Trans., 1975, PAS-94, No.1, 72-80

SYUSYUKIN, A.I.: 'Longitudinal edge effect in linear induction devices with changing phase velocity and magnetic waveguide width along its length', <u>Probl. Tekh. Elektr</u>. (USSR), 1975, No.51, 118-22. In Russian

WAVRE, N. and JUFER, M.: 'Choice of linear induction motor parameters', <u>Bull. Assoc. Suisse Electr</u>. (Switzerland), 1975, 66, No.10, 530-40. In French

Index

[Bold type denotes major topics]

accountant 93, 122, 180
Adkins, Dr B. 45, 91, 212
AEG 191
Aérotrain 130, 136, 149, 160
   Société de l' 137, 150, 151
   Suburban 149, 150
airgap 3, 5, 44, 115, 194, 213, 218, 220, 232, 233
   windings 219
Alcan Industries 113
ALIP 44, 89
Allan, T. 20, 21, 30
alternator 3, 46
Ampère, A. M. 6, 35
analogue 6, 36, 87, 90, 95
Andrée, H. 185
angled-field motor 86, 117
Apps 18, 19, 30
APT 149, 174
arch motor 114, 115, 117
Armstrong, D. S. 92, 135, 136, 226, 228
Astroglide 192, 193

Babbage, C. 41, 56, 229
Bachelet, E. 38–41, 52, 61–4, 68, 79, 96, 97, 125, 137
back-to-back 71–5
Baily, Prof. W. 27, 28
Ball, R. D. 44
Barthalon, M. 143, 153, 156
Barwell, Prof. F. T. 42, 126–8, 142, 143, 174, 187

Beams, J. W. 104
Bedford, Peer and Tonks 100, 101, 125
Beeching, Lord 128
Behrend, B. A. 76
Bertin, J. 130, 133, 136, 137, 143, 149, 150, 152, 160
Birkeland, K. 36, 38
Birmingham Airport 226
Bliss, D. S. 145, 148
Bolton, Dr H. R. 112, 148, 169
Boucherot, P. 2, 21, 36–8, 114
Bowers, Dr B. 13–15, 19
British Rail 40, 42, 130, 135, 136, 141, 167, 168, 177–9, 197, 198, 211, 226
   research labs 129
British Transport Commission 111, 127
Brown, Dr R. 122
Brown Boveri 191
Brush dynamo 24, 25, 30
Budd Co. 226
Budig, P-K. J. 143

Cabinentaxi 192
Centener, P. 34
Chariot d'essais 151
chemist 29
China Lake 183
Chirgwin, K. M. 183, 185
Churchill, Sir W. S. 40, 63
claw-pole motor 220, 221
Cockerell, Sir C. 133, 137
Colquhoun and Partners 179
communication 48, 121, 180, 226

INDEX

commutator motor 3, 8, 11, 12, 19, 44, 59, 225
concedance 123
cone of attraction 101, 165
core steel 76
counter-rotating fields 75
courage 229, 230
Crompton, Col. R. E. B. 25, 27, 30
Crompton-Brunton alternator 28, 30
cryogenic 140, 169, 171, 177, 178, 180, 189, 192, 195, 222, 225, 22
curiosity 229
current collection 181, **184**

Dakeyne, E. and J. 19
Davey, A. W. 162, 165
Davidson, R. 32
Davies, Prof. E. J. 219
de Méritens & Cie, A. 30
deep-bar rotor 37
Demag 192
Department of the Environment 168
Department of Trade and Industry 167, 175, 176, 180
Department of Transportation, Washington D.C. 121, 140, 152, 153, 157, 178-80, 182, 187, 188
design 2, 93, 230, 231
diamagnetic 98, 99
Dickenson 23, 30
disc dynamo 3, 7, 8, 9, 10, 21
disc rig 132
Donnaureid 190
double-cage rotor 37
double-sided sandwich motor 38, 93, 142, 200, 205
dual 36
Duke of Edinburgh 128, 129

Earith 138, 139, 145, **172**
Earnshaw, S. 98
  theorem 98, 99, 105, 144
Eastham, Prof. J. F. 153, 161, 162, 163, 169, 174, 200
eccentric engine 14-17, 30
E-core 200, 201, 202, 216, 217
edge effect 114, 119, 120, 130
Edwards, Prof. S. F. 176
EEC 190
efficiency 29, 44, 87, 88, 92, 98, 108, 120, 135, 188
Eisselt, G. 66, 79
electric hammer 31
electromagnetic gun 33

electromagnetic joint 130, 131, 136, 202, **209**, 210, 211
Electropult 45-7, 52, 69, 85, 111, 125
Ellison, Prof. A. J. 176
Emsland 226
energy machine 120
Erlangen **191**, 192, 226
evolution 2, 93, 230, 231

Fairbanks-Morse 8
Fairweather, W. 79
falling upwards 207
Faraday, M. 2, 3, 6-11, 21, 23, 25, 30, 31, 35, 36, 41, 77, 93, 125
fashion 5, 38, 44, 52, 69, **84**, 92, 93
fashiongraph 87-9
Federal Railroad Administration 186
feedback amplifier 42, 98, 113, 177
Fellows, T. G. 134, 145, 147, 148, 160, 163, 166, 169, 209
Ferranti, S. Z. de 26, 30
Ferraris, G. 75, 76, 90
ferromagnetic 4, 9, 28
first cost 88
Fitzgerald 26
Fleming, Prof. J. A. 35, 36, 40, 64, 97
FLIP 44, 89
force machine 120
Ford, Prof. Sir H. 175, 176
Ford Motor Co. 152, 158, 171, 178
Forman, J. 66, 80
Fox Talbot, W. H. 19, 21, 32
Freeman, Prof. E. M. 108-10, 142, 190
Froment 18, 19, 26, 30

Gamow, R. J. and Harris, J. F. 37
Garrett Corporation 121, 140, 145, 153, 156-8, 179, 181, 182, 184
Geary, P. J. 100
General Motors 159
generalised machine theory 45, 91
generator 3, 107
Gometz 151, 156, 160
goodness (factor) 5, 36, 60, 77, 87, 102, 129, 202, 208
Gorton (Loco Works) 127, 209
  experimental machine 128-30, 169, 174, 179
Gourdon, G. 62, 79
Gramme, Z. T. 23, 25, 30
  ring winding 23-5, 27, 30, 144, 153, 156, 157, 186, 190, 209
Greatorex, Dr N. 206

INDEX

Greig, Prof. J. 110
Grondhal, K. T. 66, 80

Haberhauer, K. 68, 81
Hart, G. F. 178, 179
Hawker Siddeley 137, 169, 171, 179, 190
Hearder 32
Heaviside, O. 108, 231
helical motor 120
Hennessey, D. 134
Herbert Morris (Cranes) 157
Heseltine, M. 167
Hesmondhalgh, Dr D. E. 36
Hjorth, S. 32
Hodges, P. 66, 80
Holmes, F. H. 21, 30
Hopewell, F. B. 64-6, 79, 82
Hopkinson, J. 26
Hovercraft Development Ltd 137, 153
Hovershow 66 137, 138
hysteresis motor 3, 58, 220

Imperial College 108, 109, 132, 136, 141, 143, 149, 153, 162, 173-5, 194, 206, 217
Inglis, C. C. 127

Jacobi, Prof. 8, 32
Jacquard, J. M. 55, 56, 61, 79, 229
Japan Air Lines 172, 189, 197
Japanese National Railways 159, 189, 197, 226, 227
Japolsky, Dr N. 41, 43, **108-10**
 hammer 42, 43
Jasicek, A. and Polnauer, F. 64-6, 79
Jayawant, Prof. B. V. 34, 177
jet engine 38
Johnson and Johnson Ltd 67, 80
Jones, Dr S. 149
jumping ring 35, 39, 59, 66, 97, 104

Kapp, Dr G. 25
Karapetoff, V. 76
Kelvin, Lord 26
Kemper, H. 100, 125
King's College, London 109
Kings Norton, Lord 179
Kolm, Prof. H. H. 158
Komet 194
Krauss-Maffei 140, 143, 149, 153, 169, 171, 172, 178, 187-9, 197, 226
Kron, G. 45

Ladd, W. 21, 24, 30
lamination 45
Lamme Medal 93
Landspeed **178**, 179
 University Consultants 179
Lawrenson, Prof. P. J. 232
laws of induction 31
Laws of Nature 21
Le Moteur Linéaire 152, 157, 180, 186
Lenihan, Dr J. M. A. 1
Leonardo da Vinci 38
levitation melting 102
Linear Motor Programme Committee 177
Linear Motors Ltd 134, 157, 162, 172, 179, 224
Linier, C. and Latieule, Y. 67, 81
Lintrol Systems 157
liquid metal 42, 103
 pump 43, 44, 66, 92, 122, 225
 stirring 45, 66, 122, 204
Llewelyn, J. D. 32
Lodge, Sir O. 26
logmotor 120
loom 53, 56-8, 69
 Lancashire 54
Lovell, W. V. 101, 102, 165
Lowe, J. 112, 113

Maglev 100, 149, 152, 169, 172, 178, 180, 183, 187, 189, 194, 197-200, 222, 226, 227
Magnarail 113
Magneplane 158, 187
magnetic circuit 4, 6, 8, 9, 30, 33, 45, 64, 77, 78, 100, 141, 144, 147
 equivalent circuit 13, 123
 impedance 123
 inductance 123
 machine 5, 6, 37, 38
 pull 37, 38, 57, 64, 66, 73, 93, 111, 114, 147, 205
Magnetic Reynolds Number 78
Magnetic River 101, 163, 164, 174, 176, 178, 181, 187, 194, 195, 200, 201, 209, 210, 213, 227
Manchester 48, 62, 85, 91, 101, 106, 110, 112, 121, 134, 149, 153, 156, 161, 198, 212
 College of Technology 123
 University 69, 126
Massachusetts Institute of Technology 116, 158, 180, 187
Maxim, Sir H. 64

Maxwell, J. C. 99, 230, 231
  equations 93
Mayor of Pittsburgh 52
M-Bahn 225, 227
Meeus, J. 57-9, 61, 79
Merlin Gerin 149, 150, 151, 157, 177, 180
Mershon, R. D. 34
Messerschmitt-Bölkow-Blohm 140, 153, 155, 189, 192, 197, 198, 226
metallurgist 29
'mixed μ' system 227
Motor Industry Research Association 121, 153-5
Muck, O. 103

National Aeronautics and Space Administration 133
National Research Development Corporation 133, 134, 166-8, 173, 179
Neave, A. 176
niobium-tin 8, 99, 149
Nix, Dr G. F. 70, 105, 112

Okress, E. C. 104, 105
Ontario Urban Transportation Development Corporation 187
  Queen's University, Kingston 188
open magnetic circuit 117
open-sided motor 122
Otis Elevator Co. 193

Pacinotti, A. 23, 30
Palmer, A. 167, 176
parallel magnetic circuit 123
Park, R. H. 45, 91
Parsons, C. A. & Co. Ltd 30
Partridge, G. W. 26
permanent magnet 5, 9, 16, 19, 21, 59, 60, 67, 82, 98, 105, 221, 225, 227
permeability 4
Personal Rapid Transit systems 193
Peyton, Rt Hon. J. 174
Physical Society 27
Pilsen and Joel 25, 30
pinch effect 45, 104
Pixii, A. H. 3, 6, 7, 10, 11, 19, 30, 46
Powell, J. R. and Danby, G. T. 227
power factor 92, 108, 120, 135, 144, 186-8, 200, 220, 233
power machine 120
power/weight ratio 88, 120
Pyke-Harris 28, 30

rack and pinion motor 34
radar 38
radiation pressure 98
ratchet 36, 37
reciprocating machine 14, 31-4, 36, 37
reliability 44, 69, 88, 120
reluctance 4-6, 20, 66, 67, 215, 232
  motor 3, 13, 28, 30, 33, 38, 57-9, 62, 82, 89, 220, 222, 232
repulsion motor 59
Reynolds, Prof. O. 196, 202, 212
Rhodes, Prof. R. G. 177
rocket sled 183
Rohr Corporation 152, 157, 159, 179, 180, 186
Romag 152
Royal Aircraft Establishment 48, 49
Royal Institution 35, 179
Russell, R. L. and Norsworthy, K. H. 66, 198, 209
Rutherford Laboratory 176

Sabonnadière, J-C. 180
Sadler, G. V. 110, 120, 134
Scelzo, G. P. 192
Schräge motor 77
Science Museum 6, 8, 13, 15, 17-20, 22, 24, 26, 27, 69, 73
Science Research Council 130, 176, 192
scrap metal sorting 225
Select Committee 126, 152, 157, 167, 169, 173-8
self-oscillating motor 70, 71-5, 81, 82
series connection 122
shaded pole 35
shading ring 105
shape 8, 26, 52, 67, 77, 123
sheet-rotor motor 38, 39, 59, 64
short primary 44
Shturman, G. I. 85, 116, 120
  and Aronov, R. L. 85, 115, 117
shuttle propulsion 53-9, 61, 64-7
Siemens 21-3, 25, 30, 191
single-phase motor 74, 101
SIP 44
skewed slots 46
skin effect 44
slots 4, 136
Smithsonian Institution 6
Snow, Sir F. 113
Souter Point lighthouse 21, 22, 30
SPAR Aerospace Products Ltd 171-3
Speedover Transport Ltd 137

spherical motor 44, 76, 87, 116, 118, 120, 205, 212-14
Spring, Dr K. 177, 178
stability 200, 206
Stanford Research Institute 152, 158, 178
Stöhrer, E. 7, 8, 23, 30
Sturgeon, W. 11, 13, 14, 30
superconductor 8, 9
superimposed skewed windings 215, 216
Sussex University 152, 177
synchronous motor 3, 58, 61, 67, 86, 180, 189, 194, **220**, 223, 226, 227

Taylor, W. H. 14, 19, 30
teeth 4, 76, 136, 221
Tesla, N. 2, 27, 30, 53, 105
theory 230, 231
Thomson, A. 26
Thomson, E. 35
Thomson and Houston 26, 27, 30
Tomlinson, *Cyclopaedia of Useful Arts* 31-3
topology 2, 3, **7-12**, 16, 25, 26, 34, 52, 57, 91, 92, 136, 148, 197, 209, 210, 215, 218, 222, 225, 226, 228
Toronto Urban Transit Scheme **170**
track joints **130**, 131, 136, 202, 209-11
Tracked Hovercraft 109, 126, **134**, 136-41, 145, 148, 149, 158-60, 166, 167, 169-72, 175-9, 190, 197, 198, 202, 205, 213, 217
transferance 123
Transpo 72 **160**, 164, 166, 169
Transportation Test Center, Pueblo 181, 182, 186, 187
Transrapid 153, 155, 156, 194, 197, 226
transverse flux 13, 36, 92, 136, 146, 157, 166, 172, 181, 185, 192, 197, 199, 202, 213, 216-18, 227
    motor **203-5**
travelling field 20, 33, 44, 59, 60
Tridim 152
Trombetta, P. 31, 34, 35

tubular motor 33, 35, 44, 59, 66, 68, 83, 202, 203

unbalanced magnetic pull 38, 76, 143
unbalanced magnetic push 93, 143
Universities Steering Committee 178, 192
URBA system 143, 153, 156

Van Depoele, C. 35
vernier motor 89
Victorri, M. 180
Von Miller, O. 28
Von Vago, P. J. 68, 80

Warwick University 177
Watkinson, Rt Hon. H. 113
Wedgwood Benn, Rt Hon. A. 153
Weh, Prof. H. 192
Weil, A. 68, 83
West, Prof. J. C. 34, 149, 152
West German Ministry of Technology 186
Westinghouse Corporation 45-7, 111, 125
Wheatstone, Sir C. **13-17**, 19, 21, 30, 32
Wilde, H. 23, 25
    alternator 24, 30
Williams, Prof. Sir F. C. 26, 74, 76, 82, 110, 116-18, 120, 122, 134, 162
Wilson, P. B. 66, 80
wisdom 229
Woolrich 8, 9, 11, 30
Wordingham, C. H. 26
World War I 38
World War II 38, 45, 69, 107, 108, 221

Xi-core 201, 202

Young, Y. L., Jun. 66, 80

Zehden, A. 36-8, 41, 125
zig-zag track conductor 223
    synchronous motor 227

INDEX

spherical motor, 48, 76, 82, 116, 118, 120, 205, 212–14
Spring, Dr S. 177, 178
stability, 200, 206
Stanford Research Institute, 156, 158, 198
Stöcher, E. 7, 8, 9, 10
Sturgeon, W. 11, 13, 14, 30
superconductor, 8, 9
superimposed skewed windings, 214, 216
Sussex University, 180, 177
synchronous motor, 3, 58, 61, 67, 86, 150, 185, 194, 220, 224, 226, 227

Taylor, W. B. 14, 19, 30
teeth, 4, 10, 156, 221
Tesla, N. 2, 20, 30, 52, 105, theory, 210, 221
Thomson, A. 26
Thomson, J. 35
Thomson and Houston 20, 21, 30
Tennissen, Corporation of Israel Arts 31–3
topology, 3, 5, 9, 13–14, 16, 19, 20, 21, 22, 25, 41, 92, 136, 148, 194, 205, 210, 215, 218, 222, 225, 226, 228
Toronto Urban Transit Scheme, 170
track joints, 130, 131, 150, 202, 209–11
Tracked Hovercraft 104, 106, 134, 136–41, 145, 148, 149, 158–60, 165, 167, 168–72, 175–9, 190, 191, 198, 202, 205, 213, 217
transference, 122
Transpo 72, 160, 166, 166, 169
Transportation Test Center, Pueblo, 181, 184, 186, 187
Transrapid 152, 153, 150, 190, 197, 220
transverse flux, 13, 30, 92, 136, 140, 147, 165, 172, 151, 155, 182, 187, 199, 202, 213, 215–18, 227
motor, 203–5
travelling field, 20, 83, 84, 59, 60
Traider, 152
Trombetta, F. 31, 38, 25

tubular motor, 34, 35, 44, 59, 66, 68, 82, 202, 203

unbalanced magnetic pull, 28, 76, 142
unbalanced magnetic push, 93, 143
Universities Steering Committee, 178, 192
URBA system, 147, 153, 156

Van Depoele, C. 35
vernier motor, 59
victorri, M. 180
Von Miller, O. 28
von Vago, P.J. 58, 80

Warwick University, 177
Watkinson, Rt Hon. H. 113
Weagwood Benn, Rt Hon. A. 153
Weh, Prof. H. 192
Weil, A. 58, 82
West, Prof. J. C. 28, 149, 132
West German Ministry of Technology 186
Westinghouse Corporation, 45–7, 111, 127
Wheatstone, Sir C. 13–17, 19, 21, 30, 32
Wilde, H. 22, 25
alternator 24, 30
Williams, Prof. SEC F. C. 25, 74, 76, 82, 110, 126–18, 120–1, 72, 184, 167
Wilson, F. R. 46, 90
vacuum 229
woolrich, S. 9, 14, 30
Worthington, C.h. 20
World War I, 38
World War II 38, 45, 63, 107, 108, 221

Xi-Gore, 201, 202

Young, V L, Junior 60

Zehden, A. 36–8, 41, 75
on track construction 202
synchronous motor 37